The Economic Value of Landscapes

Landscape has been ill-served by 20th century economics. Within existing economic fields – such as ecological economics and the more traditional environmental (and resource) economics – some scholarly attention has been given to the economic aspects of landscapes. This attention, however, appears to be fragmented and lacks an overarching relational framework, or even a consistent and systematic analysis. Moreover, the discipline of ecological economics often regards the landscape as an ecosystem, or as an aggregation of ecosystems, thereby throwing the concept of landscape back into the realm of landscape ecology.

This book aims to explore the avenue of landscape economics and provides the building blocks (from different scientific disciplines) for an economic analysis of landscapes. What exactly constitutes and determines the value of a landscape? It focuses on the value of landscapes in its broadest sense, thereby covering a variety of topics including stakeholder involvement in landscape design, landscape governance and landscape perceptions from different countries. Merely saying that landscapes have value or are important is not sufficient – not when resources are scarce and have alternative uses. Measuring and quantifying the economic value of changes in landscapes would help ensure that landscape management decisions are both (economically) rational and sound.

A prominent feature of the upcoming field of landscape economics is that it goes beyond traditional neo-classical orthodoxy as various contributions in this book illustrate. It has characteristics that do not fit well with the standard presumptions of the neo-classical model. The aesthetic phenomenon and the holistic perception of landscapes have the potential of creating inconsistencies in an economic model of instrumental rationality that is based on independence and easily demarcated objects. This book integrates perspectives and practices from a variety of disciplines and specialized professions, such as ecology, sociology, governance studies, and psychology.

C. Martijn van der Heide is an Economic Researcher at LEI, Wageningen UR and *Lector* (Reader) 'Integrated Nature and Landscape Management' at the University of Applied Sciences Van Hall Larenstein, in Velp, the Netherlands.

Wim J.M. Heijman is Professor of Regional Economics at Wageningen University, the Netherlands.

Routledge studies in ecological economics

1. **Sustainability Networks**
   Cognitive tools for expert collaboration in social-ecological systems
   Janne Hukkinen

2. **Drivers of Environmental Change in Uplands**
   Aletta Bonn, Tim Allot, Klaus Hubaceck and Jon Stewart

3. **Resilience, Reciprocity and Ecological Economics**
   Northwest coast sustainability
   Ronald L. Trosper

4. **Environment and Employment**
   A reconciliation
   Philip Lawn

5. **Philosophical Basics of Ecology and Economy**
   Malte Faber and Reiner Manstetten

6. **Carbon Responsibility and Embodied Emissions**
   Theory and measurement
   João F. D. Rodrigues, Alexandra P.S. Marques and Tiago M. D. Domingos

7. **Environmental Social Accounting Matrices**
   Theory and applications
   Pablo Martínez de Anguita and John E. Wagner

8. **Greening the Economy**
   Integrating economics and ecology to make effective change
   Bob Williams

9. **Sustainable Development**
   Capabilities, needs, and well-being
   Edited by Felix Rauschmayer, Ines Omann and Johannes Frühmann

10. **The Planet in 2050**
    The Lund discourse of the future
    Edited by Jill Jäger and Sarah Cornell

11. **Bioeconomics**
    Edited by Mauro Bonaiuti

12. **Socioeconomic and Environmental Impacts on Agriculture in the New Europe**
    Post-communist transition and accession to the European union
    S. Serban Scrieciu

13. **Waste and Recycling: Theory and Empirics**
    Takayoshi Shinkuma and Shusuke Managi

14. **Global Ecology and Unequal Exchange**
    Fetishism in a zero-sum world
    Alf Hornborg

15. **The Metabolic Pattern of Societies**
    Where economists fall short
    Mario Giampietro, Kozo Mayumi and Alevgül H. Sorman

16. **Energy Security for the EU in the 21st Century Markets, Geopolitics and Corridors**
    Edited by José María Marín-Quemada, Javier García-Verdugo and Gonzalo Escribano

17. **Hybrid Economic–Environmental Accounts**
    Edited by Valeria Costantini, Massimiliano Mazzanti and Anna Montini

18. **Ecology and Power**
    Struggles over land and material resources in the past, present and future
    Edited by Alf Hornborg, Brett Clark and Kenneth Hermele

19. **Economic Theory and Sustainable Development**
    What can we preserve for future generations?
    Vincent Martinet

20. **Paving the Road to Sustainable Transport: Governance and Innovation in Low-Carbon Vehicles**
    Edited by Måns Nilsson, Karl Hillman, Annika Rickne and Thomas Magnusson

21. **Creating a Sustainable Economy**
    An institutional and evolutionary approach to environmental policy
    Edited by Gerardo Marletto

22. **The Economics of Climate Change and the Change of Climate in Economics**
    Kevin Maréchal

23. **Environmental Finance and Development**
    Sanja Tišma, Ana Maria Boromisa and Ana Pavièiæ Kaselj

24. **Beyond Reductionism**
    A passion for interdisciplinarity
    Edited by Katharine Farrell, Tommaso Luzzati and Sybille van den Hove

25. **The Business Case for Sustainable Finance**
    Edited by Iveta Cherneva

26. **The Economic Value of Landscapes**
    Edited by C. Martijn van der Heide and Wim J.M. Heijman

The Economic Value of Landscapes

Edited by
C. Martijn van der Heide and
Wim J.M. Heijman

Routledge
Taylor & Francis Group
LONDON AND NEW YORK

First published 2013
by Routledge
2 Park Square, Milton Park, Abingdon, Oxfordshire OX14 4RN

Simultaneously published in the USA and Canada
by Routledge
711 Third Avenue, New York, NY 10017

First issued in paperback 2014

Routledge is an imprint of the Taylor & Francis Group, an informa business

British Library Cataloguing in Publication Data
A catalogue record for this book is available from the British Library

Library of Congress Cataloging in Publication Data
A catalog record for this book has been requested

ISBN 978-0-415-56328-4 (hbk)
ISBN 978-1-138-80823-2 (pbk)
ISBN 978-0-203-07637-8 (ebk)

Typeset in Times New Roman
by Cenveo Publisher Services

Contents

Figures

Tables

Preface

Landscapes are never finished, they are always a work in progress. Every time you blink and open your eyes, the landscape has changed. How many of us do not sigh pensively that the landscape of our childhood has vanished? Landscapes are places where memories are stored and recollected; they provide an anchor of authenticity and identity in turbulent times of modernization and globalization. All of our historic city- and townscapes, polders, reclaimed land and rural areas are the indelible stamp of our national identity and are loaded with cultural memories. Everywhere you stand in the landscape, you stand within the thoughts of "passed" people, especially in a country as densely populated as the Netherlands. As such, landscapes are rich in reminiscence and anecdote – they form, according to the Dutch poet and novelist Willem van Toorn, a "pictorial book of our memory". But when landscapes change, these *lieux de mémoires* – sites of collective memory – may disintegrate and disappear, together with nature and heritage resources. As landscapes are always evolving and susceptible to modifications, preferences for and appraised values of landscapes may change as well. And conversely, landscapes are changing in response to changing values; many of the current urban and rural landscapes are the result of human preferences and creations.

This book focuses on the economic value of landscapes and provides the building blocks for an economic analysis of landscapes. It is intended to facilitate and promote the use and further development of economics within the spatial context of landscapes. Here, we consider economics as much more than money, inflation, supply and demand. We view economics as the discipline that tries to explain how to fulfil people's unlimited needs and aspirations under scarce resource constraints. This volume reviews and presents our current understanding of the linkages between economics and landscape, and disentangles the complex relations between the two. It pulls together bits and pieces from a wide assortment of analyses directed at a theoretical and practical exploration of "landscape economics". Besides methodological and conceptual considerations, illustrative case studies looking at particular landscape valuation issues in Israel, Japan, the Netherlands, and Spain are presented.

The Economic Value of Landscapes is a topic that is located near the boundary of the natural and social sciences. As such, this book is meant for scholars of diverse interests, who wish to work on interdisciplinary problems or whose work crosses interdisciplinary lines. Moreover, the book is intended to assist planners, designers and policy makers in making informed decisions about landscape design and planning. They have to make choices between different, and usually competing, uses of the land. These choices will have impact on the character and quality of the landscape. Having insight into the economic value of landscape can support practitioners inside and outside government to allocate scare resources to achieve specific landscape goals.

This book is the product of great work by many contributors whose efforts are highly appreciated. Editing the book was challenging in countless ways, yet very rewarding, thanks to the support and patience of the contributors and the Routledge staff. The book's editing has been enhanced by our practical experience of investigating a range of landscape economics issues gained through projects financed by the strategic research programme (*Kennisbasis*) of the Dutch Ministry of Economic Affairs, Agriculture and Innovation, and through research conducted for the Netherlands Environmental Assessment Agency (PBL) that has been commissioned by the Statutory Research Tasks Unit for Nature & the Environment (*WOT Natuur & Milieu*). Our insights into the topic have also been enhanced over the years by our discussions with colleagues at Wageningen University, LEI Wageningen UR, University of Applied Sciences Van Hall Larenstein, and the Ministry of Economic Affairs, Agriculture and Innovation.

Some of the chapters were presented at or grew out of an international conference on Landscape Economics, held in Vienna in July 2009, organized by the European Consortium for Landscape Economics (CEEP). We offer our deep appreciation to the participants of this conference for their contributions that have enabled us to produce this volume. Furthermore, we are particularly indebted to Henk Folmer, for organizing a Special Session on the book during the European Regional Science Association Congress (ERSA) in August 2010 (Jönköping, Sweden). We also thank Adam Walker for his help in reading through some of the chapters and making a number of useful suggestions. We also appreciate the assistance provided by Ms. Apurupa Mallik for her participation in the preparation of the final manuscript. Finally, we owe a special debt to the Routledge staff, and especially to Robert Langham and Simon Holt, for their enduring support and patience in the past three years.

Martijn van der Heide and Wim Heijman
Leiden and Wageningen, July 2012

Contributors

Marc Antrop obtained his PhD in geography at the University of Ghent (Belgium). Currently he is Emeritus Professor at the Department of Geography, within the research unit Landscape Research, University of Ghent and member of the National Committee of Geography (Academy of Sciences), Vice-President of the Royal Committee for Protection of Monuments and Landscapes in Flanders and member of the Belgian Scientific Committee of ICOMOS-IFLA. In 2003 he received the Distinguished Scholarship Award of the International Association for Landscape Ecology and in 2007 was proclaimed doctor honoris causa at the University of Tartu (Estonia). His research focuses on landscapes and is holistic, integrating landscape typology and evaluation, landscape ecology, historical geography, landscape perception and landscape architecture.

Olivier Aznar obtained his PhD degree in 2002 with a thesis entitled *Environmental Services and Rural Areas*. Since 2002, he has been working as a research scholar at Cemagref/Irstea, Metafort Unit, Clermont-Ferrand (France), where his main areas of research are environmental economics, services economics and landscape economics. He is working on different aspects of environmental services: supply, demand and institutions.

Arianne T. de Blaeij is an environmental economist and since 2006 has been working as researcher and project leader at LEI, part of Wageningen UR (the Netherlands), within the organizational unit Regional Economy and Land Use. She is a specialist in monetary valuation of and payment mechanisms for conservation of and improving ecosystem services, and in cost–benefit analysis. She obtained her PhD in economics at VU University, Amsterdam in 2003.

Arjen Buijs is a sociologist who obtained a PhD degree in 2009 with a thesis entitled *Public Natures: Social Representations of Nature and Local Practices*. He currently combines positions as Assistant Professor at the Forest and Nature Policy group of Wageningen University (the Netherlands) and as Senior Researcher at Alterra Green World Research. His research interests include public participation and social conflicts in natural resource management, landscape perceptions and cultural diversity in social representations of nature.

Jasper Dekkers is Assistant Professor in spatial informatics and spatial economics at the Department of Spatial Economics (VU University Amsterdam, the Netherlands). He is coordinator of the master of science programme in geographical information sciences (UNIGIS) at the SPINlab – Centre for Research and Education on Spatial Information of VU University, Amsterdam. He studied business economics and spatial economics at VU University and graduated in 2001 with a thesis on location-based services and the mobile market, which won an award from the Geospatial Information & Technology Association (GITA), the Netherlands. His PhD dissertation, completed in 2010, focused on the explanation and modelling of land and house prices, in particular at the urban fringe, and on the use of these explanatory models for future land-use modelling.

Dafna M. DiSegni is an Assistant Professor at the Department of Natural Resources and Environmental Management at the University of Haifa, and adjunct faculty at the Tel Aviv University in Israel. She received her PhD in natural resources and environmental economics in 2004 from the University of California at Davis, USA. Her research within the field of resources and environmental economics focuses on market design, trade and regulatory economics. She has recently extended her research and publications on valuation theory, including studies on: the macroeconomic value of ecosystem services and the impact of climate change on these values; the value of information on water pollution for useful management of water drilling in Israel; and the development of best management practices for sustainable orchards and agricultural landscapes for soil, water and environment conservation.

Hans Farjon works as a Senior Policy Researcher at the Netherlands Environmental Assessment Agency, where his main fields of interest are landscape ecology, landscape perception and spatial planning. He has extensive experience with development and implementation of indicators, monitoring systems and models for use in landscape assessments of national policy schemes.

David Soto Fernández holds a PhD in contemporary history from the University of Santiago de Compostela (Spain) with a thesis defended in 2003 and entitled *Productive Transformations in Contemporary Galician Agriculture. From Organic Agriculture to the Green Revolution (1752—1986)*. In 2003, he became Associate Professor of Contemporary History at the University Pablo de Olavide of Seville. His main research focuses on the analysis of contemporary agrarian history from an environmental perspective, with particular attention to the study of social metabolism of agricultural systems.

Sylvie Ferrari is an Assistant Professor at the University of Bordeaux (France) and has been a researcher at the GREThA (Research Unit in Theoretical and Applied Economics – CNRS affiliated research unit) since 2005. Her teaching is primarily related to ecological economics and sustainable development. Her research areas concern: environmental sustainability of production, environmental ethics and sustainable development, and sustainable management of water resources. From 2002 to 2005, she was a visiting researcher at IRSTEA (National Research Institute of Science and Technology for Environment and Agriculture) with a research project focusing on the biophysical analysis of agricultural production processes in the presence of landscape amenities, and their implications for public policy.

María D. Domínguez García holds a PhD in social sciences from Wageningen University (the Netherlands) with a thesis entitled *The Way You Do It Matters. A Case Study: Farming Economically in Galician Dairy Agroecosystems in the Context of a Cooperative*. Since 2007, she has worked as a lecturer and researcher at Vigo University (Spain). Her research is in ecological economics, agro-ecology and rural sociology. She has wide experience with the farming styles approach. Since 2010, she has focused on governance and learning processes related to territorial rural development through research applied to the Galician context in several FP7 (Seventh Framework Programme) projects.

David Haim is an applied economist. He holds a Research Associate position in the Department of Forest Engineering, Resources and Management in the College of Forestry, Oregon State University (USA). His research focuses on the economics of land-use and climate change. Currently, he is investigating afforestation integrity under carbon markets, exploring climate change impacts on land-use changes in the USA and analyzing the role of carbon sequestration as a greenhouse gas mitigation strategy. He received his PhD in applied economics from Oregon State University in July 2011.

Wim J.M. Heijman received MSc degrees, in economics and human geography, respectively, from Tilburg University and the University of Utrecht in the Netherlands. He received his PhD degree from Wageningen University. In 2000 he was appointed Professor of Regional Economics at Wageningen University. In the last four years he has been specializing in a new area of research: landscape economics. Research in this field is inspired by the European Landscape Convention, which aims to preserve valuable landscapes, especially in areas surrounding heavily populated urban areas.

Ian Hodge is Professor of Rural Economy in the Department of Land Economy and Fellow of Hughes Hall at the University of Cambridge (UK). He was Head of Department between 2002 and 2011. His primary research interests are in rural environmental governance, property institutions and rural development. He has previously worked at the Universities of Queensland in Australia and Newcastle upon Tyne in the UK. He has a PhD from the University of London and a BSc from the University of Reading.

Iddo Kan is a Lecturer at the Department of Agricultural Economics and Management of the Hebrew University of Jerusalem (Israel). His research activities are related to the scientific areas of agricultural economics, environmental and resource economics, regional economics and political economics.

The common factor among most of his studies is the economic analysis of agricultural land-use. He investigates this topic in relation to irrigation and drainage management, solid-waste management, assessment of climate-change impacts, rural landscape preservation, evaluation of externalities, siting of noxious facilities and water pricing in political decision-making systems. His expertise is in integrating natural processes into economic analyses with the objective of characterizing management strategies and policies under optimal and/or equilibrium conditions.

Eric Koomen is Assistant Professor at the Department of Spatial Economics of VU University, Amsterdam in the Netherlands. His research interests include urban development, land-use change, spatial modelling and policy support. He has edited or co-authored three books and published numerous papers in scientific and applied journals and books. He teaches courses on land-use change and assessing the landscape in the Earth and Economics programme, and is tutor for the Urban Environment Lab at Amsterdam University College and for GIS and Environmental Impact Assessment in the UNIGIS MSc programme. Furthermore, he supervises BSc and MSc theses and PhD research. Eric graduated in physical geography in 1992 and since then has worked, among others, for the Dutch Ministry of Transport and Water Management where he focussed on the use of GIS for environmental impact assessment. In recent years Eric also worked part-time at Geodan where he managed projects to inform regional authorities about likely spatial developments, their potential impacts and possible policy alternatives.

Christian Lippert is a Research Scholar and Lecturer at the Institute of Farm Management (Section Production Theory and Resource Economics) of Universität Hohenheim (Germany). He teaches farm planning and environmental and natural resource economics. His main research areas are institutional economics applied to agri-environmental and food safety issues, and agricultural economics with respect to climate change.

Nora Mehnen is an environmental scientist. She obtained her degree in 2005 at the University of Vechta (Germany) with a diploma thesis entitled *The Regional-Economic Value of National Park Tourism in the new Harz National Park, Germany*. Her interests include environmental policy, protected areas, (economic) valuation, regional (sustainable) development, governance and (environmental) decision-making. Currently, she is completing her dissertation research on governance of and in protected landscapes at the Department of Cultural Geography, Faculty of Spatial Sciences of the University of Groningen.

Ingo Mose was trained as a geographer and has held a professorship at the Institute of Biology and Environmental Sciences at Carl von Ossietzky University Oldenburg, Germany, since 2005. He is also attached to ZENARiO – the Centre of Sustainable Spatial Development in Oldenburg. His research interests cover protected areas, regional governance, rural development and sustainable tourism. Ingo has held a number of visiting professorships across Europe, including at the University of Vienna and Kingston University, London. Currently he also acts as a spokesman for the German Rural Geographers Working Group.

Pierre v. Mouche studied theoretical physics and mathematics at the University of Nijmegen (the Netherlands). He received his PhD degree from the University of Utrecht, under the supervision of Hans Duistermaat, with a thesis about the discrete Mathieu operator. Since 1989, he has held a position in economics at Wageningen University. His scientific interest concerns mathematics, and in particular mathematical economics.

Kenji Okubo obtained his PhD degree with a thesis entitled *An Econometric Analysis of Environmental Value and Preservation of Grassland in Aso Area, Kumamoto Prefecture, Japan* in March 2002. Since 2004, he has worked as an Assistant Professor in the Department of International Bio-Business Studies of Tokyo University of Agriculture, Japan, where his main areas of research are econometric analysis of multifunctionality of agriculture, rural tourism and ecological marketing.

Nico B.P. Polman is a Senior Researcher responsible for investigations regarding institutional economics, scaling and governance, as well as the impact of agriculture on rural land-use, including aspects related to regional development and rural policy. He is employed at LEI, part of Wageningen UR (the Netherlands), within the organizational unit Regional Economy and Land Use. He has experience in coordinating research projects for the Dutch government and has participated in a number of European projects.

Colin Price has master's and doctoral degrees from Oxford University (UK) in, respectively, forestry, and land-use economics. He taught land economics at Oxford Brookes University and Oxford University, before moving to Bangor University where he became Professor of Environmental and Forestry Economics. He has written three books, including *Landscape Economics* (Macmillan, 1978) and about 200 papers on his special topics of forestry economics, cost–benefit analysis and environmental evaluation, specialising in landscape valuation. He is now a freelance academic.

Mickey Rapaport-Rom is a PhD student at the Department of Natural Resources and Environmental Management in the University of Haifa (Israel). Her thesis is on the topic of water policy in the agricultural sector in Israel and its impact on efficient allocation and the spatial distribution of different types of water. She obtained her MA in 2007 with a thesis entitled *Economic Assessment of Climate Change Impacts on the Vegetative Agriculture in Israel*. Since 2004, Mickey has worked as a researcher at the Natural Resources and Environmental Research Center (NRERC), where she studies the area of agriculture and water economics. She has published several papers in peer-reviewed publications, and has been awarded an Advanced Studies scholarship and a Jewish National Fund research grant.

Stijn J. Reinhard is with the Agricultural Economics Research Institute (LEI) at Wageningen UR (the Netherlands) and Head of the Department of Regional Economy and Land Use. He received the American Agricultural Economics Association best dissertation award for his work on the econometric analysis of the environmental and economic efficiency of dairy farms. He has led projects for multiple clients and published on impact analysis and cost–benefit analysis of policy plans, e.g. Natura 2000, EU Soil Strategy, EU Water Framework Directive and EU Marine Strategy Framework Directive. Stijn has organized stakeholder participation to gather relevant information as well as to increase support for the analysis. Currently he leads the work package on spatial econometrics of the EU project SPARD (Spatial Analysis of Rural Development).

Jan Rouwendal is a Professor at the Department of Spatial Economics of VU University, Amsterdam (the Netherlands). His research interests include the economic evaluation of cultural heritage, the economic analysis of spatial planning, and the relation between housing, urban amenities and labour markets. He is a Research Fellow of Tinbergen Institute and the Amsterdam School of Real Estate.

Marjanne Sevenant received a PhD degree in geography (Ghent University, Belgium), with a dissertation entitled *Variation in Landscape Perception and Preference: Experiences from Case Studies in Rural and Urban Landscapes Observed by Different Groups of Respondents*. In 1998–2010, she worked as a researcher at Ghent University on different projects in landscape perception and preference, landscape ecology, historical geography and environmental impact assessment. Currently, she is an Environmental and Social Specialist at IFC (International Finance Corporation/World Bank Group) in Washington DC, providing advisory services to private-sector clients on environmental, social and governance issues.

Mordechai Shechter currently serves as the Founding Dean of the School of Sustainability at the Interdisciplinary Center Herzliya (IDC), Israel. He is Professor Emeritus of the Department of Economics and of the Department of Natural Resource & Environmental Management, at the University of Haifa. He was the founder (1985) and Director (until 2010) of the Natural Resource & Environmental Research Center. He served as the University Rector (Provost), Dean of Research, Head of the Economics and of the Natural Resource and Environmental Management Departments, and Chair of the Senior Faculty Association. In 2000–2005 he served as President of Tel-Hai Academic College in the Upper Galilee region in northern Israel. He also headed Israel's National Parks and Nature Reserves Authority Council, the Israel Information Centre for Climate Change Adaptation and the Board of the City of Haifa's Museums.

He holds a PhD in Agricultural Economics from Iowa State University (1968), and has taught and published in the areas of environmental and natural resource economics and management.

Frans J. Sijtsma is Assistant Professor at the Faculty of Spatial Sciences of the University of Groningen (the Netherlands), and his main areas of research are regional economic development in relation to nature, landscape and human well-being. He holds a PhD in economics, and his thesis was entitled *Project Evaluation, Sustainability and Accountability – Combining Cost–Benefit Analysis (CBA) and Multi-Criteria Analysis (MCA)*. He publishes and teaches on subjects relating to these topics.

Louis H.G. Slangen is Associate Professor of Agricultural Economics and Rural Policy at Wageningen University (the Netherlands). His teaching, research and supervision of MSc and PhD theses are in the field of resource economics, economics of land-use, new institutional economics and economic organisation theory. He has published about 35 articles in refereed journals, 15 chapters in books and 10 books. He has long and extensive experience in various types of research projects for a number of Dutch Ministries and the European Union.

Dirk Strijker is Professor in Rural Development (the Mansholt-chair) in the Department of Cultural Geography within the Faculty of Spatial Sciences of the University of Groningen (the Netherlands). He worked at the Faculty of Economics at the University of Groningen (1979–1983), and at the Agricultural Economics Research Institute (LEI) in The Hague (1983–1988). He has published on various subjects, especially on the economic aspects of agricultural policy, the relations between agriculture, nature and rural development, and on spatial subjects related to agriculture and agricultural policy. He is also a columnist for Dutch newspapers (*Dagblad van het Noorden* (until 2009) and *De Boerderij* (since 2010)).

Carolina Tagliafierro has a master's degree in agricultural economics and policy and a doctorate in agro-forestry resources management with a dissertation in landscape economics (both at the University of Naples "Federico II", Italy). Between 1998 and 2006 she worked as a research assistant at the Department of Agricultural Economics of "Federico II" in consumer preference analysis, economic valuation of environmental resources, and evaluation of plans and programs for agricultural and environmental resources. She is completing her PhD in environmental economics at the Queen's University of Belfast (UK) with a dissertation entitled *An Integrated Approach to Landscape Economic Valuation*, in which a transdisciplinary methodology is implemented that integrates non-market valuation methodologies, landscape ecology and GIS-based techniques.

C. Martijn van der Heide obtained his PhD degree with a thesis titled *An Economic Analysis of Nature Policy* in 2005. Since 2004, he has worked as a research scholar at LEI Wageningen UR, The Hague (the Netherlands), where his main areas of research are environmental cost–benefit analysis, economics (and valuation) of biodiversity conservation, ecological economics and landscape economics. In June 2010, Martijn was appointed as *Lector* (Reader) in Integrated Nature and Landscape Management at the University of Applied Sciences Van Hall Larenstein, in Velp (the Netherlands).

J. Willemijn Weijschede-van der Straaten holds a PhD that combined the research areas of cost–benefits of open space and the location choice of households. She worked on both topics at the VU University in Amsterdam (the Netherlands) and at the Central Bureau for Economic Policy Analysis (CPB). In 2010 she finished her dissertation and continued working at the CPB. Her research areas include policy analyses of mobility, housing and environment.

Veerle Van Eetvelde holds a PhD in geography (University of Ghent, Belgium) and a master's degree in urbanism and spatial planning. Her research focuses on a holistic landscape approach, landscape analyses and characterization using GIS, landscape ecology and historical geography. Since 2008, she has been a Lecturer in the Department of Geography, within the research unit Landscape Research, University of Ghent. Currently she is Secretary General of the European chapter of the International Association for Landscape Ecology IALE–Europe and Vice President of UNISCAPE (a Network of Universities dedicated to the implementation of the European Landscape Convention) and a member of the Royal Committee for Protection of Monuments and Landscapes in Flanders.

Arjen van Hinsberg works as a Senior Policy Researcher at the Netherlands Environmental Assessment Agency. He holds a PhD in ecology, and his main field of interest is the relationship between environmental conditions and biodiversity, and the way policy actions influence this relationship. He has extensive experience with development and implementation of indicators, on the national and international scale, for use in biodiversity assessments. He is coordinator of the Dutch National Focal Centre on effects of air pollution for the UN–ECE convention on long range transboundary air pollution.

Peter van Kampen obtained his master's degree on the subject of internet-based decision support systems at the University of Groningen (the Netherlands). As a spin-off of the thesis, several ideas were put into practice, recently resulting in his company De Ontwikkelfabriek, where he and his colleagues are dedicated to inventing, creating and implementing IT tools to support academic research. Peter is also an applied research project manager for the Faculty of Economics and Business at the University of Groningen and Chairman and coordinator of all Groningen-based Science Shops.

Sandy van Tol studied environmental sciences at the Hogeschool Utrecht (the Netherlands). At the National Institute for Public Health and the Environment (RIVM), she contributed to emission calculations and estimates for Dutch agriculture. Since 2000, she has worked at the Netherlands Environmental Assessment Agency, where she supports and conducts research on biodiversity. In particular, Sandy is working in and contributing to several projects that deal with environmental and biodiversity assessments.

Frank Witlox holds a PhD in urban planning (Eindhoven University of Technology, the Netherlands), a master's degree in applied economics and a master's degree in maritime sciences (both from the University of Antwerp). Currently, he is Professor of Economic Geography at the Department of Geography of Ghent University. He is also a visiting professor at ITMMA (Institute of Transport and Maritime Management Antwerp), and an Associate Director of GaWC (Globalization and World Cities, Loughborough University). Since 2010, he has been the Director of the Doctoral School of Natural Sciences (Ghent University). His research focuses on transport economics and geography, economic geography, spatial modelling techniques, (city) logistics, world cities and globalization, and land-use change and development.

JunJie Wu holds the Emery N. Castle Chair in Resource and Rural Economics at Oregon State University (USA). He also is a University Fellow at Resources for the Future, Washington DC, and holds a Chang-Jiang Visiting Professorship at Renmin University of China, Beijing. He received his PhD from the University of Connecticut. His research interests include the optimal design of agri-environmental policy and spatial modeling of land-use change and its economic and environmental impacts. He has published more than 70 refereed journal articles and received several awards for research, including the 2002 American Agricultural Economics Association Quality of Research Discovery Award and the 2004 and 2009 Western Agricultural Economics Association Outstanding Published Research Award.

Wenchao Xu received his PhD from the Department of Agricultural and Resource Economics of Oregon State University (USA) in 2010. His fields of interest include applied microeconomics, environmental economics and resource economics. His dissertation explored the interactions between wildfire risk, natural amenities and land development. He is currently working as a post-doctoral researcher on the Idaho NSF EPSCoR "Water Resources in a Changing Climate" project at Boise State University. His current research addresses the impacts of climatic and water supply conditions on agricultural land-use decisions under the institutional governance of water appropriation.

1 Landscape and economics

Perceptions and perspectives

C. Martijn van der Heide and Wim J.M. Heijman

Introduction

Human alterations to the landscape are as old as humanity itself. In fact, landscape – as the word itself says – is land-shaped, reflecting social and economic needs of a particular society at a given moment (Antrop 2006). Over the last 10,000 years, human activities have brought about major changes to the world's landscapes. These changes occur as a result of the expansion of settlements, the creation of fields and pastures for agriculture, the clearing of forests and the draining of marshes and wetlands etc. Some theoreticians claim that any landscape formed and influenced by man is cultural. According to Taylor (2002: 94) 'cultural landscapes can be thought of as any landscape bearing the impact of human activity, historic or prehistoric.'[1] Cultural landscapes are in contrast to natural landscapes, which are spaces not (yet) changed by human hands. By Taylor's definition, most landscapes are cultural, shaped over time but with varying degrees of modification. In Europe, for instance, every landscape can be considered cultural. Rural as well as urban, central, marginal or remote landscapes, but also industrial, rural or urban wastelands are covered by this term – only a few of them do not reflect the immediate impact of human interventions.

Besides the physical interaction with the landscape, mankind 'responds emotionally and aesthetically to it and this response can be seen in the work of artists, musicians, writers and poets' (Garrod and Willis 1999: 239).[2] The concept of landscape as an aesthetic whole that could affect society evolved mainly during the Renaissance and in the baroque gardens of French Absolutism, including the royal Gardens of Versailles (covering 800 hectares of land). Moreover, the English landscape parks of the 18th century laid the foundation of, and prompted the attempt to create urban public parks and the municipal green belts throughout Europe in the 19th century. The motivation behind the creation of these green open spaces, however, was not aesthetic, but mostly pragmatic and related to issues of public health and the rapid increase in urban population.[3] As such, it is important to realize that landscapes, be they natural or man-made, are shaping the world and (cultural) society. Aesthetic appreciation of landscapes can be found in, for example, the symphonic poem *The Moldau*, by the classical Czech composer Smetana, which describes the course of the river Vltava through the

Bohemian landscape.[4] Aesthetics of landscapes are also reflected in landscape paintings (or not, in the case of a lousy or messy painter). As Tress and Tress show (2001), landscape painting emerged as an independent painting genre from the 15th century onwards and began to occupy a preeminent place in art production. During the 17th and 18th centuries, when landscape painting became a popular and widespread style of art, the paintings were collected by a growing middle class who loved them not just aesthetically, but also because they reflected a national identity.[5] Moreover, landscape plays a compelling role in literature (see, for example, the Special Issue on "Landscape through literature" in the *Naturopa* magazine (Council of Europe 2005)).

A fundamental characteristic of all landscapes is that they are never finished. In fact, landscapes are continuously changing and evolving through natural and human induced processes and activities (Antrop 2005). On the other hand, current landscape changes are increasingly seen as a threat, 'because they are characterised by the loss of diversity, coherence and identity of the existing landscapes' (Balej et al. 2010: 108). Antrop (2005) argues that there are too many landscape changes everywhere and that current landscape dynamics, both in speed and scale, are considered to be too fast when compared to traditional land management practices that lasted for centuries. Indeed, against a backdrop of global climate change and dwindling natural resources, landscapes change rapidly due to population growth, changes in lifestyles and the competing demands for land for local and global food and fuel security. These rapid changes and the sometimes chaotic landscape dynamics make people feel the urgency of preserving (traditional) landscape values, or at least of managing and developing landscapes sustainably. Although the concept of sustainable landscapes seems to be a *contradictio in terminis* (Antrop 2006), it is clear that short term changes need to be within the physical limits of the landscape system to provide future generations with economic goods and services.

This brings us to the topic of this book, namely the value of landscapes. Many choices involve values – that is, judgments about what it is good to do or to be. It is perhaps worth insisting at this point that when we refer in this book to the *economic* value of landscapes, we are aware of the fact that economics is not a value in and of itself. Economics is only a way of weighing one value against another. It addresses problems of scarcity, trying to explain how to fulfill people's unlimited needs and aspirations under scarce resource constraints. Without scarcity – for example, the Garden of Eden, where all the scarcity has disappeared – there are no economic problems which force people to make choices among available alternatives (Sowell 2007). So, measuring and quantifying the economic value of changes in landscapes would help ensure that landscape management decisions are both (economically) rational and sound. Of course, there may be an air of unease with linking the welfare-centered motivations of the "narrowly self-interested" *Homo Economicus* with public goods such as landscapes that provides benefits to (local) society as a whole. That is, how to pursue and achieve the public good while also glorifying the values of utilitarianism and individualism? However, this agitation might be based on the perception that economics stands only for anthropocentric, welfarist and egoistic interests – a perception that probably has been exacerbated by the 2008 credit crisis and its concomitant cash crunch, and which tends to be narrowly confined to economics in neo-liberal terms. However, merely saying that landscapes have value or are important is not sufficient – not when resources are scarce and have alternative uses. Moreover, a prominent feature of the upcoming field of landscape economics is that it goes beyond traditional neo-classical orthodoxy as various contributions in this book illustrate. It integrates perspectives and practices from a variety of disciplines and specialized professions, such as ecology, sociology, governance studies, and psychology.

The purpose of this chapter is twofold. First, it briefly provides perceptions and perspectives on the concept of landscape and its economic dimension. The emphasis here is on "briefly" because much of these perceptions and perspectives are discussed in greater detail in the various chapters of this book. The second purpose of this chapter is to describe the objectives of this book and provide an overview of the contents. These two purposes form the basis for the succeeding three sections, which deal respectively with (i) the definition of landscape; (ii) the theoretical background on the economic analysis of landscapes; and (iii) the key objectives and structure of the book.

The term "landscape"

It is not particularly original – but therefore not less true – to point out that there are many interpretations and connotations of the term "landscape" (see, for example, Tress and Tress 2001; Antrop 2006; Daugstad 2008). The disparity in definitions is partly due to the fact that changing landscapes result in changing concepts, but also because the notion of landscape has often been the focal point, rationale, or even unifying entity for geography, archeology, ecology, and spatial planning and design (landscape architecture). Since the various aspects of the landscape are approached through these different (and apparently distant) disciplines, the term "landscape" has become somewhat of an umbrella term that covers and includes many different things. After all, the specific disciplines have developed their own theories and applications related to, in name, one and the same study object, namely landscape. Needless to say the many different interpretations of the term "landscape" make it difficult to communicate clearly, and even more challenging to identify and implement consistent landscape policy measures.

In everyday or colloquial language, "landscape" is a very familiar term. In fact, as *lantschap* the term was already known in the Netherlands in the 13th century. According to Tress and Tress (2001: 144) the Dutch term *lantschap* 'described an administrative entity, a certain area, or was used as a synonym for land and one's native country.' Also in other Germanic languages, landscape was traditionally synonymous with "region" and "territory". The meaning of the term "landscape" changed over time, however. The same authors show that at the end of the 16th century, landscape was coming more and more to have a perceptual meaning. They tell us that this was particularly due to the Dutch landscape painters, because when their paintings reached England, the term "landscape" gradually came to refer to 'a piece of art, a painted scene' (Tress and Tress 2001: 144).

In the transition from the 17th to the 18th century, landscape was introduced into the practice of science by, especially, geographers. It was the German geographer and naturalist Alexander von Humboldt (1769–1859) who defined landscape concisely but holistically as 'the total character of a region' (the quote is from Farina 2000: 12). This definition implies landscape is seen as a complete entity perceived by humans and having a distinct identity. Since then, numerous landscape concepts have abounded in the scientific literature, each of which served the specific purposes of a specific discipline. An anthology of some (more or less) recent definitions is as follows.

- Forman and Godron (1986) consider 'landscape as a heterogeneous land area composed of a cluster of interacting ecosystems that is repeated in similar form throughout' (cited in Farina 1998: 2).
- Stiles (1994: 144), in his search for a unifying "landscape theory", writes that the term "landscape" is 'often used to include outdoor spaces at all scales, from a single private garden through neighborhoods, parks and public gardens, green space networks, the whole urban landscape, rural cultural landscapes through to the environment as a whole in the present, past and future.'
- According to the definition used in the European Landscape Convention adopted by the Council of Europe (2000: Article 1), landscape 'means an area, perceived by people, whose character is the result of the action and interaction of natural and/or human factors.'
- Bell (2001, quoted in Karjalainen 2006: 17) defines landscape from a landscape architectonic perspective as 'a part of environment that we can engage with a given time.'
- Terkenli (2001: 199) explores landscape through the perspective of the discipline of geography and considers the landscape as 'a perceived segment of earth's surface' and as 'a way of seeing, rather than as a scene or image.'
- Burel and Baudry (2003: 43) argue that, from an ecological viewpoint, landscape 'is a level of organization of ecological systems that is higher than the ecosystem level. It is characterized essentially by its heterogeneity and its dynamics, partly governed by human activities. It exists independently of perception.'
- Antrop (2006: 188) sees the landscape as more than a physical entity. 'The concept of landscape encompasses more than an area of land with a certain use or function. I consider landscape as a synthetic and integrating concept that refers both to a material-physical reality, originating from a continuous dynamic interaction between natural processes and human activity, and to the immaterial existential values and symbols of which the landscape is the signifier.'
- Opdam et al. (2006: 323) "take" the landscape as a 'geographical unit characterized by a specific pattern of ecosystem types, formed by interaction of geographical, ecological and human-induced forces.'

- Wylie (2007: 6) points out that 'contemporary English dictionaries commonly define landscape in something like the following terms: "that portion of land or scenery which the eye can view at one." Most then go on to note that the term landscape may refer to a picture or image of the land, as well as the land itself.'

The definition of Forman and Godron considers landscapes from a rather ecocentric perspective. This "organism-centered" definition of a landscape implies that a landscape could range spatially in absolute scale from an area smaller than a single forest to national and beyond (for example, entire ecoregions). Also Burel and Baudry (2003) and Opdam et al. (2006) make an explicit and direct link between landscape (characteristics) and ecological properties. All other aspects, including man's use (tangible or intangible) are of secondary consideration. When so regarded, the deterioration of ecological qualities, say species richness and species composition, automatically means a degradation of the landscape (and conversely: landscape changes can lead to habitat loss and alteration).

Most of the other above-mentioned definitions are anthropocentric in nature, meaning that they refer in one way or another to the utility or satisfaction that individuals gain from a "holistic spatial entity". The definition of the European Landscape Convention, which is based on a comprehensive, integrated understanding of landscape and makes no distinction between natural or cultural landscape, is used by most of the authors contributing to this book as the starting point of their treatise. Indeed, the basic idea behind the economic value of landscapes is anthropocentric; it is to identify (and possibly maximize) benefits that are accruable to man. Because an anthropocentric orientation emphasizes the utilitarian aspects of landscape, it is generally associated with mainstream (neoclassical) economics (Van Kooten 2000).

Having said this, it is essential to be aware of the fact that a geographical area may be visualized by its landscape, but the latter is not the area itself. Besides landscape, an area has more characteristics, e.g. the quality and quantity of the land it comprises and climatic conditions. The land may be seen as the substrate and main determinant of land-use and its resulting landscape. For example, an area with sandy soils may generate a landscape completely different from an area with clay soils. In this approach, landscape represents the visual aspect of the land-use of a geographical area, containing physical components such as trees, grass, houses and roads, but also land type diversity. As such, the visual exploration of landscapes is clearly linked to these components – they are the visible landscape elements – or to put it more prosaically, to the "landscape structure" (the role of landscape structure in visual perception is described more fully in, *inter alia*, Dramstad et al. 2001). But the effect of a landscape is not only visual, however, as smell, sound, touch and even taste play a major role in the appreciation of landscapes. After all, people also perceive landscapes that are noisy and smelly, for example when going through an area with intensive pig farming.

Although this book devotes less attention to urban landscapes, the focus is certainly not on special natural features, landscapes of extraordinary beauty or special ecological areas – quite the contrary.[6] The book deals with various different types of landscape, varying from the rural landscape to recreational green spaces, and from national ecological networks to built-up areas.

Values, externalities and public goods

A vast amount has been written about values, externalities and public goods in order to understand environmental commodities (such as biodiversity and nature) that lie outside the market nexus, and we are not going to try and sum it up here (see, for example, Garrod and Willis 1999; Bateman et al. 2002; Hanley and Barbier 2009). However, when related to landscapes it is wise to briefly touch upon a couple of key issue, namely (i) past research efforts and the current state of landscape in economic analysis; (ii) valuing landscapes; and (iii) the public good dimensions of landscape.

The emergence of landscape in modern economic thought

The disciplines of geography, archeology, ecology, and spatial planning and design color the way in which people view and perceive the landscape around them. Ideas of economics, however, have recently emerged as a new force and have come to exercise a profound influence over the way in which landscape is valued and treated. 'Not only the professed economic system of a society, but the hidden assumptions of economics and the value systems that it enshrines, are central to understanding the modern view of the relationship between humans and the natural world,' writes Ponting (2007: 129). This is also particularly true for landscapes, although it has not always been seen this way.

In Western Europe until the 1970s, landscape itself was not considered to be of particular economic value and because of that it was not studied explicitly by economists. Much that Dasgupta (2008: 1) has said about nature – 'nature did not appear much in twentieth century economics and it doesn't do so in current economic modelling' – holds also for landscapes. As already noted above, the driving force behind changing landscapes in which successive generations and different societies have lived has been the need to feed, clothe and house the increasing population. The landscape of the past could then best be considered an unintended consequence of economic activities such as agriculture, forestry or industry. During that time, landscape used to be an externality (positive or negative) to the "real" economy (Heijman 2007). In other words, landscapes were created as "incidental" co-benefits (or co-disbenefits) to activities in economic sectors or product markets. Of course there are exceptions to this general picture. "Designed" and "deliberate" landscapes of private parks and gardens have always been an important element in the life of the wealthier part of society. But generally speaking, "landscape has no owner" (*a paisagem não tem dono*) as a mural in huge letters on a Tagus riverside wall in Lisbon says. The interest for "public" landscapes fits into the pattern of a growing interest for the quality of life in general, which is generated by growing incomes and the change in needs connected to it.

One could say that as far as the economist's interest in landscape is concerned, the history of the interest for the environment at large is repeated. Starting in the 1960s, economists have become more and more aware of the damage done to the environment by economic activities. They also realized that the benefits of these activities should be weighed against their environmental costs. Therefore, to a high degree, environmental economics is concerned with the measurement of environmental damage in monetary terms, which allows public decision-making taking into account the environmental costs (Boardman et al. 2006). As landscape values also clearly need to be taken into account in decision making, landscape economists can support debates on how to pursue landscape policy goals amid competing claims. As we will elaborate upon later, monetary valuations are easy to communicate to stakeholders and the wider public and can feed relatively smoothly into policy formulation. Valuation does enable making a case to stakeholders and decision makers of the size of trade-offs between landscape attributes that are likely to be involved. That is, policy makers can make better choices if the valuation issue is made as explicit as possible. However, not all the values lend themselves easily to quantification; that is, there are still several values provided by landscapes upon which it is difficult to place a financial figure, particularly the more aesthetic values. We will come to that in the following sub-section.

Valuing landscapes

The field of landscape economics can be said to have started with the classic work of Colin Price in 1978. In the preface of his seminal book, simply entitled *Landscape Economics*, Price writes (1978: xi): 'As for economists, the strong subjective component of landscape value has dissuaded them from applying their science in the field.' The book focuses especially on landscape values and valuation techniques and tries to identify common grounds between aesthetics and economics. As such, if Price writes about the value of landscape, he goes beyond the instrumental value of its spatial (and mostly ecological) components, such as trees, hedgerows or ponds. The specific value of a landscape is not restricted to this instrumental value, but also contains aesthetic value. We mentioned earlier that the experience of landscapes – how they are looked at and interpreted – and inspiration drawn from them has for centuries been the subject of countless works of (painted) art, music and literature. Here, landscapes show their intrinsically aesthetic and emotional value. For landscapes, this category of value is much more important than for, say, nature or biodiversity. However, many economists have a problem with the concept of aesthetic value because it is regarded as being non-instrumental. That is, the aesthetic value of a landscape is an end in itself, not a means to some other end (Brouwer and Van der Heide 2009). This value refers to the problem of "incommensurability", which relates to 'an intelligible choice between feasible options, where there is no appropriate value in terms of which the options might be compared as "better", "worse", or even approximately equal' (Holland 2002: 23). Landscape, therefore, creates special challenges for economic theory. It has characteristics that do not fit well with the standard presumptions of the neo-classical model. The aesthetic phenomenon and the holistic perception of landscapes have the potential of creating inconsistencies in an economic model of instrumental rationality that is based on independence and easily demarcated objects.

The valuation of landscapes and the damage done to them is an important element in landscape economics. All kinds of methods that are known from environmental economics can also be applied in the realm of landscape economics. The practice of monetary valuation of the environment – that gained considerable momentum in the late 1980s, but started decades earlier with the travel cost method to measure the positive welfare effects from the provision of national parks and other recreational facilities – is older than the scholarship in landscape economics. However, it is good to realize that when compared to natural assets, such as biodiversity, landscape is arguably characterized by more intangible outcomes (as opposed to material) in terms of capturable returns related to conservation and sustainable use success and also lacks a single standardised metric. In the words of Garrod and Willis:

> Landscape is perhaps the most complex environmental good to value: it can be made up of an infinite number of configurations in terms of scale, shape, relief, vegetation cover, colour, and man-made features, all of which interact with each other in distinctive permutations. Landscapes also have seasonal effects which make them more attractive in some seasons than in others (e.g. New England in the fall).
>
> (Garrod and Willis 1999: 259)

This makes landscape valuation a subjective exercise as the value to an individual can include benefits that are quite intangible to measure (for a detailed treatise on subjectivity and objectivity in landscape economics, see Chapter 3 of this book). It is the absence of this metric and a price signal of scarcity that presents perhaps the biggest challenge to landscape economics.

Measuring the loss of *economic* value of landscape – which links landscape with wellbeing – provides a compelling motive for investment in its conservation, sustainable use or improvements to ensure the landscape reflects a favourable balance between collective preferences, although this statement is ridden with controversy among some (economic) practitioners (see, for example, Bromley and Paavola 2002). Nevertheless, landscape economics involves the careful measuring and balancing of costs and benefits of particular (spatial) activities so that actors (be they economic, political or otherwise) get a clear signal when to stop expanding such an activity.[7] After all, one person's willingness or unwillingness to support landscape conservation – his or her choice of landscape preferences – will directly influence what others can "consume" or perceive. As such, the value of landscape is anchored in a human perspective – it subscribes to an anthropocentric value orientation – and can be interpreted as the result of an interaction between who attaches value, humans and the exact object of valuation.

But first, of course, we have to agree on what landscape is and what should be conserved or developed.

Various researchers have estimated the value of landscape (attributes). McVittie et al. (2009), for example, present summary values of European landscape valuation studies in terms of different landscape features attributable to agriculture. The average willingness-to-pay estimates vary widely depending on the landscape attributes that were valued (such as dry stone walls, hedgerows or heather moorland). Madureira et al. (2007) provide a comparable overview, but only for case-studies from France and Portugal. Garrod and Willis (1999) review several studies which have attempted to value the landscape of the Yorkshire Dales in Northern England. Various contributions to this book add to and enrich the existing literature on landscape values.

Albeit landscape valuation methods are now highly mature, not every researcher declares his or her faith in economic landscape valuation – especially when expressed in monetary terms (see also Chapter 8 of this book). For example, economist Heal (2000: 29) emphasizes that 'valuation is neither necessary nor sufficient for conservation. We conserve much that we do not value, and do not conserve much that we value.' And although Heal's statement applies to valuing ecosystem services, questioning the supposed pragmatism of valuation studies while standing clear from ideological statements is also highly relevant for the analysis and protection of landscapes.

Moreover, landscape valuation is not sufficient in itself. The question is whether the measurement of (monetary) landscape values can be regarded as a prerequisite for better landscape management decisions. Generally speaking, there are three major contributions that landscape economists can make to improve landscape decision making. First, landscape valuation studies can be used for awareness-raising and advocacy: by demonstrating that the economic value of landscape has previously been underestimated, it can be argued that the landscape considerations should be integrated into public policy. Putting landscape changes in monetary terms is a pretty effective way to spread the message. After all, there are few people who would deny that money speaks louder than words.

Second, landscape valuation studies can assist policy makers in making fair and transparent decisions. Assessing the value of landscape can make a worthwhile contribution as it helps policy makers to make informed decisions about spatial planning and development where there are alternative ways of allocating scarce financial and other resources. By placing a value on landscape change, it can be weighed up against other (monetized) costs and benefits to assess the relative net benefit of the project that causes the landscape change.[8]

The third contribution comprises the fact that landscape economics involves an important design element. Landscape economists could become important partners for landscape architects in designing rural and urban landscapes. Especially by way of choice experiments the usual top-down way of landscape planning could be replaced by a bottom-up approach, in which every stakeholder could have his say in the design of the "optimum landscape", which is based upon the maximum of the available local information (Van der Heide et al. 2008).[9]

The use of available local information is generally lacking in present-day central planning practices.

The public good character of landscape and its amenities

What kind of a good is "landscape"? This is an important question, now the economic value of landscapes has been recognized. Generally speaking, the landscape itself is non-excludable and non-rival in nature, which makes it a public good. Nobody "owns" the landscape. However, landscape *elements*, such as hedgerows, trees, grass strips, ditches, or walls, may be publicly, commonly, or privately owned. As far as they are publicly owned, the government may design the landscape in such a way that is best for society. This is not the case if the landscape elements are commonly or privately owned. The preservation of, say, a common forest may require central regulation not only from a resource point of view but also from the viewpoint of landscape. From a theoretical perspective, it can be shown that if the landscape elements are a common pool resource – it is difficult or costly to exclude or limit users from using them, while one person's consumption reduces resource availability for others – two solutions are possible: 'either socialism or the privatism of free enterprise' (Hardin 1998: 682). External authorities are then presumably needed to impose rules and regulations on users, or to create private property rights and to allocate them to key users and beneficiaries (Ostrom 1999).

Landscape-elements can also be privately owned. In the countryside, for example, farmers can be regarded as important stewards or managers of the landscape. These landowners are needed in conserving and maintaining the landscape. However, it can be wondered whether unregulated private ownership will provide an "efficient" use of the landscape. After all, because the benefits from landscape are not valued in market exchanges, private landowners tend to undersupply them. A farmer reaps the benefits of his harvest but has no means of capturing the benefits his arable land may provide in aesthetic terms and to society at large. His relationship with landscape is primarily functional. Without a mechanism to capture the non-market values of landscape, the farmer will under provide non-market based assets. In these cases, restrictions on land-use changes, or internalisations of externalities (that is, an external "bad" that is imposed on the landscape by the farmer) through taxes and subsidies may be suitable approaches to achieve a socially optimum of provision of the landscape. Therefore, it is necessary that, apart from central regulations, farmers will be paid for the service they provide to maintain the landscape. The EU modulation policy will hopefully prove to be an effective and efficient tool in matching the private objectives of landowners with the public aim of preserving the landscape.

Due to the public good character of landscape, the effective governance of landscape management is complex and challenging. Complexity is reinforced by the notion that current public policy rarely has effects on landscape in its entirety, but rather at the scale of landscape attributes or elements. Landscape management is more cost-efficient when it is integrated across sectors, which strengthens ownership for landscape policies. The public good character of landscape is reflected implicitly or explicitly in many of the contributions to this book. Chapters 5 and 6 address specific governance issues associated with landscapes.

Key objectives and the structure of the book

Within existing economic fields – such as ecological economics and the more traditional environmental (and resource) economics – some scholarly attention has been given to the economic aspects of landscapes. This attention, however, appears to be fragmented and lacks an overarching relational framework, or even a consistent and systematic analysis. Moreover, the discipline of ecological economics often regards the landscape as an ecosystem, or as an aggregation of ecosystems, thereby throwing the concept of landscape back into the realm of landscape ecology. This book aims at exploring the avenue of landscape economics and provides the building blocks (from different scientific disciplines) for an economic analysis of landscapes. It focuses on the value of landscapes in its broadest sense, thereby covering a variety of topics including stakeholder involvement in landscape design, landscape governance and landscape perceptions from different countries.

The book consists of four parts, each part is addressed individually in the succeeding four sub-sections. In each sub-section, we turn to highlight certain of the chapters in that specific part.

Part I: Rise to the challenge – an awakening to awareness

The first part of the book includes five chapters. In Chapter 2, "Setting a framework for valuing the multifunctional landscape and its multiple perceptions", Marc Antrop, Marjanne Sevenant, Carolina Tagliafierro, Veerle Van Eetvelde and Frank Witlox describe how landscape has become a theme in spatial and environmental planning policy and in nature and heritage management, and how specific applied research resulted from this. Landscape became an integrating concept in transdisciplinary projects and landscape research, encompassing a growing set of disciplines: geographers, historians, planners, archaeologists, landscape ecologists and landscape architects. According to the authors, the European Landscape Convention has stimulated landscape characterization and research in landscape perception and preference. This increases the need for robust and reliable indicators that allow assessing landscape changes and measuring the effects of these changes on the value/valuing of landscapes, which stimulated the research based on landscape metrics and the need for monitoring. Many new concepts have recently been introduced, such as multifunctional landscapes, landscape and ecosystem services, landscape as a resource and sustainable landscapes; others shifted in meaning and importance, such as scale and landscape dynamics. Specific approaches and methods from different disciplines inspired one another mutually. Economic landscape valuation and landscape economics are the latest rising trend. Antrop and his co-authors discuss how different concepts are interpreted and applied in economic valuation of landscapes and how methods can be transformed to that end.

Chapter 3, by Colin Price, is entitled "Subjectivity and objectivity in landscape evaluation: an old topic revisited". Price critically examines landscape appraisals. Controversies over the economic valuation of landscape have some roots in an earlier debate about landscape assessment, in which subjective and holistic evaluation contended with objective and components-based appraisal. Subjectivity can occur in many senses, and it is important that subjective elements are identified within what purport to be objective appraisals. Economists have in recent times attempted to find holistic values through contingent valuation, and by statistically relating purchase decisions to objectively measurable components of landscape. To do so in fact requires many covert subjective judgments, and yet results in values which may not be practically useful. It can be argued that what is needed is not replacement of subjective judgment, but an enrichment and expansion of this judgment through protocols that make it more representative, more quantified, and more explicitly comparable with other values.

In Chapter 4, titled "From private values to social value of landscapes", Dafna DiSegni argues that economic literature has devoted much attention to individual valuation of landscapes, providing direct goods and services, consumed over time and space (including benefits from agricultural landscape, recreational activities in forested area, open spaces, beaches, etc.). However, the social value of landscape is determined not only by the aggregate of individuals' benefits from the landscape services, but should also cover the provision of additional public goods and ecosystem services (such as carbon sequestration), not directly valued by the individuals. The objective of this chapter is to position the individual benefits within the larger framework of social benefits from a given landscape. DiSegni concludes that valuation should clearly distinguish between private goods, public goods, and unobserved goods, where the latter often provides "supporting ecosystem services". At the stage of social value computation, it is essential to consider the integration of these three categories of goods and services with the process of aggregation: (i) the benefits of goods revealed through market signals; (ii) goods that have no market signals, but are known to possess an individual value; and (iii) goods which are not considered by individuals because of an unawareness of the existence of the goods/services or due to strategic ignorance of their existence.

Ian Hodge continues with Chapter 5: "The governance of rural landscape: property, complexity and policy". In his view, landscapes represent major sources of value that are primarily privately owned but with values that are predominantly in the public realm. Governance thus involves a mix of private property and public policy, operated through a variety of institutions. At the same time, landscapes, especially with regard to their aspect as a major element of an ecological system, represent a complex good, one that is characterized by many elements that interact in ill-defined and often unpredictable ways. Policy thus needs to represent the public interests in landscape management decisions, but in a context where there are strong elements of uncertainty and ignorance.

The final chapter in this part, Chapter 6 by Nora Mehnen, Ingo Mose and Dirk Strijker, is entitled “Governance of protected landscapes and its implications for economic evaluation”. The authors state that the past years have shown increasingly rapid advances in the field of nature policy, especially into the direction of bottom-up decision-making and non-hierarchical structures. According to them, a paradigmatic shift occurs from a static-preservation to a dynamic-development oriented approach. Related to this, the classical structure of governmental decision-making in nature conservation is top-down. The modern bottom-up processes are designed to resolve conflicts through cooperation. These developments have an impact on economic evaluation. Strict rules and strict objectives defined by the state are easier to evaluate than fuzzy rules from bottom-up. The more actors are involved, the more different and maybe conflicting interests are involved, and the more complex the economic evaluation becomes. This chapter focuses on the economic evaluation of protected landscapes, which are characterized by a combination of different functions and actors. That is, these landscapes are typically based on the interactions of people and nature over time, and this chapter sheds some new light on how the increasing complexity of these interactions requires a multifaceted and nuanced economic evaluation approach.

Part II: Reorientations and reflections – building blocks of landscape economics

This part comprises three chapters. Chapter 7, by Wim Heijman and Pierre v. Mouche, is entitled “A procedure for determining an optimal landscape and its monetary value”. The authors present a possible way to carry out a cost–benefit analysis of landscapes on the basis of stated consumer preferences of landscape characteristics. The idea implies that the net benefits of landscapes are computed through a two-step valuation procedure. This chapter is the briefest of the book and entirely mathematical in nature.

In Chapter 8, entitled “Evaluation of landscape impacts – enriching the economist’s toolbox with the HotSpotIndex”, Frans Sijtsma, Hans Farjon, Sandy van Tol, Peter van Kampen, Arjen Buijs and Arjen van Hinsberg examine the evaluation of landscape impacts from the perspective of the scenic beauty discourse. They state that integrated land development projects combining various types of land-use interventions pose serious challenges to Cost–Benefit Analysis (CBA). That is, Sijtsma and his co-authors describe the unexpected main weaknesses in the actual use of theoretically available CBA tools. Moreover, they briefly review the tools that landscape researchers have to offer to the CBA analyst in order to evaluate landscape change. These methods of non-economist landscape researchers certainly have strengths which economists’ tools lack, but they also have their own drawbacks when applied within CBA. To bypass these problems, the authors introduce the so-called Hotspotmonitor: a new tool used to gather landscape preferences in a standardized, systematic and spatially-precise way. They examine its ability to assess landscape impacts (within a CBA) in the upgrade of the Dutch city of Almere.

The final chapter in Part II, Chapter 9 by Iddo Kan, David Haim, Mickey Rapaport-Rom and Mordechai Shechter, is entitled: "Rural landscape and optimal agricultural land-use". This chapter theoretically and empirically analyzes the effectiveness of a regional-scale agricultural policy to encourage changes in land allocation among crops, as a mechanism to increase social welfare through substitution of production profits with landscape amenity values. The empirical analysis is based on a positive-mathematical-programming model, which enables smooth variations of land allocation among crops. The model is applied to the heavily populated northern part of Israel, using a quadratic landscape amenity-value function that discriminates among the landscape contribution of three groups of crops. As the amenity services provided by the agricultural area exhibit a decreasing return to scale, the efficiency of the policy is found to be dependent on the size of the region in which it is implemented.

Part III: Worldwide applications and detailed case studies – integration of practices

Part III consists of three chapters. In Chapter 10, "An economic evaluation of the grassland landscape in Aso Kuju National Park, Japan", Kenji Okubo analyzes the economic value of landscape in Aso-Kuju National Park. Over 18 million people visit the Aso area in Japan each year. The Aso area has the largest caldera (a volcanic feature usually formed by the collapse of land following a volcanic eruption) in the world. Also, there is a magnificent view of a central cone and somma that are covered by grassland. In addition, these landscapes with grassland have been maintained by livestock farmers – the Aso area is one of the largest livestock farming areas in Japan – who put cattle to grass with right of common. With the help of correspondence analysis and the travel cost method, the author aims at determining visitors' purposes in visiting the area and the economic value of the landscape.

Chapter 11, by María D. Domínguez García and David Soto Fernández is entitled "From an 'integrated' to a 'dismantled' landscape". According to these two authors, landscape is subject to continuous transformation: visually, socio-economically, ecologically and culturally. They approach changing landscapes by focussing on changes in the so-called "social metabolism". To describe and illustrate this approach, García and Fernández focus on the transformation of a traditional, integrated landscape to a dismantled one in the region of Galicia in the Northwest of Spain. The transformation implies a change from a sustainable landscape in which the socio-economic, ecological, cultural and symbolic realms are strongly related and interconnected to another less sustainable one in which part or all of that interconnection has been lost.

In Chapter 12, entitled "What can hedonic analysis tell us about the value of landscapes?", Jan Rouwendal and J. Willemijn Weijschede-van der Straaten state that policy evaluation of preserving landscape implies the comparison of costs and benefits of various scenarios. In many cases, some costs or benefits refer to non-market goods, which implies that the required willingness-to-pay measures cannot be obtained from prices. Hedonic analysis is probably the best known technique used by economists to deal with this issue. Its main advantages are that there is a good theoretical basis, and the valuations are market-based, implying that revealed (instead of stated) preferences can be used. Moreover, the popularity of the technique ensures that there is a frame of reference against which new studies can be placed. This motivates the question in the title of this chapter: could this technique that has been fruitfully applied in many other areas also be used in landscape policy? Their answer is mainly positive. They develop a formal model for protecting a valuable landscape, which results in a cost–benefit rule that can be made operational through hedonic price analysis. To illustrate this, the model is applied to the optimal provision of open space in cities.

Part III of the book is concluded by Chapter 13, written by Jasper Dekkers and Eric Koomen. Their chapter is entitled "The monetary value of open space in urban areas: evidence from a Dutch house price analysis". It attempts to determine a value for the non-built-up landscape, generally referred to as open space, in urban areas in a systematic and integrated way and by using an econometric approach. The authors attach a monetary value to open space using two separate hedonic house price analyses at the local housing market level in the Randstad region, the strongly urbanized western part of the Netherlands. From this analysis, it appears that the impact of open space on property price depends on the type of open space and other local conditions.

Part IV: Outlook for landscape economists – burgeoning perspectives on recreation, agriculture and urban agglomeration

Part IV contains three chapters. In Chapter 14, entitled "The importance of landscapes for recreational firms", Nico Polman, Arianne de Blaeij, Louis Slangen and Stijn Reinhard analyze economic benefits of the National Ecological Network (NEN) for recreational firms. In the Netherlands, the NEN is a network of existing nature areas, nature development areas, connecting zones, and agricultural areas with potential for agri-environmental management. The objective of the NEN is to protect and enhance nature areas and landscape structures. In this chapter the authors analyze whether firm size, represented by the number of workers, depends on the landscape composition of the NEN zones. To estimate the relevance of the NEN for recreational firms, a landscape index is developed for individual firms, including the distance of the firm to and the size of the NEN in the firm's surroundings. This "Landscape Reilly-index" was computed for about 29,000 recreational firms. The results indicate that recreational firms in the surroundings of the NEN employ relatively more workers than other firms. To be more specific, the empirical analysis shows that recreational firms located in the neighborhood of marshes, forests, coastal areas or grasslands have more workers than those that do not have such types of nature in their proximity. However, the effect is small. The size of the effect depends on the size of the NEN and the distance to the recreational firm, and on the specific type of nature. A smaller size and/or a larger distance has a decreasing effect on employment by recreational firms.

Chapter 15, by Sylvie Ferrari, Christian Lippert and Olivier Aznar, is entitled "Agricultural policies and rural landscape: Some insights from theoretical and empirical literature". The authors determine the conditions that ensure the provision of environmental services by rural landscapes that have been shaped in part by agricultural activities. Moreover, they analyze the economic values associated with these services and thereby account for the existence of specific social demand. Within this context, the conditions of public intervention capable of ensuring rural landscapes' provision of environmental services – and of deriving value from their non-market functions – is studied both theoretically and practically. Ferrari and her co-authors conclude that until now, public policy based analyses of rural landscapes have taken place within a static framework. However, research into the dynamics underlying rural landscapes is a crucial stage in the development of long-term policies aimed at landscape preservation and a regulated provision of environmental services. A long-term management of rural landscapes cannot be organized without accounting for the interactions between agricultural and natural systems, no matter which spatial scale is considered.

The book is concluded by Chapter 16: "Amenity-driven migration and the spatial distribution of economic activity", written by Wenchao Xu and JunJie Wu. They state that the past three decades have seen an increasing demand for (locational) amenities. With rising income and decreasing transportation costs, forests, lakes and other forms of open space are more valued as sources of amenities that support quality of life than solely as production factors. This chapter examines the effects of amenities (or locational amenities – the desirable features of a place, location-specific in nature) on interregional migration and development patterns. The authors assume households maximize their utility by choosing their consumption bundles and residential location subject to a budget constraint. Monopolistic competitive firms maximize their profit by employing labor and supplying products. The model takes account of the effect of agglomeration on interregional migration. The results indicate that with intermediate transport cost, residents tend to move to large cities to take advantage of job opportunities, lower living cost and varieties of goods, which in turn makes a metropolitan area even larger. Amenities tend to exacerbate the agglomeration effects by attracting more residents.

Notes

1 Because landscape is by its very definition "land shaped", it is, to be precise, a pleonasm to speak of a cultural landscape.

2 There are two paradigms of landscape aesthetics, as Lothian (1999) shows: the objectivist and the subjectivist. The objectivist, or physical, paradigm presupposes that the quality of the landscape is an intrinsic, physical attribute that can be assessed by applying criteria. The subjectivist, or psychological, paradigm considers landscape aesthetics as solely a human construct and as a product of the mind – the eye of the beholder.

3 Interest in landscape (and nature) in the 19th century basically boils down to two points (Balej et al. 2010: 110): first, 'there is a romantic zeal for untouched wilderness, in which the countryside is a place to escape society' and second, 'townspeople show a growing interest in the countryside as a place for relaxation after the daily work activities.'

4 Other comparable examples are the 20th-century English composer Vaughan Williams who more or less romanticized the British landscape and his contemporary colleague Delius who has been characterized as a "composer of landscapes". For much of the time they found their creative inspiration from the landscapes that they knew and loved.
5 The reciprocity between landscape and painting is beautifully described and illustrated by Wylie (2007), who starts his book on the cultural geographical aspects of landscape with Paul Cezanne's paintings of Mont Saint-Victoire (Provence, southern France).
6 One of the greatest changes in the experience of landscapes has been caused by the rise of the city and urban living in the last two centuries. Cities are the most artificial landscape produced by humans. For a thorough treatment of urban landscape dynamics and their impact on ecological systems, see Czamanski et al. (2008).
7 The branch of microeconomics is characterized by such a "when to stop" rule, namely when the marginal cost equals the marginal benefit. This is in sharp contrast to conventional macroeconomics, the study of the economy as a whole, which has no analogous "when to stop" rule (Daly 2007).
8 Related to this second contribution is the income generating aspect of a landscape. The landscape may be the basis of important private regional industries, such as rural tourism, direct sales from farms to the public, care farms et cetera. Also income and jobs generated by public services such as maintaining and the keeping of the landscape and nature conservation could contribute to regional development. An important role for the landscape economist is therefore developing and applying the tools for evaluating the economic effects of landscape policies.
9 Economists often portray themselves as specialists in optimization. The optimal amount of landscape is defined as the point at which marginal landscape costs equal marginal costs of maintaining landscape qualities. Although there are scientists who tone down the emphasis on optimality (see, for example, Vatn 2002 and Pannell 2006), we fully recognize that the "optimum" landscape will certainly depend on more than money. In general, economics recognizes the importance of different preferences – whether they are material or spiritual, aesthetic or instrumental – and even moral choices. Optimization can be a multidimensional and capacious concept.

References

Antrop, M. (2005) 'Why landscapes of the past are important for the future', *Landscape and Urban Planning*, 70(1–2): 21–34.

Antrop, M. (2006) 'Sustainable landscapes: Contradiction, fiction or utopia?' *Landscape and Urban Planning*, 75(3–4): 187–197.

Balej, M., Raška, P., Anděl, J. and Chvátalová, A. (2010) 'Memory of a landscape – a constituent of regional identity and planning?' In: J. Anděl, I. Bičík, P. Dostál, Z. Lipský and S.G. Shahneshin (eds), *Landscape Modelling; Geographical Space, Transformation and Future Scenarios*, pp. 107–121, Dordrecht: Springer.

Bateman, I.J., Carson, R.T., Day, B., Hanemann, M., Hanley, N., Hett, T., Jones-Lee, M., Loomes, G., Mourato, S., Özdemiroğlu, E., Pearce, D.W., Sugden, R. and Swanson, J. (2002) *Economic Valuation with Stated Preference Techniques: A Manual*. Cheltenham, UK, and Northampton, MA: Edward Elgar.

Boardman, A.E., Greenberg, D.H., Vining, A.R. and Weimer, D.L. (2006) *Cost–Benefit Analysis: Concepts and Practice*. Upper Saddle River, NJ: Pearson-Prentice Hall.

Bromley, D.W. and Paavola, J. (eds) (2002) *Economics, Ethics, and Environmental Policy: Contested Choices*. Oxford: Blackwell Publishers Ltd.

Brouwer, F. and Heide, C.M. van der (2009) 'Conclusions and prospects'. In: F. Brouwer and C.M. van der Heide (eds), *Multifunctional Rural Land Management – Economics and Policies*, pp. 335–347, London: Earthscan.

Burel, F. and Baudry, J. (2003) *Landscape Ecology: Concepts, Methods and Applications*. Enfield (NH), USA: Science Publishers, Inc.

Council of Europe (2000) *European Landscape Convention*. Florence, 20.X.2000. http://conventions.coe.int/Treaty/en/Treaties/Html/176.htm.

Council of Europe (2005) 'Special Issue on "Landscape through literature"', *Naturopa*, 103: 1–97.

Czamanski, D., Benenson, I., Malkinson, D., Marinov, M., Roth, R. and Wittenberg, L. (2008) 'Urban sprawl and ecosystems – Can nature survive?' *International Review of Environmental and Resource Economics*, 2: 321–366.

Daly, H.E. (2007) *Ecological Economics and Sustainable Development; Selected Essays of Herman Daly*. Cheltenham, UK, and Northampton, MA: Edward Elgar.

Dasgupta, P. (2008) 'Nature in economics', *Environmental and Resource Economics*, 39(1): 1–7.

Daugstad, K. (2008) 'Negotiating landscape in rural tourism', *Annals of Tourism Research*, 35(2): 402–426.

Dramstad, W.E., Fry, G., Fjellstad, W.J., Skar, B., Helliksen, W., Sollund, N.-L.B., Tveit, M.S., Geelmuyden, A.K. and Framstad, E. (2001) 'Integrating landscape-based values – Norwegian monitoring of agricultural landscapes', *Landscape and Urban Planning*, 57(3–4): 257–268.

Farina, A. (1998) *Principles and Methods in Landscape Ecology*. Dordrecht: Kluwer Academic Publishers.

Farina, A. (2000) *Landscape Ecology in Action*. Dordrecht: Kluwer Academic Publishers.

Garrod, G. and Willis, K.G. (1999) *Economic Valuation of the Environment: Methods and Case Studies*, Cheltenham, UK, and Northampton, MA: Edward Elgar.

Hanley, N. and Barbier, E.B. (2009) *Pricing Nature: Cost–Benefit Analysis and Environmental Policy*. Cheltenham, UK, and Northampton, MA: Edward Elgar.

Hardin, G. (1998) 'Extensions of "the tragedy of the commons"', *Science*, 280: 682–683.

Heal, G. (2000) 'Valuing ecosystem services', *Ecosystems*, 3(1): 24–30.

Heide, C.M. van der, Blaeij, A.T. de and Heijman, W.J.M. (2008) *Economic Aspects in Landscape Decision-making: A Participatory Planning Tool based on a Representative Approach*. Wageningen: Mansholt Working Papers, no. 41.

Heijman, W.J.M. (2007) *Regional Externalities*. Berlin: Springer.

Holland, A. (2002) 'Are choices tradeoffs?' In: D.W. Bromley and J. Paavola (eds), *Economics, Ethics, and Environmental Policy: Contested Choices*. pp. 17–34, Oxford: Blackwell Publishers Ltd.

Karjalainen, E. (2006) *The Visual Preferences for Forest Regeneration and Field Afforestation – Four Case Studies in Finland*. The Finnish Society of Forest Science: Academic Dissertation, Dissertationes Forestales 31.

Kooten, G.C. van (2000) 'Biodiversity and ethics: religion, science and economics'. In: G.C. van Kooten, E.H. Bulte and A.R.E. Sinclair (eds), *Conserving Nature's Diversity; Insights from Biology, Ethics and Economics*. pp. 143–159, Aldershot: Ashgate.

Lothian, A. (1999) 'Landscape and the philosophy of aesthetics: Is landscape quality inherent in the landscape or in the eye of the beholder?' *Landscape and Urban Planning*, 44(4): 177–198.

Madureira, L., Rambonilaza, T. and Karpinski, I. (2007) 'Review of methods and evidence for economic valuation of agricultural non-commodity outputs and suggestions to facilitate its application to broader decisional contexts', *Agriculture, Ecosystems and Environment*, 120(1): 5–20.

McVittie, A., Moran, D. and Thompson, S. (2009) *A Review of Literature on the Value of Public Goods from Agriculture and the Production Impacts of the Single Farm Payment Scheme*. SAC: Land Economy and Environment Research Group, Report Prepared for the Scottish Government's Rural and Environment Research and Analysis Directorate (RERAD/004/09).

Opdam, P., Steingröver, E. and Rooij, S. van (2006) 'Ecological networks: A spatial concept for multi-actor planning of sustainable landscapes', *Landscape and Urban Planning*, 75(3–4): 322–332.

Ostrom, E. (1999) 'Coping with tragedies of the commons', *Annual Review of Political Science*, 2: 493–535.

Pannell, D.J. (2006) 'Flat earth economics: The far-reaching consequences of flat payoff functions in economic decision making', *Review of Agricultural Economics*, 28(4): 553–566.

Ponting, C. (2007) *A New Green History of the World; The Environment and the Collapse of Great Civilizations*. London: Penguin Books.

Price, C. (1978) *Landscape Economics*. London: Macmillan.

Sowell, T. (2007) *Basic Economics: A Common Sense Guide to the Economy*, Third Edition. New York: Basic Books.

Stiles, R. (1994) 'Landscape theory: A missing link between landscape planning and landscape design?' *Landscape and Urban Planning*, 30(3): 139–149.

Taylor, P.D. (2002) 'Fragmentation and cultural landscapes: Tightening the relationship between human beings and the environment', *Landscape and Urban Planning*, 58(2–4): 93–99.

Terkenli, T.S. (2001) 'Towards a theory of the landscape: The Aegean landscape as a cultural image', *Landscape and Urban Planning*, 57(3–4): 197–208.

Tress, B. and Tress, G. (2001) 'Capitalising on multiplicity: A transdisciplinary systems approach to landscape research', *Landscape and Urban Planning*, 57(3–4): 143–157.

Vatn, A. (2002) 'Efficient or fair: Ethical paradoxes in environmental policy'. In: D.W. Bromley and J. Paavola (eds), *Economics, Ethics, and Environmental Policy: Contested Choices*. pp. 148–163, Oxford: Blackwell Publishers Ltd.

Wylie, J. (2007) *Landscape*. Abingdon, Oxon: Routledge.

Part I

Rise to the challenge – an awakening to awareness

2 Setting a framework for valuing the multifunctional landscape and its multiple perceptions

Marc Antrop, Marjanne Sevenant, Carolina Tagliafierro, Veerle Van Eetvelde and Frank Witlox

Introduction

In this chapter, we propose a framework for the economic valuation of multifunctional landscapes with their inherent multiple perceptions and preferences. It is an attempt to integrate the diverse approaches in landscape science with new emerging trends of economic valuation of landscape. The first sections of this chapter review basic and new concepts and paradigms in landscape research which are also significant in landscape planning and policy. The European Landscape Convention (Council of Europe 2000) is used as a framework for this discussion. Many disciplines are involved and there are multiple approaches to landscape. Consequently, landscape research and its applications are very diverse. Paradigms, definitions, concepts and methods need to be compared and linked to achieve the necessary integration. This demands an inter- and transdisciplinary approach. From this review, possible bridges to the economic valuation of landscape are presented, and illustrated with an example in the last section.

The meaning of landscape since the European landscape convention

Traditional cultural landscapes have become lost and disturbed, and the speed and magnitude of the ongoing changes is still growing. Since the 1990s, this has caused a growing popularity of landscape in policy debates and scientific conferences. The conclusion of Chapter 8 in the First Assessment of the European Environment by the EEA, the Dobříš Assessment, which was devoted to landscape, sums up the importance of landscape:

> The richness and diversity of rural landscapes in Europe is a distinctive feature of the continent. There is probably nowhere else where the signs of human interaction with nature in landscape are so varied, contrasting and localised.
>
> Despite the immense scale of socio-economic changes that have accompanied this century's wave of industrialisation and urbanisation in many parts

> of Europe, much of this diversity remains, giving distinctive character to countries, regions and local areas.
>
> (http://www.eea.europa.eu/publications/92-826-5409-5/page008new.html)

This report was a signal for the Council of Europe to take action as well, resulting in the European Landscape Convention (ELC) (Council of Europe 2000). This Convention initiated more research and action programs related to the landscape in most European countries than ever before. This is remarkable as the Convention has no legal basis, such as an EU-directive, and no financial means are provided. The Convention also introduces a series of formal definitions related to the landscape, as well as a series of recommendations, which give a common and international basis for action. Article 1 of the ELC defines landscape as: 'an area, as perceived by people, whose character is the result of the action and interaction of natural and/or human factors.' (Council of Europe 2000: Art. 1). This definition has not only reached consensus between the ministers of the member countries of the Council of Europe, but is also supported by positive recommendations of the committees on diversity and landscape (CO-DBP) and cultural heritage (CC-PAT).

The general measures proposed by the European Landscape Convention include the recognition of landscapes in law as an essential component of people's surroundings, as an expression of the diversity of their shared cultural and natural heritage and a foundation of their identity. Consequently, the integration of landscape in all kinds of policies was proposed. The specific measures include awareness-raising, training and education, identification and assessment of landscapes (i.e. landscape character assessment) and defining landscape quality objectives.

Opened for signature on October 20, 2000, it entered into force on March 1, 2004 after 10 member states had ratified the Convention. In November 2009, the Convention was ratified by 30 of the 47 member states, and six more signed it.

Multiple meanings: overview of basic and new concepts

Land and landscape

Although some criticism has been formulated that the definition of landscape in the ELC is too broad and vague, it is innovative in many aspects. Essentially, the landscape is:

- seen as a spatial entity, having a variable extent and scale, and having territorial properties;
- perceived and experienced by humans, and consequently is relative to the observer;
- holistic, expressed by its character, which also defines its identity;
- dynamic, changes being an inherent property of it;

- the result of continuous interaction between natural processes and human activities.

From these properties, landscape can be conceived as a dynamic holistic phenomenon, consisting of scale dependent entities that are hierarchically structured. This makes landscape fundamentally different from the concept "land" (Zonneveld 1995), which is considered as a piece of terrain, bounded in space and bordered and very often owned by someone or some institution. Land refers to (private) property that can be used more or less freely by its owner. Human impact upon the landscape mainly acts indirectly through land-use. The concept of landscape as a tangible area or region was the object of the traditional regional geography in the early twentieth century that aimed to identify and delineate regions, based on their particular character as a result of the interaction between the physical environment and the society living there (Johnston 1986; Muir, 1999).

Landscape is holistic: holons and the importance of scale and context

The basic paradigm of holism is that the whole is more (or different) than the sum of its parts. The concept of holism was introduced in landscape geography through the recognition of the *Totalcharakter* of landscape, which – according to Zonneveld (1995: 12) – is attributed to von Humboldt. This *Totalcharakter* was raked up again in landscape science as a positivistic and experimental research approach, in particular in its applications in (eco)systems theory and *Gestalt*-psychology of perception (Antrop 2005a; Antrop and Van Eetvelde 2000; Naveh and Lieberman 1993; Sevenant 2009). A *Gestalt* of a phenomenon is the whole that is more than the sum of its composing parts. The premise behind holism explains e.g. the complexity of factors influencing experimental landscape preference studies (Sevenant and Antrop 2009, 2010; Sevenant 2010).

Holism also became a basic paradigm in landscape ecology, where the concept of a "holon"as a building block of the Total Human Ecosystem (THE) was introduced (Naveh and Lieberman 1993). Holons are considered as open (sub)systems that are hierarchically structured, thereby forming the complex landscape. Scale and context are two important factors that define the meaning and functioning of holons and are considered as basic parameters in setting up a research project. Thus, the holistic nature of landscape both refers to landscape as a system with its organization in patterns, structures and processes as well as the perceptions of it.

All is landscape

Another innovative aspect of the ELC is that there is no focus on specific types of landscapes. In fact, the scope (Art. 2) states that the Convention 'applies to the entire territory of the Parties and covers natural, rural, urban and peri-urban areas. It includes land, inland water and marine areas. It concerns landscapes that might be considered outstanding as well as everyday or degraded landscapes.' This broadened the attention to landscape that was previously restricted to mainly selected and particularly valuable landscapes such as natural ones (IUCN) or cultural ones (UNESCO World Heritage).

Perceptions of landscape

Landscape differs from the concept of "land" through its relation to the human observer, as clearly formulated in the European Landscape Convention. Perception shows up in the definition of landscape (Art. 1); preference is referred to in "aspirations of the public" in the definition of landscape quality objectives (Art. 1).

In the context of landscape research, perception and preference are two different concepts referring to distinct processes that have been intermingled in literature. They are often brought up in one and the same breath both by experts and non-experts. This confusion is already obvious from the designation of research paradigms in perception and preference research, where the "perception-based approach" (e.g. Daniel 2001) and the "public preference model" (e.g. Scott and Benson 2002) refer to the same paradigm.

Kaplan and Kaplan (1995) acknowledge that perception and preference are closely related. They argue that perception is a key element in preference in that the measurement of preference allows investigation of the perceptual process. Nevertheless, despite their converging meanings, both concepts need to be defined separately as they are used in literature (e.g. Antrop 2007; Kaplan and Kaplan 1995).

Bell (2004), for example, defines perception as follows: 'The activity carried out by the brain by which we interpret what the senses (mainly sight for most people) receive. It is not merely a factual reporting but tends to be referenced to associations and expectations already present in the mind of the beholder' (Bell 2004: 185). From this definition, two steps are prominent in perception: the senses and the brain. Bell resumes this in the physiology of perception versus the psychology of perception. He puts forward vision as the most important sense for perception for two reasons: vision is the sense that provides most information (87% of perception, Bell 2004) and it is the sense in which we think. That what can be seen is defined by physiology: the mechanisms of sight, the structure of the eye, how it receives light, and the limitations thereby. These are influenced by the circumstances at a particular time such as distance, movement, attention, fatigue etc. All aspects of the physiological perception can be measured in an objective way. The psychology behind (visual) perception is more subjective: whereas the functionality of the brain is more or less similar for all people, the brain determines what we see and how we see things in the world through processing the information and constructing images from it.

Preference is still more subjective. Zajonc and Markus (1982: 124) describe preference as follows: 'Preferences are themselves primarily affectively based behavioural phenomena. A preference for X over Y is a tendency of the organism to approach X more often and more vigorously than Y. ... [P]references are conceptualized as the subjective counterparts of object utilities and values.'

In this sense, landscape preference can be set along an ordinal scale, denoting ranked order. Preference is described as an interaction between cognition and affect (Kaplan and Talbot 1988; Zajonc and Markus 1982). Cognition can be understood as 'a non-sensory process that transforms sensory input and produces or recruits representations' (Zajonc and Markus 1982: 127). Kaplan (1988) distinguishes between two sorts of affective reactions – pleasure/pain and interest – and between two facets of cognition – content and process. The link to affect then is sometimes through the contents of cognition and sometimes cognitive processes can have affective implications.

An alternative view of preference is that it is an indicator of aesthetic judgment, induced by properties of the stimulus rather than by cognition (Berlyne 1960). This alternative (or extended) view is the reason why preference is easily confused with aesthetics in literature (e.g. Kaplan and Kaplan 1995).

In the realm of landscape perception and preference research, the term "preference" has been added to varying adjectives, indicating either slightly or substantially differing ideas: e.g. "environmental" preferences (e.g. Johnson et al. 2004; Herzog and Kropscott 2004), "landscape perceptual" preference (e.g. Kaplan et al. 2006), "landscape" preferences (e.g. Dearden 1985; Fry et al. 2009; Hagerhall 2000, 2001; Purcell et al. 2001), "aesthetic" preference (e.g. Jessel 2006; Ulrich 1983), and "visual" preference (e.g. Strumse 1994). Sometimes, preference has been replaced by "evaluation" (e.g. van den Berg et al. 1998). The question phrasings used in surveys to assess landscape preference are similarly varied.

It is obvious that perception is not the mere technical aspects of observation (or sensation) and that preference is not the mere liking or disliking of something. Perception and preference can imply very different psychological mechanisms notwithstanding their often confounding use in literature on landscape research. Yet, it is possibly the object under observation that causes this mix-up between them: the landscape. Kaplan and Kaplan most sensibly resume how perception and preference are linked, and where the confusion between them stems from:

> Perception is quite obviously important to survival. ... But being able to perceive what is safe and what is dangerous is not enough. If the information an organism requires through the power of perception is to aid in its survival, it is essential that it not only perceive what is safe but also prefer it. ... If this were the case, one would expect that what is basic to perception must also be important to preference.
>
> (Kaplan and Kaplan 1995: 41)

Landscape as a social and mental construct – meaning and values

If landscapes are subject to perceptions and preferences then landscape becomes a rather relative concept. As soon as landscape is perceived and interpreted by human observers, landscape becomes a mental and social construct with varying values and meanings appended to it. This idea was developed by human geographers such as Lowenthal and Jackson, and by phenomenological geographers such as Buttimer, Relph, Seamon and Tuan. The relativistic idea culminated in the definition of landscape by Cosgrove and Daniels as "a way of seeing" (Cosgrove 1984; Cosgrove and Daniels 1988; Daniels 1993, 1999) and in landscape as "a work of the mind", its scenery being 'built up as much from strata of memory as from layers of rock' (Schama 1995: 6–7). Landscapes are not solely tangible material, they are a product of observation and interpretation, giving birth to the concepts of "mindscapes" (e.g. Lörzing 2001; Cosgrove 2003; Jacobs 2006), and "cognitive" or "mental" maps (e.g. Tuan 1975). This makes landscape intimately related with concepts such as identity, character and place making, which also carry many symbolic meanings.

Landscapes as continuously dynamic entities

The continuous interaction between spatial structures and processes

A basic paradigm of landscape ecology is that spatial patterns and ecological processes are continuously modifying each other. Forman and Godron (1986: 3) phrase it like this: 'An endless feedback loop. Past functioning has produced today's structure; today's structure produces today's functioning; today's functioning will produce future structure.' Also the definition in the ELC states that the tangible and perceivable landscape is the result of continuous interaction between natural processes and human activities (Art. 1). Consequently, change is an inherent property of landscape and time an additional factor – besides space – to understand landscape.

Landscape genesis

The diversity of natural landscapes in the world is the result of geological, geomorphological, ecological and climatic processes, which have changed through the long history of the planet. The human impact is rather recent and leaves only permanent marks since the Neolithic (New Stone era, beginning about 10,200 BC). Gradually, natural landscapes became cultural. The broad diversity of cultural landscapes is the result of the interaction between natural processes and human activities characterized by successive adaptations to human needs – mainly due to the increase of the population – and growing technological means. Places and regions are characterized by a unique history and different landscape paths or trajectories can be defined (Käyhkö and Skånes 2006; Antrop et al. 2007). The magnitude, pace and extent of the human impact on landscapes increased exponentially since the eighteenth century with consecutive and revolutionary innovations in industry, agriculture and society (Antrop 2000, 2003, 2005b).

Actual trends of landscape transformation

The actual driving forces of landscape change are well known (Antrop and Van Eetvelde 2008; Bürgi et al. 2004). Three of these driving forces are interrelated and economically driven. First, an area gets disclosed by transportation infrastructure, which initiates development. This leads to industrialization and/or urbanization in this core area, creating powerful attraction on the hinterland. Gradually, markets and economy have become dispersed, inducing networking between these core areas. Globalization as a background driving force became increasingly important. Nevertheless, the global processes are translated into a multitude of local processes (Antrop 2008).

The general trend of transformation results in polarization of geographical space, creating rather small areas of intensification and vast areas of extensification of land-use. Population density in growing urban regions means an increase of the infrastructure density of various kinds, multifunctional land-uses, as well as growing networking between these places. The vast hinterland gradually becomes abandoned and land-use becomes extensive and often leads to rewilding and reforestation. Examples of this are found in many remote rural areas throughout Europe.

Besides these economical driving forces, there are natural causes, which are becoming increasingly important, as they are often unpredictable and catastrophic. Natural hazards cause disasters, which can cause severe human and economic loss. This is particularly important as the core areas of intensification are often situated in high-risk zones (coastal zones, river deltas and valleys, volcanic and tectonic active zones, etc.).

Keeping up with the changes

Landscape changes have reached such a pace that it is getting difficult to study them. Simultaneous data collection of all aspects has become impossible. New methods for monitoring and dealing with the uncertainty involved are urgently needed.

Very often, stratified sampling is used for monitoring changes (Brandt et al. 2002; Dramstad et al. 2001) and changes are expressed by indicators. These mainly relate to land cover changes and changing landscape patterns. Indicators are however indirect measures of ecological processes or human activities. The relationship between the indicator and the process is not always proven or causal (Li and Wu 2007). This causes the significance for policy evaluation to remain difficult and uncertain (Dramstad and Sogge 2003; Parris 2004).

Other more qualitative forms of landscape monitoring use re-photographing of the landscape from selected sites showing landscape changes in a series of photographs taken from the same spot (Puschmann and Dramstad 2002).

Actors of change

Landscape is the appearance of a land, characterized by a specific form of organization by humans. The real action takes place in the land, in particular in the land-use. The basic question here is: who has the competence and authority to make land-use changes? Hägerstrand (1995) defined different forms or scales of competence. The most important one is the *territorial competence* as Hägerstrand calls it, and belongs to the land owner (private or public) who is the only one who is capable of making real, material and tangible changes. The second form is the *spatial competence,* which has an indirect, more regulating power on land-use changes. It belongs to municipalities, local management boards and planning authorities, or higher administrative and policy levels. The spatial competence may be organized in two ways: (i) as functional specialization, which is reflected in a sector-based policy and authority, and (ii) as geographical integration of the competences in a territory (Hägerstrand 1995).

One of the main problems in contemporary landscape planning and management is the large number of actors that have territorial competence. Many of the traditional landscapes in the countryside are the result of initiatives of a rather restricted number of land owners over time (Muir 2003). Democratization in the twentieth century led to an increase of small landownership, resulting in many non-concerted small land-use changes. In accumulation, they cause a more chaotic change of the landscape, which is not always clearly planned. It is referred to as "autonomous development" (Antrop 1998). Liberalism and free-market ideologies stimulate these processes. The different competences also determine the involvement of various stakeholders. Selman (2006) discussed the different valuing of and behaviour of insiders and outsiders. Landscapes are read differently by outsiders, who have not been living in the landscape and can relatively deliberately choose which aspects of landscape to enjoy. Selman observes that the inside population is nowadays in a transitional status, some of them living from the land, others only residing there, and yet others being recent incomers or passers-by without any social and emotional ties to the area (Selman 2006).

Multifunctional landscapes and sustainability

Multifunctionality and natural and human capital

Producing an operational definition of "multifunctional landscapes" is not simple (Brandt and Vejre 2004). In general, multifunctionality refers to different (potential) uses of the same piece of land. In particular, it concerns the possibility of simultaneous or consecutive uses of the land without making fundamental changes to the structure or morphology of the landscape. So, the transition of rural land to urban built-up land is not to be considered as multifunctional, but more as a functional change resulting in a new landscape type. Moreover, multifunctionality is also related to the scale of observation (Antrop 2004).

Functionality has to do with usefulness for some kind of purpose and therefore also deals with assigning values. In the context of landscape, functionality refers first and foremost to the suitability of the land for a specific use. Second, it refers to how something works, for example the ecological functioning of a landscape. Common to both definitions is a causal relationship, between elements in the landscape or between the landscape and its user. The notion of value is implicit in both meanings, although different values can be contradictory. A high value for ecological functioning might correspond to no value at all for land development. Also the achievement of the function might change the set of values, as land development can destroy the functional ecological value. Multifunctionality also has to do with combining different values, which can be complementary or conflicting.

Related to this assigning of values, Haines-Young and Potschin (Haines-Young 2000; Haines-Young and Potschin 2004; Potschin and Haines-Young 2003) have introduced the natural capital paradigm into questions about sustainable multifunctional landscapes. They state that in terms of this natural capital paradigm, 'landscape multifunctionality arises according to the way in which different people, or groups in society, value the different outputs from an area' (Haines-Young and Potschin 2004: 181–192). Yet, according to them, it is not easy to define value in the context of multifunctional landscapes. They ascribe this to the fact that people value outputs differently, not only between but also within single interest groups of people. As an example, they refer to the possible incompatibility between recreational and biodiversity goals.

The authors claim that 'multifunctionality is not a property of ecological systems per se but a result of the interaction and linkage between society and the environment' (Haines-Young and Potschin 2004: 182). The importance given to (ecological) functions can change through time. The properties and functions we select for designing cultural landscapes depends on our value systems. Ideally, multifunctional landscapes should consist of configurations that sustain the outputs that society appreciates and (highly) values. Haines-Young and Potschin (2004) consider the development of tools to define ecologically viable futures as one of the major challenges of actual landscape ecology. They state that there is no one single optimal configuration for a sustainable landscape but rather a set of alternatives that are 'sustainable in terms of continuing outputs of goods and services we currently value' (Haines-Young and Potschin 2004: 189). This set of alternative spatial configurations fluctuates as biophysical parameters change either naturally or due to human impact, together with the evolution of landscape.

Landscape functions – landscape goods and services

The recent focus on ecosystem services has brought attention to the economic importance of natural ecosystems. This concept indeed aims at encompassing the different perspectives on benefits associated with ecosystems: the ecological, the resource economics and the political perspective (Brown et al. 2007). In Daily's definition (1997: 3), ecosystem services are 'the conditions and processes through which natural ecosystems' support human life.They produce ecosystem goods, the material products, along with 'many intangible aesthetic and cultural benefits as well' (see also Daily et al. 2000: 395). However, Costanza et al. (1997) distinguish the "ecosystem services" from the "ecosystem functions" (physical and biological cycles) that produce ecosystem services. This distinction between "ecosystem

Table 2.1 Functions, goods and services of natural capital

Functions	*Goods and services provided*
Regulation	Maintenance of essential ecological processes and life support systems. These are important for example for climate regulation, disturbance prevention, water regulation and control, soil productivity and erosion, nutrient regulation and pollution control
Habitat	Providing habitat (suitable living space) for wild plant and animal species
Production	Provision of natural resources, such as food, raw materials, genetic and medical resources, but also ornamental resources
Information	Providing opportunities for cognitive development. This includes aesthetic information related to the landscape scenery, recreation, cultural and artistic information, historic and spiritual information and scientific and education information

Source: De Groot et al. (2002).

services" and "ecosystem functions" is finalized in de Groot et al. (2002) to translate the ecological complexity into a limited number of ecosystem functions, which, in turn, provide the goods and services that are valued by humans. Table 2.1 summarizes functions, goods and services of natural capital according to de Groot et al. (2002).

Integrating landscape economics in landscape science

Landscape science is an umbrella term to cover all disciplines involved in landscape studies (Naveh 2005). Traditionally, these include geography, history, landscape architecture, landscape ecology and (landscape) archaeology. Recently, landscape economics has contributed to landscape research.

Landscape economics has emerged as a branch of environmental economics with a focus on the specific issues of landscape and to provide economic tools for identifying efficient levels of landscape protection and provision. The term landscape economics was first used by Price in 1978 (Santos 1998), and gained more attention, because of the peculiarity of landscape valuation. Landscape economics holds true to the theoretical framework and analytical methodologies from its roots in environmental economics. In turn, landscape economics is based on neoclassical economics, which refers to the market for the optimal allocation of resources among economic agents and activities. In the markets, rational individuals express their preferences making choices and prices signal resources' scarcity (Pearce and Turner 1990). When prices do not give an accurate signal for the full range of benefits provided or there is no market at all, a "market failure" occurs and private benefits and costs diverge from social benefits and costs (Hanley et al. 1997). This is the case for several ecosystem services and goods. Nonmarket valuation helps to correct this situation, assigning an economic value to non-marketed goods and services reflecting individuals' preferences. It uses trade-offs between conservation and development to assess landscape value (Hanley et al. 1997; Santos 1998). Economists estimate the monetary value reflected by such trade-offs as accurately as possible. The fundamental idea is that demand and supply functions can be interpreted as behavioural relations and repositories of values themselves (Samuelson 1948). Economics is, then, not just the study of markets but more generally the study of human preferences and behaviour (Hanemann 1984) as expressed through the market. The theory of nonmarket valuation has made significant advances over the past three decades, increasing the optimism of economists about its use as a viable tool to assist decision making.

Inter- and transdisciplinarity

Integrating landscape economics in landscape science needs an interdisciplinary approach and, as stakeholders may also be involved, a transdisciplinary approach as well. Although terms such as "interdisciplinary" and "transdisciplinary" are frequently used in research proposals and projects, a lot of confusion exists about their real meaning. Tress et al. (2004) give the following definitions in relation to landscape research:

- *disciplinary* studies take place within the bounds of the recognized academic disciplines;
- *multidisciplinary* means that different academic disciplines study one and the same subject but each one with their disciplinary goals;
- *interdisciplinary* studies involve several unrelated academic disciplines in a problem-oriented research goal; and
- *transdisciplinary* studies integrate both academic researchers from unrelated disciplines and user-group participants (stakeholders, laymen) to reach a common goal.

Integrated landscape studies group interdisciplinary and transdisciplinary approaches together (Tress et al. 2004). To achieve "real" integration, bridges should be built between different disciplines concerning concepts, definitions, methods, and specific goals (Tress et al. 2003, 2007). A common language needs to be developed and benefits of combining different methods should be explored.

Landscape classification and evaluation

Landscape identification and assessment

The specific measures formulated in the ELC (Art. 6) include the identification and assessment of the landscapes throughout the territory, analyzing their characteristics, as well as the forces and pressures transforming them and taking notes of the changes. Once identified, landscapes should be assessed (see below), taking into account the particular values assigned to them by the interested parties and the population concerned. All over Europe, new landscape inventories and identifications are being elaborated in various forms and at different scales and by using different methods. Some of the methods concern the collection of maps (landscape atlases) often linked to Geographical Information Systems (GIS) databases, while others are more descriptive and result in well-illustrated landscape catalogues or landscape biographies.

The ELC does not suggest a specific methodology for supporting and implementing this landscape identification and assessment. It rather stimulates the party countries to co-operate and exchange experiences. This is achieved through numerous conferences, workshops and seminars. The guidelines for the ELC elaborate on the terms "identification" and "assessment" as well as on "landscape quality objectives". To start with the latter term, the "landscape quality objectives" for a specific landscape are defined as: 'the formulation by the competent public authorities of the aspirations of the public with regard to the landscape features of their surroundings.' (Art. 1.c). The main idea of these guidelines is to obtain knowledge and understanding of how landscapes are characterized as a result of interacting natural processes and human activities. This knowledge is not purely academic but problem and goal oriented.

The term "identification" refers to the description of the specific characteristics that make one landscape different from another. However, this description is oriented according to the purpose of the inventory. Sometimes, these characteristics are referred to as landscape qualities, however without a value judgment.

The term "assessment" refers to a successive phase of evaluating the meaning and significance of the landscape qualities identified. This results in the formulation of the earlier mentioned landscape quality objectives (Council of Europe 2008). Both "identification" and "assessment" will be explored in further detail in the following two sub-sections.

Landscape typology and chorology

Landscape identification is based upon making and often mapping of appropriate landscape classifications. Two types of landscape classification can be recognized (Van Eetvelde and Antrop 2009).

1. A *landscape typology* is a systematic classification of landscape types based on attributes that describe properties of interest, such as land-use, scenic properties, cultural characteristics or history. Landscape types are defined by unique relations between natural components (such as geology, soil, morphology, land cover) and human components (such as settlement and field patterns, land-use, building and farming styles). They often reflect a specific landscape history or are formed by specific processes. Landscape types are generic in nature: they may occur in different areas and in different geographical contexts. For example, the landscape type "polder" can have different origins and can occur in different regions.
2. *Landscape chorology* focuses on the spatial patterns formed by different landscape types to form unique spatial arrangements with a distinct identity. They are often unique, which is reflected by a proper name given to the area. Landscape chorology is part of defining geographical regions. It is a

hierarchical spatial classification at different scale levels. For example, the Hercynian mountains of the Massif Central in France are composed of volcanic landscapes (Auvergne), granitic landscapes (Limousin) and limestone landscapes (Causses). Each of these consists of different landscapes types, such as limestone plateaus and canyons for the Causses, or volcanos and basaltic flows for the Auvergne.

Methodologically, two approaches are possible for classifying landscapes: the holistic and the parametric methods. The principle of the holistic method is to start with a hierarchical chorological subdivision of an area (Figure 2.1). Very often it is based on detailed documents giving a synoptic view of the landscape in a bird's eye perspective, such as aerial photographs. The procedure is very similar to unsupervised image classification where the observed patterns are delineated and then identified, and is highly based on the *Gestalt*-abilities of our perception in interpreting complex patterns. The holistic method starts with building a spatial framework that becomes gradually filled when more detailed information becomes available. It is typically a process of zooming in on the landscape.

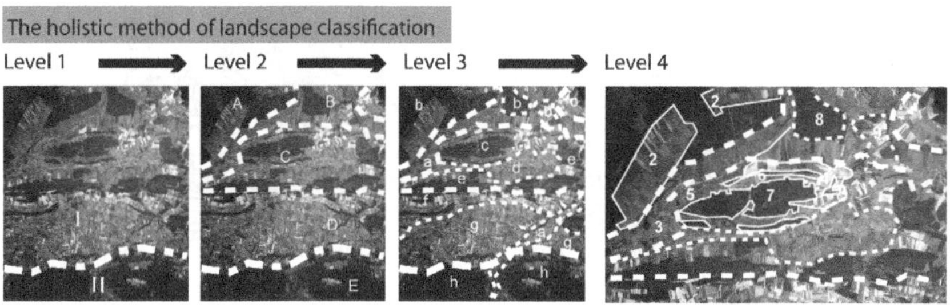

- stepwise interpretation of synoptic documents and delineating of hierarchical landscape units.
- first and second level are based on the most important characteristics (1 + 2)
- third and fourth level divide units into smaller units based on more detailed characteristics (3, 4).
- each spatial landscape unit is described by different attributes in a database.

hierarchy of spatial landscape units

Levels				Attributes		
1	2	3	4	soil	slope	land use
I	A	b	1			
			2			
	B	a	3			
		b	8			
			9			
	C	c	4			
			5			
			6			
			7			
		d				
		e				
	D	a				
		f				
		g				
	...	...				
II	E	a				
		h				
	...	...				

Figure 2.1 The holistic method of landscape classification.

Source: Van Eetvelde and Antrop (2009).

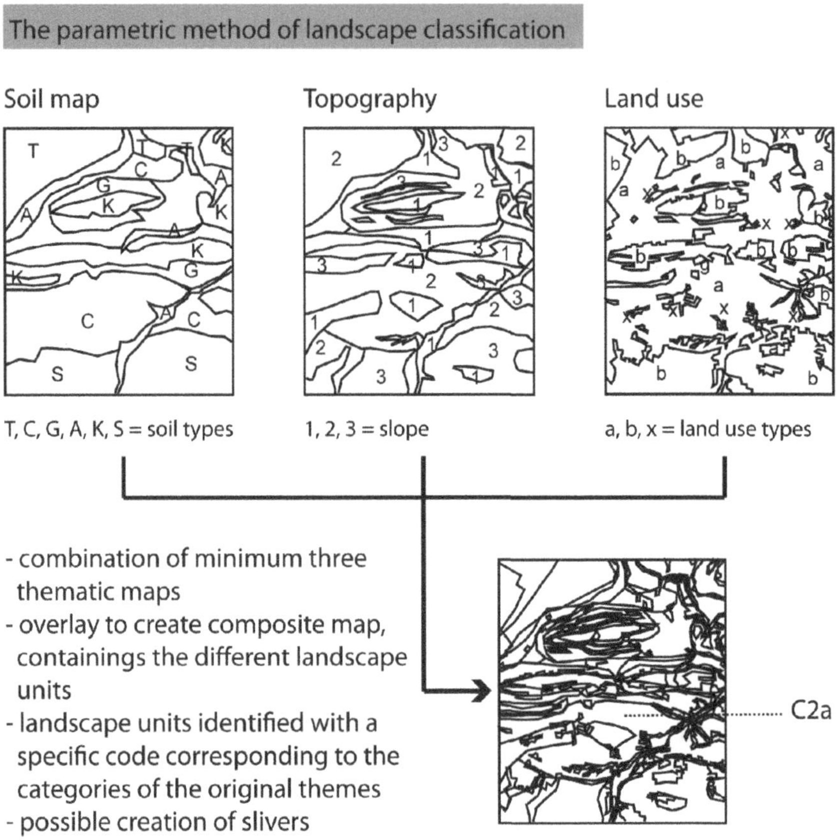

Figure 2.2 The parametric method of landscape classification.

Source: Van Eetvelde and Antrop (2009).

The parametric method starts from overlaying a set of thematic maps, forming a composite map where the overlay polygons define the landscape units or patches and the combined themes describe the landscape types (Figure 2.2). This technique became very popular when GIS and digital maps became available. The process is rather one of zooming out on the landscape, making aggregations and generalizations.

Older landscape typologies are based on classifications of geographical regions and are often holistic and generic in nature. More recent typologies are based on GIS-overlay of digital thematic maps, using a spatial and statistical analysis to define landscape types and employing more parametric methods. Because of the hierarchical structure of the holistic nature of landscapes, landscape classifications are also made at successive embedded spatial scales that reflect this structure (Van Eetvelde and Antrop 2009).

Landscape Character Assessment

The renewed interest in the concept of landscape character stimulated the development of methods for holistic landscape characterization, such as the Landscape Character Assessment (LCA) of the English Countryside Agency or the Historic Landscape Characterization (HLC). Landscape character became a central concept in the Landscape Character Assessment in the UK and is defined as a distinct, recognizable and consistent pattern of elements in the landscape that makes one landscape different from another, rather than better or worse (Swanwick 2002). The LCA distinguishes landscape character types and landscape character areas. Landscape character types are distinct types of landscape, which are relatively homogeneous in character. They share broadly similar combinations of geology, topography, drainage patterns, vegetation and historical land-use and settlement pattern. Landscape character areas are defined as single unique areas and are the discrete geographical areas of a particular landscape type (Swanwick 2004). Thus, the LCA comprises a holistic landscape typology and chorology (see above).

The HLC is very similar to LCA but adds the time dimension. Starting from the present, landscape units are identified and traced back to the past. Thus new concepts such as "time depth" and "landscape paths" or "trajectories" were introduced (Fairclough and Macinnes 2003; Clark et al. 2004; Turner 2006).

Assigning landscape values

Landscape values in landscape management and planning

Landscape values only exist when people, users, assign them to specific landscape qualities which they consider important and significant (Antrop 2004). The specific measures of the ELC (formulated in Art. 6.d) proposes to define "landscape quality objectives" for the landscapes that were identified and assessed, after public consultation and after establishing procedures for the participation of the general public, local and regional authorities, and other parties (Art. 5.c). However, the ELC does not provide a list of landscape qualities nor proposes a detailed methodology how to proceed. Nevertheless it is clear that the competent public authorities are responsible for organizing the public's participation, in collecting their aspirations and translating them into policy actions for each of the identified landscapes.

The holon-concept as discussed above, allows the individual evaluation of the composing parts (elements, objects) of the landscape. As such, landscape evaluation is a process of assigning values to landscape properties. However, this is only meaningful in relation to their spatial context and to the surrounding landscape features (Antrop 2008). This is referred to as the intrinsic and contextual value (for a further explanation, see Antrop 2004). Value categories can be defined and these often receive a legal status for protecting landscapes. These categories are however broad and generic and cannot be assessed by one criterion alone, so a series of criteria is needed to assess different aspects of their composed value. Many of these criteria can be seen as indicators of properties that cannot be measured directly. Criteria can be expressed in very different ways according to their operational definition. Illustrative for this is the ongoing quest in the literature to find descriptive quantifiable (mainly visual) indicators to be linked with landscape evaluation and landscape character to serve as a response to the indicators that already have gained theoretical ground in most other environmental fields. A very valuable example is the description by Tveit et al. (2006) of nine visual key concepts, based on an extended review of the literature of landscape research. These concepts are meant to form a framework for landscape assessment and the compilation of landscape character. The authors distinguish between stewardship, coherence, disturbance, historicity, visual scale, imageability, complexity, naturalness and ephemera (Tveit et al. 2006). Ode et al. (2008) investigate the existence of relationships between the quantitative indicators selected to represent these concepts and the individuals' aesthetic preferences, suggesting that landscape characteristics described by these concepts can help explain why people prefer/like or do not prefer/dislike some landscapes.

The scale of measurement used defines the possibilities of combining criteria, for example in a multicriteria evaluation procedure. The last step in the procedure consists in defining the evaluation scale. Elements can be very valuable for locals but can be meaningless at the scale of the World Heritage for example.

Table 2.2 summarizes the components and variables that could be taken into account when evaluating landscapes. Table 2.3 links landscape assets or qualities to functions, which have to be related successively with utilities or values and specific (multifunctional) use(s), taking the acceptability for the predefined management goals at the appropriate scale level into account.

How to assess and express landscape values?

Valuing means an assessment of landscape qualities, described by criteria that need to be operational and well-defined. This can help to make comparisons between different landscape units or conditions, e.g. in scenarios. Also, the outcome is represented on a map, and the development of GIS-techniques makes the total process much easier. Different methods are available depending on the measurement scale used (ordinal, interval, ratio). Most often, several criteria need to be measured and combined. In a nominal scale, the number of categories of assessment is restricted. Early studies of land evaluation, e.g. by FAO, used categories such as suitable, conditional suitable and unsuitable. Ordinal scales of measurement allow ranking. When using an ordinal scale of more than five classes, the scale is very often considered as being continuous and scores or ratings are used for each of the criteria and these scores or ratings are treated as quantitative measurements of interval or ratio scale. For decision-making support, the scores of different criteria are combined using multicriteria-evaluation (MCE) techniques. Therefore the criteria need to standardized and weighted, and a combination

Table 2.2 Components of landscape evaluation

Value categories of a landscape (element or spatial unit)	*Criteria for evaluation*	*Expression of value (scale of measurement)*	*Scale of importance or significance to assess the value*
Intrinsic values: - Natural - Historical, cultural, social - Aesthetical - Symbolic … **Context value:** - Location - Spatial association - Vertical or horizontal relation …	**Related to the content:** - Rarity - Authenticity - Information content - Time depth - Coherence - Diversity, heterogeneity, … **Related to perception and preference:** - Legibility - Identity, character - Order, variation, contrast - Atmosphere, mystery, … - About utility - Accessibility - Potential uses - Stewardship - Tangible or not? … **→ Marketable or not?**	**Qualitative:** - Nominal **Quantitative:** - Ordinal (ranking) - Interval - Ratio **→ Monetary or not?**	Local Regional National International Universal (world)

model or function needs to be chosen (Eastman et al. 1993). Commonly used is a linear weighted combination of standardized scores in the general form of:

$$S_u = \Sigma w_i X'_i \cdot \Pi c_j, \qquad (2.1)$$

where

S_u is the combined score for a given land unit u (or cell in a raster model);
X'_i is the standardized score for criterion i, and w_i its weight ($0 \leq w_i \leq 1$); their product is summed;

Table 2.3 Linkages between landscape qualities, functions, values and specific (multifunctional) use(s)*

Assets or landscape qualities		*Functions*		*Potential utility/value for…*	*Principles to evaluate allowable use or multifunctionality*
Natural Historical Cultural Social Aesthetical Symbolic …	→ **link to:** →	Space Information Products (goods) Services, amenities …	→ **implement at appropriate scale level for:** →	Agriculture Housing Recreation Heritage …	Accessibility Sustainability Type of multifunctionality Spatial combination Temporal combination Synergy Combined effects (positive, negative) …

*Assets or landscape qualities have to be linked to one or more functions, which then have to be implemented for specific (multifunctional) use(s), acceptable for the management goals at an appropriate scale level.

Source: Adapted from Antrop (2004).

c_k are the constraints (expressed by the Boolean values 0 or 1); and their product Π defines whether the land unit u can be taken into consideration.

Most crucial is the weighting procedure. Eastman et al. (1993) proposed the pairwise comparison method by Saaty (1977), which allows the assignment of weight in an interaction participation procedure by all stakeholder groups involved. This method can be used to formulate indicators for complex landscape qualities of one location in the landscape. However, spatial patterns in landscapes require other methods, such as the use of landscape metrics (McGarigal et al. 2002; Li and Wu 2007; Uuemaa et al. 2009).

The economic value of landscape

Landscape values in landscape economics

Landscape is often found in the list of public goods associated with agriculture, forestry and nature. The economic value of landscape is for example clearly expressed in the reform of the Common Agricultural Policy within the EU. As no rational decision process on natural resources management can be abstracted from their value (Lockwood 1997), policy makers had to consider other means to assess landscape values.

In the broader context of ecosystem services, the benefits of nature to humans can be analyzed from various perspectives: physiological, psychological and economic. In this way, the connections between assessment (or evaluation) and valuation are more straightforward. Yet, though linked to each other and often conflated in literature, "evaluation" and "valuation" cannot be treated as synonyms (Daniel 2001). Evaluation is the process of scoring or rating the quality of landscape while valuation assigns an economic (i.e. monetary) value to landscape or its attributes. Over the last two decades, evaluation research has focused on objective assessment criteria, to minimize undue subjectivity. However, there is generally little reference to how individual preferences can be measured and integrated (Moran 2006). This has left a void for the evaluation of trade-offs between benefits associated with different landscape states (Santos 1998), that have most recently been filled by environmental economics; that is, the transition between evaluation to valuation.

The economic value of the landscape depends on its components as well as on how these components are perceived by consumer, with respect to his/her preferences (Hasund 1998). In economics, definition of value is based on the ideals of consumer's rationality and sovereignty: each individual consistently knows his/her preferences (rationality) and is best able to make choices based on them (sovereignty). Rational choice applies to landscape as well: individuals express their preferences through willingness to pay (WTP) for an improvement or willingness to accept (WTA) as a compensation for degradation. WTA and WTP, then, represent measures of economic value for landscape (Hanley et al. 1997; Earnhart 2001; Brander and Koetse 2007;

Hoyos 2010). Economic value here refers to the Total Economic Value (TEV) of the environmental resources, that encompasses "use" and "non-use" values (Krutilla 1967; Pearce and Turner 1990).

The idea of assigning landscape a monetary value is still perceived by some as immoral, maybe because of the fear that 'what can be sold, should be sold' (Price 1994: 1). However, decisions involving landscape by definition assigns (at least implicitly) monetary values to the implied landscape benefits. Conservation costs are well documented and also socially conspicuous, while benefits are mainly "intangible" and systematically understated. However, "intangible" does not mean these benefits are unreal in welfare terms (Santos 1999). The willingness to pay/accept concept automatically gives a monetary indicator of preferences. As long as another meaningful unit in benefit and cost terms is not available, money remains the best indicator we have (Pearce and Turner 1990). Monetary estimates of the values of ecosystem services, even if inexact, may be far better than a complete lack of such estimates (Brown et al. 2007).

Placing monetary values on landscape has never been an easy task because of its complexity and holistic nature. In particular, major difficulties are related to landscape representation in the economic models. What individuals perceive of the landscape is difficult to measure empirically (Michael et al. 2000). Economists have used a number of measurable attributes as proxies for different landscape frames, spanning from percentage of the total land area that a category of open space occupies within a specified neighborhood (Irwin 2002) to perceptual categories of landscape built through the factor analysis of a 1–6 scale scoring by respondents of several rural landscape images (Howley 2011). Now, the integration of landscape economics within the broader context of landscape science offers the opportunity of more actively interacting with the other landscape disciplines. This broadening perspective is not totally novel. Already in the very early stages of the development of environmental economics, Ciriacy-Wantrup acknowledged the importance of integrating developments in social psychology and the newly emerging academic field of survey research 'to put [welfare economics] on a more realistic foundation' (Hanemann 1994: 21). The time has come to benefit from the information provided by many other landscape disciplines. Some of this information allows quantitative analyses of the bio-physical components (landscape ecology) and also allows for a cross-over with environmental psychology and landscape preferences that share the same interest in the interrelation between landscape and people's perception.

The integration with such disciplines will facilitate the adoption of the holistic approach in landscape economic analyses, which entails both the landscape biophysical components and individuals' landscape perception and preferences. Indeed, this has recently become a focal point in environmental economics which has witnessed an increasing use of GIS to link the economic assessment to the geographical context (Grabaum and Meyer 1998; Palang et al. 2000; Soini 2001; Bastian et al. 2002; Bateman et al. 2002; González and León 2003; Kong et al. 2007). However, the adoption of an explicit spatial analysis in the practice of economic valuation has been quite slow and only relatively few studies have incorporated spatial results from GIS analysis in econometric models (Geoghegan et al. 1997; Irwin 2002; Hilal et al. 2009).

The second consideration concerns the choice of appropriate valuation methodologies that can exploit the information made available by other disciplines. Stated preferences valuation methods, in particular Choice Experiments, offer such potentials. Their attribute-based analysis of individuals' preferences (Adamowicz et al. 1994, 1998) can complement the component-based landscape analysis. However, a wide and often sharp debate has accompanied this since the very beginning of non-market valuation of landscape, and in particular stated preference methods. In particular, two aspects are controversial.

The first is the general scepticism towards the "monetization" of "intangibles". The concept of "non-use values" encompasses the "intangibles" value and provides one of the building bridges between economists and environmentalists as it introduces the altruistic dimension in the economic analysis of individuals' attitudes and preferences towards the environment (Pearce and Turner 1990). In this respect, the focus on stated preference methodologies for landscape valuation is justifiable: unlike revealed preference methods, stated preference methods do provide a behavioural trace for the economists to glean information about landscape non-use values (Portney 1994). The focus is also on Choice Experiments because they represent the only viable alternative when the interest is in exploring the change in many (spatial) attributes.

The second, specific for stated preference, relates to the criticism that individuals state their choice in a hypothetical scenario instead of revealing it through actual behaviour. A major research effort in the field has focused on improving experiments and choice tasks so to make them closer to the real market situation (Adamowicz et al. 1998). Moreover, since the pioneering work of Adamowicz et al. (1994), Choice Experiments have proved particularly versatile in open space and recreational sites valuation.

The debate around these controversial issues has led to substantial methodological improvement of non-market valuation methods, thereby increasing their applications. Many empirical results suggest that stated preference methods are reliable, or, when not, offer solutions (Randall 1994; Carlsson and Martinsson 2001; Bedate et al. 2012). Their role in cost–benefit analyses has been widely acknowledged by now as the only available way to account for costs and benefits related to non-market and non-use values.

Towards an integrated method for landscape economic evaluation

An integrated method for landscape economic valuation develops along two main phases. The first phase analyses the biophysical and spatial components of the landscape, using geographical and landscape ecological methods to identify landscape types, makes an appropriate landscape characterization and quantifies landscape attributes to be used as criteria further on. This results in a GIS-based database, mapping and visualization. The results of this are used for making a stratified sampling of locations for which terrestrial photographs are taken and viewsheds are determined. Georeferencing of the photographs is undertaken by linking their locations to the maps representing the landscape attributes. In the second phase, this sampling is used in a Choice Experiment to elicit people's preferences for these landscape views and relate them to the characterizing criteria previously defined. This integrated valuation method was applied in a case study on the Peninsula of Sorrento (Italy) (Tagliafierro et al. forthcoming). Four successive steps within the two phases can be identified and are summarized as follows.

Step 1: Landscape classification

The first step consists of compiling a landscape classification, which combines several landscape components, such as terrain properties like elevation, slope degree and slope aspect, which can be assessed from digital terrain models, and components as parent material, soil conditions and land cover or land-use. Depending on the data available, a holistic or parametric method can be used (Van Eetvelde and Antrop 2009). The result is a landscape characterization where landscape types are mapped and their attributes described in a related GIS-database. Additional attributes, such as cultural ones, resulting from ancillary data sources or analysis, can be added to these landscape types. Finally, landscape attributes are quantified, using landscape metrics (McGarigal et al. 2002) and visual indicators (Ode et al. 2008; Tveit 2009).

Step 2: Sampling design, terrestrial photographs and viewshed analysis

Based on the landscape types as strata, a sampling design covering the whole study area is set for localizing representative terrestrial photographs. This sampling design is used to define visual indicators and as visual stimuli in the survey (step 3). For the selected locations, terrestrial photographs are taken and georeferenced in the field using GPS tracking. A viewshed analysis for the horizontal angle of view of each photograph is conducted in a GIS-environment. This viewshed corresponds with the representation of the landscape on the terrestrial photographs.

Step 3: Visual indicators and inquiring peoples preference

Based on the photographs and viewsheds, a set of visual and perceivable landscape attributes are selected, as suggested in literature on landscape preferences (Palmer 2004; Ode et al. 2008, 2009; Tveit 2009). For each viewshed, these visual and perceivable properties are quantified, and the resulting quantifications relate to properties of the terrestrial photographs. This set of perceivable landscape attributes, formulated as quantitative variables, is used to represent different frames in the economic valuation models and make the link between the spatial attributes defined previously and the results from the Choice Experiments. An efficient experimental design is used to form choice sets in an efficient way, for example by maximizing an efficiency criterion or minimizing an error criterion (Sándor and Wedel 2002; Campbell 2007). These choice sets take into account that no photomontages or photo-simulations are used as stimuli. Also, they must contain variables measured on a continuous scale to allow quantitative analysis (Tagliafierro et al. forthcoming).

Step 4: Landscape economic valuation

Several Choice Experiments questions are used to assess trade-offs between the landscape visual indicators and their marginal values through eliciting people's preferences for landscape types (Louviere et al. 2000; Hensher et al. 2005). In each of the Choice Experiments, respondents are asked to choose between alternatives, formed by combinations of different attribute levels, and represented by the photographs of the actual landscape. Following standard practice in Choice Experiments studies, different respondents receive different landscape photographs. Answers to the Choice Experiments questions are then analyzed within the Random Utility Model framework (McFadden 1974; Hanemann 1984). This analysis allows estimating the parameters that show the effect each landscape attribute has on the probability of choosing a particular attribute combination representing a certain landscape type. These estimates are then used to calculate the willingness to pay for landscape attributes (Tagliafierro et al. forthcoming). The results from the case study on the Peninsula of Sorrento shows that most attributes are significantly different from zero, suggesting that respondents do consider the selected attributes when choosing their most preferred alternatives (Tagliafierro forthcoming).

Conclusions and future research

This overview shows the possibilities and conditions of integrating geographical and landscape ecological classification, analysis and evaluation methods with economic valuation methods for expressing landscape values in monetary terms. Yet, interdisciplinarity, and even more transdisciplinarity, is challenging: it requires that researchers from the different participating disciplines accept each others' paradigms and establish a common language. For landscape scientists, it is useful to reconsider their traditional reluctance towards landscape "monetization" and explore the potentials it can generate in terms of landscape planning and management. If landscape economics aspires to become established as a recognized branch in environmental economics, it should develop more specific valuation procedures, which acknowledge that landscape complexity requires a more active integration with the other landscape disciplines using their approaches in a correct and proper manner to achieve integration.

To achieve this we suggest working with case studies in very different settings: different landscapes, different target groups and planning goals. Only practice will stimulate the integration of concepts and methods.

References

Adamowicz, A., Louviere, J. and Williams, M. (1994) 'Combining revealed and stated preference methods for valuing environmental amenities', *Journal of Environmental Economics and Management*, 26: 271–292.

Adamowicz, W., Boxall, P., Williams, M. and Louviere, J. (1998) 'Stated preferences approaches for measuring passive use values: Choice Experiments and Contingent Valuation', *American Journal of Agricultural Economics*, 80: 64–75.

Antrop, M. (1998) 'Landscape change: Plan or chaos?' *Landscape and Urban Planning*, 41: 155–161.

Antrop, M. (2000) 'Changing patterns in the urbanized countryside of Western Europe', *Landscape Ecology*, 15: 257–270.

Antrop, M. (2003) 'Continuity and change in landscapes'. In: Ü. Mander and M. Antrop (eds), *Multifunctional Landscapes. Volume III. Continuity and Change*. pp. 1–14, Southampton, Boston: WIT Press.

Antrop, M. (2004) 'Assessing multi-scale values and multifunctionality in landscapes'. In: J. Brandt and H. Vejre (eds), *Multifunctional Landscapes. Theory, Values and History*. pp. 165–180, Southampton, Boston: WIT Press.

Antrop, M. (2005a) 'From holistic landscape synthesis to transdisciplinary landscape management'. In: B. Tress, G. Tress, G. Fry and P. Opdam (eds), *From Landscape Research to Landscape Planning: Aspects of Integration, Education and Application*. pp. 27–50, Dordrecht: Springer.

Antrop, M. (2005b) 'Why landscapes of the past are important for the future', *Landscape and Urban Planning*, 70: 21–34.

Antrop, M. (2007) *Perspectieven op het Landschap. Achtergronden om Landschappen te Lezen en te Begrijpen*. Gent: Academia Press (in Dutch).

Antrop, M. (2008) 'Landscapes at risk: about change in the European landscapes'. In: P. Dostal (ed.), *Evolution of Geographical Systems and Risk Processes in the Global Context*. pp. 57–79, Prague: Charles University, Faculty of Science.

Antrop, M., Ongenae, T., Sevenant, M. and Van Eetvelde, V. (2007) 'Cultural landscape paths in a polarised Belgium'. In: B. Pedroli, A. van Doorn, G. De Blust, M.K. Paracchini, D. Wascher and F. Bunce (eds), *Europe's Living Landscapes. Essays Exploring our Identity in the Countryside*. pp. 145–157, Wageningen: Landscape Europe, Wageningen/Zeist: KNNV Publishing.

Antrop, M. and Van Eetvelde, V. (2000) 'Holistic aspects of suburban landscapes: visual image interpretation and landscape metrics', *Landscape and Urban Planning*, 50: 43–58.

Antrop, M. and Van Eetvelde, V. (2008) 'Mechanisms in recent landscape transformation'. In: Ü. Mander, J.F. Martin-Duque and C.A. Brebbia (eds), *Geo-Environment and Landscape Evolution 2008*. pp. 183–192, Southampton, Boston: WIT Press.

Bastian, C.T., McLeod, D.M., Germino, M.J., Reiners, W.A. and Blasko, B.J. (2002) 'Environmental amenities and agricultural land values: a hedonic model using geographic information systems data', *Ecological Economics*, 40: 337–349.

Bateman, I.J., Jones, A.P., Lovett, A.A., Lake, I.R. and Day, B.H. (2002) 'Applying Geographical Information System (GIS) to environmental and resource economics', *Environmental and Resource Economics*, 22: 219–269.

Bedate, A.M., Herrero, L.C. and Sanz, J.A. (2012) 'Ex ante and ex post valuations of a cultural good. Are preferences or expectations changing?' *Journal of Environmental Planning and Management*, 55 (1): 127–140.

Bell, S. (2004) *Elements of Visual Design in the Landscape*. London: Spon Press.

Berg, A.E. van den, Vlek, C.A.J. and Coeterier, J.F. (1998) 'Group differences in the aesthetic evaluation of nature development plans: a multilevel approach', *Journal of Environmental Psychology*, 18: 141–157.

Berlyne, D.E. (1960) *Conflict, Arousal, and Curiosity*. New York: McGraw-Hill.

Brander, L.M. and Koetse, M.J. (2007) *The Value of Urban Open Space: Meta-Analyses of Contingent Valuation and Hedonic Pricing Results.* Amsterdam: Vrije Universiteit, Institute of Environmental Studies (IVM), Working Paper I.07/03. http://www.ivm.vu.nl/en/Images/FC28CE82920A02A711A184A85CD2E66B_tcm53-85983.pdf.

Brandt, J., Bunce, R.G.H., Howard, D.C. and Petit, S. (2002) 'General principles of monitoring land cover change based on two case studies in Britain and Denmark', *Landscape and Urban Planning*, 62: 37–51.

Brandt, J. and Vejre, H. (eds) (2004) *Multifunctional Landscapes. Theory, Values and History*. Southampton, Boston: WIT Press.

Brown, T.C., Bergstrom, J.C. and Loomis, J.B. (2007) 'Defining, valuing and providing ecosystem goods and services', *Journal of Natural Resources*, 47: 329–376.

Bürgi, M., Hersperger, A.M. and Schneeberger, N. (2004) 'Driving forces of landscape change – current and new directions', *Landscape Ecology*, 19: 357–369.

Campbell, D. (2007) 'Willingness to pay for rural landscape improvements: Combining mixed logit and random-effects models', *Journal of Agricultural Economics*, 58(3): 467–483.

Carlsson, F. and Martinsson, P. (2001) 'Do hypothetical and actual marginal Willingness to Pay differ in Choice Experiments?', *Journal of Environmental Economics and Management*, 41: 179–192.

Clark, J., Darlington, J. and Fairclough, G. (2004) *Using Historic Landscape Characterisation.* English Heritage and Lancashire County Council.

Cosgrove, D. (1984) *Social Formation and Symbolic Landscape.* London: Croom Helm.

Cosgrove, D. (2003) 'Landscape: ecology and semiosis'. In: H. Palang and G. Fry (eds), *Landscape Interfaces. Cultural Heritage in Changing Landscapes*. pp. 15–20, Dordrecht: Kluwer Academic Publishers.

Cosgrove, D. and Daniels, S. (eds) (1988) *The Iconography of Landscape*. Cambridge: Cambridge University Press.

Costanza, R., d'Arge, R., de Groot, R.S., Farber, S., Grasso, M., Hannon, B., Limburg, K., Naeem, S., O'Neill, R.V., Paruelo, J., Raskin, R.G., Sutton, P. and van den Belt, M. (1997) 'The value of the world's ecosystem services and natural capital', *Nature*, 387: 253–260.

Council of Europe (2000) *European Landscape Convention and Explanatory Report.* Florence: Council of Europe, Document by the Secretary General established by the General Directorate of Education, Culture, Sport and Youth, and Environment.

Council of Europe (2008) *European Landscape Convention. Guidelines for the implementation of the European Landscape Convention.* Florence: Council of Europe, Document by the Secretary General established by the General Directorate of Education, Culture, Sport and Youth, and Environment.

Daily, G.C. (1997) *Nature's Services. Societal Dependence on Natural Ecosystems.* Washington D.C., USA: Island Press.

Daily, G.C., Soderquist, T., Aniyar, S., Arrow, K., Dasgupta, P., Ehrlich, P.R., Folke, C., Jannson, A., Jansson, B.O., Kautsky, N., Levin, S., Lubchenco, J., Maler, K.-G., David, S., Starrett, D., Tilman, D. and Walker, B. (2000) 'The value of nature and the nature of value', *Science*, 289: 395–396.

Daniel, T.C. (2001) 'Whither scenic beauty? Visual landscape quality assessment in the 21st century', *Landscape and Urban Planning*, 54: 267–281.

Daniels, S. (1993) *Fields of Vision: Landscape Imagery and National Identity in England and the United States*. Princeton: Princeton University Press.

Daniels, S. (1999) *Humphry Repton. Landscape Gardening and the Geography of Georgian England.* England: Yale University Press.

Dearden, P. (1985) 'Focus: landscape aesthetics', *The Canadian Geographer/Le Géographe Canadien*, 29: 263.

Dramstad, W. and Sogge, C. (2003) 'Agricultural impacts on landscapes: Developing indicators for policy analysis', *Proceedings from the NIJOS/OECD Expert Meeting on Agricultural Landscape Indicators*, Oslo: NIJOS Report 7.

Dramstad, W.E., Fry, G., Fjellstad, W.J., Skar, B., Helliksen, W., Sollund, M.L.B., Tveit, M.S., Geelmuyden, A.K. and Framstad, E. (2001) 'Integrating landscape-based values – Norwegian monitoring of agricultural landscapes', *Landscape and Urban Planning*, 57: 257–268.

Earnhart, D. (2001) 'Combining revealed and stated preference methods to value environmental amenities at residential locations', *Land Economics*, 77(1): 12–29.

Eastman, J.R., Kyem, P.A.K., Toledano, J. and Jin, W. (1993) *GIS and Decision Making. Unitar Explorations in GIS Technology*. Geneva: Unitar.

Fairclough, G. and Macinnes, L. (2003) *Landscape Character Assessment. Guidance for England and Scotland. Topic Paper 5 – Understanding Historic Landscape Character.* The Countryside Agency, Scottish Natural Heritage. http://www.ccnetwork.org.uk/lca_topic_5.html.

Forman, R.T.T. and Godron, M. (1986) *Landscape Ecology*. New York: John Wiley.

Fry, G., Tveit, M.S., Ode, A. and Velarde, M.D. (2009) 'The ecology of visual landscapes: Exploring the conceptual common ground of visual and ecological landscape indicators', *Ecological Indicators*, 9: 933–947.

Geoghegan, J., Wainger, L. and Bockstael, N. (1997) 'Spatial landscape indices in a hedonic framework: an ecological economics analysis using GIS', *Ecological Economics*, 23: 251–264.

González, M. and León, C.J. (2003) 'Consumption process and multiple valuation of landscape attributes', *Ecological Economics*, 45: 159–169.

Grabaum, R. and Meyer, B.C. (1998) 'Multicriteria optimisation of landscapes using GIS-based functional assessments', *Landscape and Urban Planning*, 43: 21–34.

Groot, R.S. de, Wilson, M. and Boumans, R. (2002) 'A typology for the description, classification and valuation of ecosystem functions, goods and services', *Ecological Economics*, 41: 393–408.

Hagerhall, C.M. (2000) 'Clustering predictors of landscape preference in the traditional Swedish cultural landscape: prospect-refuge, mystery, age and management', *Journal of Environmental Psychology*, 20: 83–90.

Hagerhall, C.M. (2001) 'Consensus in landscape preference judgements', *Journal of Environmental Psychology*, 21: 83–92.

Hägerstrand, T. (1995) *A Look at the Political Geography of Environmental Management.* Dublin: University College Dublin, Department of Geography, LLASS Working Paper No. 17.

Haines-Young, R. (2000) 'Sustainable development and sustainable landscapes: defining a new paradigm for landscape ecology', *Fennia*, 178: 7–14.

Haines-Young, R. and Potschin, M. (2004) 'Valuing and assessing of multifunctional landscapes: an approach based upon the Natural Capital Concept'. In: J. Brandt and

H. Vejre (eds), *Multifunctional Landscapes. Theory, Values and History*. pp. 181–192, Southampton, Boston: WIT Press.

Hanemann, W.M. (1984) 'Welfare evaluations in contingent valuation experiments with discrete responses', *American Journal of Agricultural Economics*, 66: 332–341.

Hanemann, W.M. (1994) 'Valuing the environment through Contingent Valuation', *Journal of Economic Perspectives*, 8: 19–43.

Hanley, N., Shogren, J.F. and White, B. (1997) *Environmental Economics in Theory and Practice*. Oxford: University Press.

Hasund, K.P. (1998) 'Valuable landscapes and reliable estimates'. In: S. Dabbert, A. Dubgaard, L. Slangen and M. Whitby (eds), *The Economics of Landscape and Wildlife Consevation*. pp. 65–83, Wallingford: CAB International.

Hensher, D.A., Rose, J.M. and Greene, W.H. (2005) *Applied Choice Analysis. A Primer*. Cambridge: University Press.

Herzog, T.R. and Kropscott, L.S. (2004) 'Legibility, mystery, and visual access as predictors of preference and perceived danger in forest settings without pathways', *Environment and Behavior*, 36: 659–677.

Hilal, M., Brossard, T., Cavailhes, J., Joly, D., Tourneux, F.P. and Wavresky, P. (2009) 'Landscape metrics for determining landscape prices', *Proceedings from the 3rd Workshop on Landscape Economics of European Consortium on Landscape Economics (CEEP)*, May 29–30, France, Paris (Versailles).

Howley, P. (2011) 'Landscape aesthetics: assessing the general public's preferences towards rural landscapes', *Ecological Economics*, 72: 161–169.

Hoyos, D. (2010) 'The state of the art of environmental valuation with discrete choice experiments', *Ecological Economics*, 69: 1595–1603.

Irwin, G.E. (2002) 'The effects of open space on residential property values', *Land Economics*, 78: 465–480.

Jacobs, M. (2006) *The Production of Mindscapes: A Comprehensive Theory of Landscape Experience*. Wageningen University: PhD thesis.

Jessel, B. (2006) 'Elements characteristics and character – Information functions of landscapes in terms of indicators', *Ecological Indicators*, 6: 153–167.

Johnson, C.Y., Bowker, J.M. and Ken Cordell, H. (2004) 'Ethnic variation in environmental belief and behavior: an examination of the New Ecological Paradigm in a social psychological context', *Environment and Behavior*, 36: 157–186.

Johnston, R.J. (1986) *On Human Geography*. Oxford: Blackwell.

Kaplan, S. (1988) 'Where cognition and affect meet: a theoretical analysis of preference'. In: J.L. Nasar (ed.), *Environmental Aesthetics. Theory, Research and Applications*. pp. 56–63, Cambridge: University Press.

Kaplan, A., Taşkin, T. and Önenç, A. (2006) 'Assessing the visual quality of rural and urban-fringed landscapes surrounding livestock farms', *Biosystems Engineering*, 95: 437–448.

Kaplan, R. and Kaplan, S. (1995) *The Experience of Nature: A Psychological Perspective*. Michigan: Ulrich's Bookstore.

Kaplan, R. and Talbot, J.F. (1988) 'Ethnicity and preference for natural settings: A review and recent findings', *Landscape and Urban Planning*, 15: 107–117.

Käyhkö, N. and Skånes, H. (2006) 'Change trajectories and key biotopes – Assessing landscape dynamics and sustainability', *Landscape and Urban Planning*, 75: 300–321.

Kong, F., Yin, H. and Nagakoshi, N. (2007) 'Using GIS and landscape metrics in the hedonic price modeling of the amenity value of urban green space: a case study in Jinan City, China', *Landscape and Urban Planning*, 79: 240–252.

Krutilla, J.V. (1967) 'Conservation reconsidered', *American Economic Review*, 57: 777–786.

Li, H. and Wu, J. (2007) 'Landscape pattern analysis: key issues and challenges'. In: J. Wu and R. Hobbs (eds), *Key Topics in Landscape Ecology*. pp. 39–61, Cambridge: University Press.

Lockwood, M. (1997) 'Integrated value theory for natural areas', *Ecological Economics*, 20: 83–93.

Louviere, J.J., Hensher, D.A. and Swait, J.D. (2000) *Stated Choice Methods. Analysis and Applications*. Cambridge: Cambridge University Press.

Lörzing, H. (2001) *The Nature of Landscape. A Personal Quest*. Rotterdam: 010 Publishers.

McFadden, D. (1974) 'Conditional logit analysis of qualitative choice behavior'. In: P. Zarembka (ed.), *Frontiers in Econometrics*. pp. 105–142, New York: Academic Press.

McGarigal, K., Cushman, S.A., Neel, M.C. and Ene, E. (2002) *FRAGSTATS: Spatial Pattern Analysis Program for Categorical Maps*. Amherst: University of Massachusetts. http://www.umass.edu/landeco/reseach/fragstats/fragstats.html.

Michael, H.J., Boyle, K.J. and Bouchard, R. (2000) 'Does the measurement of environmental quality affect implicit prices estimated from hedonic models?' *Land Economics*, 76: 283–298.

Moran, D. (2006) *Economic Valuation and Landscape Character Assessment*. Report for the Scottish Government.

Muir, R. (1999) *Approaches to Landscape*. Hampshire: Macmillan Press.

Muir, R. (2003) 'On change in the landscape', *Landscape Research*, 28: 383–403.

Naveh, Z. (2005) 'Toward a transdisciplinary landscape science'. In: J. Wiens and M. Moss (eds), *Issues and Perspectives in Landscape Ecology*. pp. 346–354, Cambridge: University Press.

Naveh, Z. and Lieberman, A. (1993) *Landscape Ecology. Theory and Application*. New York: Springer-Verlag.

Ode, Å., Fry, G., Tveit, M., Messager, P. and Miller, D. (2009) 'Indicators of perceived naturalness as drivers of landscape preferences', *Journal of Environmental Management*, 90: 375–383.

Ode, Å., Tveit, M.S. and Fry, G. (2008) 'Capturing landscape visual character using indicators: touching base with landscape aesthetic theory', *Landscape Research*, 33: 89–117.

Palang, H., Alumäe, H. and Mander, Ü. (2000) 'Holistic aspects in landscape development: a scenario approach', *Landscape and Urban Planning*, 50: 85–94.

Palmer, J.F. (2004) 'Using spatial metrics to predict scenic perception in a changing landscape: Dennis, Massachusetts', *Landscape and Urban Planning*, 69: 201–218.

Parris, K. (2004) 'Measuring changes in agricultural landscapes as a tool for policy makers'. In: J. Brandt and H. Vejre (eds), *Multifunctional Landscapes. Theory, Values and History*. pp. 193–218, Southampton, Boston: WIT Press.

Pearce, D.W. and Turner, R.K. (1990) *Economics of Natural Resources and the Environment*. Baltimore: The John Hopkins University Press.

Portney, P.R. (1994) 'The Contingent Valuation debate: why economists should care', *The Journal of Economic Perspectives*, 8: 3–17.

Potschin, M.B. and Haines-Young, R.H. (2003) 'Improving the quality of environmental assessments using the concept of natural capital: a case study from southern Germany', *Landscape and Urban Planning*, 63: 93–180.

Price, C. (1978) *Landscape Economics*. London: MacMillan.

Price, C. (1994) 'Editorial', *Landscape Research*, 19: 1–2.

Purcell, A.T., Peron, E. and Berto, R. (2001) 'Why do preferences differ between scene types', *Environment and Behavior*, 33: 93–106.

Puschmann, O. and Dramstad, W.E. (2002) 'Documenting landscape change trough fixed angle photography', *NIJOS/OECD Workshop on Agricultural Landscape Indicators*, Norway, Oslo.

Randall, A. (1994) 'Contingent valuation: an introduction', *Landscape Research*, 19: 12–14.

Saaty, T.L. (1977) 'A scaling method for priorities in hierarchical structures', *Journal of Mathematical Psychology*, 15: 243–281.

Samuelson, P. (1948) 'Consumption theory in terms of revealed preference', *Economica*, 15: 243–253.

Sándor, Z. and Wedel, M. (2002) 'Profile construction in experimental choice designs for mixed logit models', *Marketing Science*, 21: 455–475.

Santos, J.M. (1998) *The Economic Valuation of Landscape Change: Theory and Policies for Land Use and Conservation*. Northampton, MA: Edward Elgar Publishing.

Santos, J.M. (1999) 'Valuing alternative bundles of landscape attributes: cost–benefit analysis for the selection of optimal landscapes'. In: K.G. Willis, K. Button and P. Nijkamp (eds), *Environmental Valuation, Vol. II*. pp. 361–385, Northampton, MA: Edward Elgar Publishing.

Schama, S. (1995) *Landscape and Memory*. New York: A.Knopf.

Scott, K. and Benson, F. (2002) *Public and Professional Attitudes to Landscape: Scoping Study*. University of Newcastle: Landscape Research Group, School of Architecture, Planning and Landscape.

Selman, P. (2006) *Planning at the Landscape Scale*. Oxon: Routledge.

Sevenant, M. (2009) 'Holism? Perception and preference in/of (remarkable) landscapes. Two case studies'. In: V. Van Eetvelde, M. Sevenant and L. Van De Velde (eds), *Re-Marc-able Landscapes. Marc-ante Landschappen. Liber Amicorum Marc Antrop*. pp. 94–109, Ghent: Academia Press.

Sevenant, M. (2010) *Variation in Landscape Perception and Preference. Experiences from Case Studies in Rural and Urban Landscapes Observed by Different Groups of Respondents*. Ghent University: Department of Geography, Doctoral Dissertation.

Sevenant, M. and Antrop, M. (2009) 'Cognitive attributes and aesthetic preferences in assessment and differentiation of landscapes', *Journal of Environmental Management*, 90: 2889–2899.

Sevenant, M. and Antrop, M. (2010) 'The use of latent classes to identify individual differences in the importance of landscape dimensions for aesthetic preference', *Land Use Policy*, 27: 827–842.

Soini, K. (2001) 'Exploring human dimensions of multifunctional landscapes through mapping and map-making', *Landscape and Urban Planning*, 57: 225–239.

Strumse, E. (1994) 'Environmental attributes and the prediction of visual preferences for agrarian landscapes in Western Norway', *Journal of Environmental Psychology*, 14: 293–303.

Swanwick, C. (2002) *Landscape Character Assessment. Guidance for England and Scotland*. The Countryside Agency, Scottish Natural Heritage. http://www.ccnetwork.org.uk/lca_topic.html.

Swanwick, C. (2004) 'The assessment of countryside and landscape character in England: an overview'. In: K. Bishop and A. Philipps (eds), *Countryside Planning. New Approaches to Management and Conservation*. pp. 109–124, London: Earthscan.

Tagliafierro, C. *An Integrated Approach To Landscape Economic Valuation*. Queen's University of Belfast: PhD thesis (forthcoming).

Tagliafierro, C., Longo, A., Van Eetvelde, V., Antrop, M. and Hutchinson, W.G. Landscape economic valuation by integrating landscape ecology into landscape economics, *Environmental Science and Policy* (forthcoming).

Tress, B., Tress, G. and Fry, G. (2004) 'Defining concepts and the process of knowledge production in integrative research'. In: B. Tress, G. Tress, G. Fry and P. Opdam (eds), *From Landscape Research to Landscape Planning: Aspects of Integration, Education and Application*. pp. 13–26, Dordrecht: Springer.

Tress, B., Tress, G., van der Valk, A. and Fry, G. (eds) (2003) *Interdisciplinary and Transdisciplinary Landscape Studies: Potential and Limitations*. Wageningen: Delta.

Tress, G., Tress, B. and Fry, G. (2007) 'Analysis of the barriers to integration in landscape research projects', *Land Use Policy*, 24: 374–385.

Tuan, Y.-F. (1975) 'Images and mental maps', *Annals of the Association of American Geographers*, 65: 205–213.

Turner, S. (2006) 'Historic Landscape Characterisation: a landscape archaeology for research, management and planning', *Landscape Research*, 31: 385–398.

Tveit, M., Ode, A. and Fry, G. (2006) 'Key concepts in a framework for analysing visual landscape character', *Landscape Research*, 31: 229–255.

Tveit, M.S. (2009) 'Indicators of visual scale as predictors of landscape preference; a comparison between groups', *Journal of Environmental Management*, 90: 2882–2888.

Ulrich, R.S. (1983) 'Aesthetic and affective response to natural environment'. In: I. Altman and J.F. Wohlwill (eds), *Human Behavior and Environment*. pp. 85–125, New York: Plenum Press.

Uuemaa, E., Antrop, M., Roosaare, J., Riho, M. and Mander, Ü. (2009) 'Landscape metrics and indices: an overview of their use in landscape research', *Living Reviews in Landscape Research*, 3(1), see http://www.livingreviews.org/lrlr-2009-1 (accessed 17 March 2012).

Van Eetvelde, V. and Antrop, M. (2009) 'A stepwise multi-scaled landscape typology and characterisation for trans-regional integration, applied on the federal state of Belgium', *Landscape and Urban Planning*, 91: 160–170.

Zajonc, R.B. and Markus, H. (1982) 'Affective and cognitive factors in preference', *Journal of Consumer Research*, 9: 123–131.

Zonneveld, I.S. (1995) *Land Ecology*. Amsterdam: SPB Academic Publishing bv.

3 Subjectivity and objectivity in landscape evaluation

An old topic revisited

Colin Price

Introduction

There has long been uncertainty about whether the basis for landscape evaluation should be subjective or objective (Appleton 1975; Gilg 1975; Clamp 1976; 1981; Jacques 1980; Dearden 1981; Aoki 1999). In particular, in the 1960s a vigorous debate was initiated between those who favoured a "holistic" approach to quantitative landscape evaluation – generally identified with the subjective philosophy – and those who preferred a "components" approach – usually interpreted as objective in orientation. The *holistic* philosophy was most popularly characterised by the system of the planner, K.D. Fines (1968): he devised a scoring system to help identify the least damaging visual corridor for an electricity transmission line through the South Downs Area of Outstanding Natural Beauty in south-east England (see Table 3.1).

At about the same time a geographer, David Linton (1968) published a system for landscape evaluation that relied upon two physical *components* of the landscape: the land form, and the land-use. The land form scores he gave ranged from 0 (lowlands) to 8 (mountain). Different land-uses and land-use combinations were given points, according to a scheme which actually evinced no small degree of judgement: for example "wild landscapes" scored 6, "richly varied farming landscapes" scored 5 and "continuous forests" scored −2. The importance of water bodies was recognised, scoring 2 points if an important part of the foreground, and 1 if otherwise significant. Urbanised and industrialised landscapes produce detractors which could qualify for negative points, −5 in the guidelines. Other components studies have also referred to loosely defined detractors (Penning-Rowsell and Hardy 1973).

Somewhat earlier, Rodney Helliwell (1967) had proposed another components-based system, for valuing amenity trees. It required the scoring, on a scale 1–4, seven characteristics of a tree, in itself and in relation to its environment. The characteristics were: size, useful life expectancy, importance of position in landscape, presence of other trees, relation to the setting, form, and special factors (such as historical or botanical importance). Helliwell's system differed from Linton's and Fines's in that it aimed directly to produce an economic value, by multiplying the product of scores for each of the characteristics by a monetary sum

Table 3.1 Fines's scales

Fines's descriptive category	*Fines's numerical scores*	*Harding's adaptation*
Unsightly	0–1	0–5
Undistinguished	1–2	5–10
Pleasant	2–4	10–15
Distinguished	4–8	15–20
Superb	8–16	20–25
Spectacular	16–32	25–30

determined by proposal, discussion and custom. The system has been modified a little in response to feedback, and is in everyday practical use: for example in determining due compensation for destruction of trees by developers. There is presently controversy between advocates of the Helliwell system, which focuses on the benefit of the tree, and those of the system promulgated by the Council of Tree and Landscape Appraisers (frequent updates), which is more concerned with replacement cost.

A third position in the debate was that any attempt to evaluate landscape was improper and vapid (Carlson 1977).

The controversy has had its parallels in the more recent discussion on economic valuation of landscape, the Contingent Valuation Method (CVM) resembling the holistic approach, while the Hedonic Pricing Method (HPM) shares aspects with the components approach. Choice Experiments potentially occupy an intermediate position. Again, a third standpoint is presently advocated: that monetary valuation is undesirable and unethical, not to mention impossible.

This chapter ruminates upon these debates, in particular querying whether objective methods are all that is sometimes claimed for them in terms of scientific reproducibility. It meditates on the nature of subjectivity, and uses data from a series of landscape evaluation exercises to investigate the subjectivity/objectivity distinction. After comparing the general results with those arising from the more recent application of economic valuation, it commends an approach that recognises the limitations of claimed objectivity, and gives explicit human judgement an appropriate role in deriving cash values.

The nature of subjectivity

While the components approach relied on physical categories that could be objectively measured, or marked as present or absent, the holistic approach could be, and often was, characterised as subjective: it relied on an individual's impression of the landscape's quality, whether expressed in descriptive terms or numerical ones.

Within scientific communities, the term "subjectivity" has derogatory connotations: to call data "subjective" is to condemn them as "of *less* worth" or even "worth*less*". But the word has – or is used in – a number of senses, each with rather different application to landscape evaluation. The following discussion is slightly adapted from a report to the UK Countryside Commissions (Price 1991).

"Objective", "accurate", "representative" and "impartial" on the one hand, and on the other "subjective", "approximate", "idiosyncratic" and "biased" are terms distinguished rather unclearly by usage in the literature. Consider the following statements:

1. The girth of this tree at 1.3 metres above ground level is 89.3 centimetres.
2. The canopy cross-section of this tree is about 200 m^2.
3. This tree is large.
4. This tree is magnificent.
5. Everyone agrees that this tree is magnificent.
6. On this tree's trunk I carved my girlfriend's name.
7. This tree has special significance to me.

Statement (1) is objective, accurate and representative: it refers to an external standard of measurement, and gives a figure which would be reproduced within close limits by all trained observers.

Statement (2) is objective, approximate and representative, given that the measurement method (for example, measuring the size of shadows) might produce a wider range of answers, although the standard of measurement remains external.

Statement (3) is subjective, approximate and representative: the tree is measured in relation to an *internal* and rather ill-defined standard, which nevertheless is held in common by numerous observers.

Statement (4) is also subjective, approximate and representative, if it is based on the responses of a number of people asked to classify the visual quality of the tree as "magnificent", "unexceptional" or "poor".

Statement (5) is objective, accurate and representative: anyone could observe, against the external standard of counted responses, that the entire interviewed population agreed with the statement.

Statement (6) is objective, accurate and idiosyncratic: a number of witnesses could judge its truth by the external evidence of their eyes, but it is only true as a statement by a limited number of people.

Statement (7).is subjective, approximate and idiosyncratic: a small proportion of the population can make it, by reference to an ill-defined internal standard of specialness.

"Bias" is a concept of rather different sort, denoting in an objective context a tendency (for example of a measuring instrument) to be inaccurate in a particular direction. In a subjective context it denotes deliberate distortion or unintentional misrepresentation of truth. Bias is not confined to subjective matter; but subjectivity, approximateness and idiosyncracy make it harder to detect. "Falsity" is the extreme form of bias, meaning that neither direction nor magnitude of bias is known. All seven statements are subject to bias, in the sense that mismeasurement, miscounting, and misrepresentation are universal possibilities.

Landscape evaluation is normally seen by outside commentators to have the character of statement (7), and to be very susceptible to bias. However, in reality landscape evaluation may be objective (we may record the number of people who say they like a landscape), accurate (all observers could agree on the number of people who said they liked it) and representative (a broad consensus could be found on which landscapes were most liked). In this sense landscape evaluation – or for that matter evaluation of ecosystem services – does not differ from the generality of economic evaluations, in which economists objectively record how many people are willing to pay how much for a given quantity of a specified commodity. The difference, if any, is that not everyone faces a similar price for high-quality landscape, or that (it is widely said) landscape does not have any price at all.

Much of the effort made in landscape evaluation can be seen as an attempt to move what is subjective, approximate, idiosyncratic and biased towards being objective, accurate, representative and impartial. But each element of the transition is different and involves a different modification of techniques, and it may even be that there is a trade-off between "improving" the data in one dimension, and improving it in another.

Statistical tests of subjectivity and objectivity

Fines's descriptive and numerical system of landscape evaluation became a popular topic for testing in the 1970s (e.g. Penning-Rowsell and Hardy 1973). The method can easily be applied either in the field or by use of photographic or other visual representation. Observers require little training, and numerical results are readily generated and replicated.

During the period 1978–2010, groups of undergraduate and postgraduate students of what is now Bangor University were asked to evaluate a given set of views, using a modified version of Fines's scale. In the first form of evaluation, a range of five points was assigned to each of the descriptive categories, it being reported that difficulties were created by Fines's geometrical scale (Harding, pers. comm – see Table 3.1.). Subjectivity does not cease to be subjectivity, at the moment when a number is attached to its judgements. But quantification allows statistical testing, to establish whether there are objectively significant patterns in the data. In 1990 some such analysis was initiated, but during the past 10 years a consistent form of analysis has been applied, to allow comparability between groups. In some of those years, data were collected by up to five separate groups, giving a total of 29 usable data sets for this analysis, encompassing the judgements of some hundreds of evaluators.

Each replication of the exercise involved views drawn from the same set of 14. Rarely, problems such as those caused by the outbreak of foot and mouth disease in the UK in 2001 or by temporary disputes over rights of way meant that the complete set could not be evaluated.

Second, a modified version of Linton's system was applied to the same views. Land form was expressed as "flat" (1 point), "undulating" (2 points), "steep"

(3 points), "hilly" (4 points) or "mountainous" (5 points). To avoid the matters of judgement that are apparent in Linton's scheme, land-use was reflected in terms of its variety, the following categories being recorded as [significantly] present or absent: cultivated land, broadleaved trees, coniferous trees, "wild" (semi-natural) vegetation, (attractive) urban use; each use present scored 1 point. Water was either absent (0 points), present (1 point) or significant (2 points). Detractors (whatever were judged as such) scored −1 point, irrespective of their number or size.

Third, Helliwell's system was applied to a defined set of trees. Helliwell's system has been modified somewhat over the years, but an early form was used throughout, to allow comparison over a long time period.

Subjective scores for views

After a brief introduction to the systems and their problems, respondents were asked to score each view *as they saw it*. Thus the raw scores were idiosyncratic, but collectively were representative of the population from which the student sample was drawn (not necessarily of the whole national population).

In all 29 data sets, differences in scores between views were very highly significant (the calculated *p*-values – the probability of the differences arising by chance – are typically one-in-an-astronomical number). The individual scores may be subjective and idiosyncratic, but one may (objectively) observe an overwhelming agreement that, in human terms at least, some views are regarded as better than others.

From the 29 data sets, there were 26 which included one particular view. In 20 cases it was identified as "the best"; in four other cases it was "second best"; and in the remaining two, "third best". (Three other views out of the 14 shared the remaining six first places.) This one, much-favoured view did not have the most grandiose mountains (the highest mountain in Wales was included in one of the other views), nor the most varied land-use, nor the most dramatic sweeps of water. What it did have was an exquisite composition. It was a "picture post-card" view. This outcome is of great importance to the later argument.

At the other end of the scale, in all 29 cases, one particular view – a scene characterised by detractors and dereliction – was classed as "unsightly" and given the lowest mean score, as indeed it had been in all the cases before formal data recording began. This consistency over many years suggests that preferences are not just the result of highly transient fashions, as has been suggested in a debate reviewed in Sheppard and Harshaw (2001).

There was less agreement on ranking of the medium-quality views, for which it seems that individual taste for a particular type of landscape (upland, rural, coastal, urban) exerts an influence.

Differences between individuals

By contrast, in 75 per cent of cases there appeared at first to be no significant difference *between* individuals in their scoring of landscape – the idiosyncratic element seemed negligible. However, when the variable tested was individuals' *deviation* from the group's mean score for each view, in 75 per cent of cases there was a consistent and significant tendency for some individuals to under-score or over-score by comparison with the group. This difference was nonetheless much less significant than the difference between scoring of views.

Despite these differences, the variation in subjective scores is not inexplicable. Rather, it demonstrates that response to landscape is a function of characteristics – perhaps objectively measurable ones – of the *observer* as well as of the *landscape*. Although no consistent attempt was made to include a formal test in these exercises, it was informally noted that the best correlations between individual scores were often between those relatively few participants who came from North Wales, where the exercises were being carried out. Conversely, if there were participants who were resident outside Europe, their scores were almost invariably the least well correlated with the group's mean. Expectedly, cultural background and customary landscape experience seems to have an influence, which is in principle susceptible to quantitative investigation.

Consistency of subjective and objective scores

The four (supposedly) objective components of landscape were assigned a number by each group. Because there was sometimes disagreement about which category a view should be placed in (where does "hilly" shade into "mountainous"?) sometimes fractional scores were given, according to the weight of opinion favouring one or other category. This consensus approach meant that no statistical tests could be done *within* a data set, but comparisons could still be made *between* the 27 data sets valid for this purpose. In each test the replicates are the scores given by each group to a particular view or particular attribute of a particular tree. The summary measures of variation are then averaged across all views and all trees.

Because the data were not strictly on a continuous cardinal scale with a normal distribution, the results should be interpreted with some caution. Nevertheless, they are interesting and revealing. Table 3.2 summarises measures of consistency between data sets, based, for each view, on the standard deviation of the 27 mean or consensus scores given by the different groups.

Subjective scores of the views' quality seem on the whole to be as consistently assessed between data sets as are assessments of landscape's objective components. Table 3.3 presents a similar result for the 29 consensus scores given to each attribute of each tree while applying Helliwell's system. (The same caveats apply to the nature of the variables as are noted for Table 3.2.)

There seems to be no major difference between arguably objective attributes such as "size" and "presence of other trees" on the one hand, and, on the other, arguably subjective ones like "suitability to the setting" or "form". This can be attributed to judgement affecting the objective attributes too. The attribute "size" might seem to be objectively measurable, but when it is measured by categories such as "very large", "large", "medium" and "small" there is a problem with internal standards: large in relation to what norm? For example, it has been quite clear from discussions that, in an enclosed setting, the tree's size was partly judged in relation to the space which it occupied. Helliwell's system now defines tree size in the objectively replicable terms of vertical-section area of crown, but there is still a subjective issue about where the boundaries are placed between size categories. While guidance is offered on all the other included categories, there remains an element of judgement. For example, a group of cabbage palms beside Bangor Pier is "unsuitable to setting" for ecological purists, but "very suitable to setting" for cultural symbolists. What defines "special" in relation to historical, botanical or other associations is inevitably subjective.

Apart from the "special factors", which were infrequently attributed and which the groups were in no position to judge, there was no overall significant difference in standard deviation or coefficients of variation across the attributes, irrespective of their degree of subjectivity: similar agreement between groups existed for all of them.

Because of the multiplicative mode of compiling monetary value, this summary measure is more variable than the individual factors. Nonetheless, there is monumental significance in the difference in value between the nine trees. By contrast and predictably, when an additive compilation was made, the variability was generally less than that for the individual factors.

Relationship of subjective score to components

For each data set, mean subjective score for each view was regressed on the four "objective" components. All except one of the valid 27 cases (groups of evaluators) produced a regression which was significant at the 5 per cent level, but only 12 were significant at the 1 per cent level. Of the individual components, land form was significant at the 5 per cent level in ten cases, land-use variety in three cases, presence of water in two cases, and detractors in five cases. Although a common set of views was being appraised, different individual components appeared to be significant in different cases. Out of the 16 logically possible combinations of significant components, seven were found in practice. In no single case were all four components significant, in only one were three significant, and in 11 cases no individual component at all was significant. The median adjusted R^2 was 0.598, which is to say, that in half of all cases at least 40 per cent of all variation in the subjective score remained unaccounted for by the four components, although these components have often been advised as the most significant determinants. While the low coefficient of variation (Table 3.2) shows that there is strong agreement on the status of water bodies as a component of individual views (absent, present, significant), it seems that in these data sets at least this physical measure has little demonstrable effect on landscape's subjective score. Moreover, in three individual regressions of subjective score on components, water had a perverse negative coefficient. These unexpected results might be attributable to collinearity between presence of mountains and absence of water in the view.

The variation in the structure of these descriptive models does not give confidence that such results can validly be transferred to other contexts.

Table 3.2 Measures of consistency in subjective and objective scores given to views

Variable	*Range across views of coefficients of variation*	*Mean coefficient of variation across views*	*Mean of [standard deviation ÷ possible range]*
Mean subjective score	0.079–0.328	0.156	0.068
Components (objective):			
• Land form	0.014–0.197	0.090	0.054
• Land-uses	0.075–0.632	0.196	0.123
• Water	0–0.368	0.132	0.047
• Detractors	0–2.367	0.787	0.313
Summed components	0.069–0.288	0.171	0.093

Table 3.3 Measures of consistency for the attributes of each tree in Helliwell's system

Attribute of a particular tree	*Range across trees of coefficients of variation*	*Mean coefficient of variation across trees*	*Mean of [standard deviation ÷ possible range]*
Its size	0–0.301	0.130	0.090
Its life expectancy	0.070–0.508	0.171	0.126
Its position	0.107–0.491	0.216	0.153
Presence of other trees	0.090–0.856	0.300	0.280
Its suitability to the setting	0.107–0.284	0.176	0.144
Its form	0.061–0.365	0.211	0.130
Special factors	0.000–0.347	0.079	0.036
Its value (a multiplicative aggregation of the characteristics above)	0.328–1.173	0.682	0.015

In sharp contrast, subjective scores showed a much more consistent statistical significance, and at a much higher level. Usually, any one individual's score was well correlated with any other individual's. In about half of all cases my own score was the best correlated with the mean of all scores. The median adjusted R^2 for the regression of mean student score on my own score was 0.897. This relationship was always significant, at any reasonable level of significance. Even in the two cases where the adjusted R^2 value indicated a higher degree of explanation of mean score by the four components than by my individual score, the relationship with my score was more significant as judged by p-value. A "representative evaluator" seems to give a better prediction of a group's score than do the objective components.

Effect of other variables

As to the variation *between* data sets, the belief was often expressed that the weather or season had affected individuals' scores. Typically, it was perceived that the order of influence, from greatest to least, was: landscape, personality, weather, time and season, mood. This is useful in that the stable factors of the view – the actual landscape – and of the individual – personality – are rated more important than the transient ones, and the data do tend to bear out this subjective perception. In recent years a composite weather score has been given by assigning +1 to "good" descriptive words, and −1 for bad ones. The perception that weather had had a moderate effect on scores, unexpectedly, was hardly borne out by the statistical evidence: at the 95 per cent confidence level, mean scores were only just significantly related to the index of "goodness" of the weather.

Subjective scoring has shown itself as robust and consistent over the years, and explicable in terms of the views scored and individuals doing the scoring. Random, it is not. Objectivity seems elusive, if the requirement is that characteristics of the landscape are uniformly, consistently and precisely measured. If the relationship between subjective and objective scores is very far from perfect, it cannot, on the basis of these data, be blamed on the waywardness of the variables that are conventionally considered subjective.

Parallels with contingent valuation versus hedonic pricing

The perceived opposition of subjectivity and objectivity in landscape evaluation has its parallels in the present debate about *economic* valuation of landscape. Expressed or stated preference takes the position of subjectivity, and revealed preference that of objectivity.

The Contingent Valuation Method (CVM) appears to be holistic in its aims and achievements. It seeks individual respondents' willingness to pay (or accept compensation) for a difference between two or more described or depicted overall states: between, for example, the landscape as it is, and the landscape as it might become, either through hostile physical exploitation, or following purposive aesthetic improvement. It stands at the end of a line of development of the holistic approach. Fines quantified what landscape architects, idiosyncratically, had heretofore judged. Exercises such as those described above add representativeness to quantification. CVM puts quantitative and representative judgement in monetary terms, so as to facilitate comparison of landscape with other values. But it does not add any objectivity. Subjectivity is equally present in the individual's judgement of the worth of the landscape: differences lie only in the terms in which that subjective valuation is expressed. Objectivity is potentially present at each stage of the process, but only in the sense that one may record what judgement it is that individuals represent themselves as making, and the number of individuals who make that judgement. It might even be argued that introducing monetary judgements, in relation to specified scenarios, increases the incentive for strategic bias.

The Hedonic Pricing Model (HPM), on the other hand, is based on the revealed preference – the observed willingness of individuals to put hand in pocket – that is so conspicuously absent from CVM. It seeks to value physical, locational and environmental attributes by recording actual payments made for houses, or for leisure trips, or for industrial or business locations, each of which offers a different package of these attributes. Then statistical analysis is deployed to associate differences in price paid with particular differences in attributes.

In character with its provenance in holistic methods, CVM of landscape, being recipient-orientated, has often focused on the characteristics of the *respondent* rather than those of the view. HPM is resource-based and has focused on the attributes of the resource, which are, perhaps, more susceptible to objective measurement. Like the components approach to landscape evaluation on which it builds, HPM is more accountable in the sense that it attempts to show due cause for willingness to pay. And, like the components approach, it has a tendency to claim more objectivity than it is rightly entitled to do, as will be discussed later on. Nonetheless, because it is based on judgements that have a real cost to the individual, its value base has rightly been considered more reliable. There is less opportunity for bias and particularly for misrepresentation of the subjective values that underlie preference. It is the strategic and other reasons to misrepresent or miscomprehend the truth, and not the underlying subjectivity of the judgement, that differs between CVM and HPM.

Once again, in the monetisation version of the evaluation debate, there is a third position, of those who revile the whole attempt to quantify – and particularly to monetise – the value of the environment: any attempt at monetary valuation is a futile, self-deluded enterprise (Sagoff 1988; Clark et al. 2000). And, by way of a final contrast, there have always been those who see monetisation as a profoundly painful but necessary expedient, if the "due regard" – so glibly talked of in policy pronouncements – is to be given to aesthetic values. I too 'would prefer to walk through a world of beauty forever unthreatened, forever intuitively appealing' (Price 1978: xi).

CVM and landscape

CVM, its deficiencies, and remedies for deficiencies have all been repeatedly reviewed, and the process need not be reprised here (Mitchell and Carson 1989; Garrod 2002). While early CVM applications focused on forest recreation (Davis 1963; Price 1970), and more recent ones have mostly concerned wildlife and habitat (open a volume of *Ecological Economics* for examples), there have been many applications over the years whose primary concern was landscape (e.g. Brookshire et al. 1976; Randall et al. 1978; Willis and Garrod 1993; Tyrväinen and Väänänen 1998; Navrud et al. 2008). Because projected changes are visual and evident, there are fewer problems in clearly representing what is to be valued than there are, say, with habitat conservation, or actions that affect global climate change or hydrological status.

One of the persistently recurring problems of CVM is embedding or part-whole bias: the situation where enhancing or preserving a particular landscape is taken as an emblem for the value of all landscapes of that type, all landscapes, or even all environmental improvement and protection. This is of major importance, because the hypothetical question is usually asked in the context of a real problem, which may be as small scale as an environmental improvement scheme for a small housing estate, or as large scale as national or even international policies towards protected areas generally. For example, MacMillan and Duff (1998) found that, implausibly, the willingness to pay for all Scotland's native Scots pine forests was only a little greater than willingness to pay for the forest in one glen. Hanley et al. (1998) suggest that follow-up questions can detect and eliminate the bias.

CVM's stock-in-trade is valuing the differential between with- and without-change scenarios. Relatively few offer a range of options from which the value of different components of landscape could be deduced. In the Yorkshire Dales National Park, Willis and Garrod (1993) did seek willingness to pay for the one among eight potential landscape treatments that the respondent favoured. Unfortunately the choice was a rather nebulous one, no differential being specified between willingness to pay for the preferred landscape treatment and willingness to pay for any other alternative: had this been established, it would have yielded an individual valuation of alternative landscape treatments. Another approach is to use post-CVM discussion to probe what features of landscape were deemed attractive by (a sample of) respondents. Structured discussion may lead to prejudgement: on the other hand, open-ended discussion tends to produce an anecdotal quality, open to subjective selection and interpretation by the researcher.

At the end of all, CVM directly seeks what decision makers desire to know: what, in monetary terms, is a difference in environmental condition worth to those who would experience it? While mediated by the objective recording of stated willingness to pay, it refers nonetheless to the idiosyncratic, holistic preferences of the respondents. Apart from its well-identified general biases and ambiguities, CVM has two particular weaknesses in relation to landscape valuation (see also other contributions in this book):

- its case-specificity and the doubt that is consequently cast on transferring benefit valuations to another context; and
- its hypothetical nature, including the problem of defining what exactly is being "purchased", as found in part-whole bias and in varied personal meanings attached to the product being offered (Price 2001).

These are two problems that HPM has apparent potential to solve.

Components: disaggregation, pricing and recomposition

Precisely because the choice made is not hypothetical, precisely because preferences must be backed up by actual payment, the product valued by HPM is well specified: it is in the consumer's interest to establish what is being legally acquired, whether in the great matter of house purchase, or lesser ones of selecting the destination or route for a journey. By buying a house with a good view, one does not buy *all good views*, and nobody would be so foolish as to suppose that they did. By contrast, in a CVM there are no very tangible ill-consequences of making the wrong assumption, about whether the hypothetical willingness-to-pay question refers to one, or to some, or to all aspects of landscape or environment. The second apparent advantage is potential transferability. Benefits recorded in one context may be transferred to another, perhaps very different context, by recomposing a value package from the value of the package's components, as measured elsewhere. Perhaps the most common, because the most readily measured, environmental variable is proximity to parks or other greenspace (Choumert et al. 2009). GIS allows a high degree of automation in treating such variables. But proximity may bring a range of effects. It provides access for physical recreation (Anthon et al. 2005). It may bring semi-natural ecosystems into view from the house (Tyrväinen and Miettinen 2000). It encourages a "better kind of neighbour" through the effects of elevated house price on *discouraging* purchase by a "worse kind of neighbour" (Price 1995). It may even bring a degree of nuisance, when parks are linked with antisocial behaviour. And, in the case of forested areas on the south side of housing in the northern boreal zone, it brings significant exclusion of winter sunlight, which may account for apparently perverse effects of distance (Tyrväinen 1997).

Very few studies have attempted, or even suggested the attempt, to relate house price or travel cost to a (subjective) assessment of holistic landscape quality (Abelson 1979; Brown and Mendelsohn 1984; Henry 1994, 1999; Bergin and Price 1994). It may be no small part of the discouragement to take such an approach, that it exposes the study to the charge of subjectivity – subjectivity on the part of the investigator or the investigator's agents, rather than that of the individuals investigated.

More common have been attempts to relate house price (or, less frequently, travel cost) to objectively measurable components of the landscape. The effect of presence, proximity, quantity or quality of trees has been particularly heavily researched (Payne and Strom 1973; Morales et al. 1976; Anderson and Cordell 1988; Garrod and Willis 1992; Hanley and Ruffell 1993; Kim and Johnson 2002). So too has the presence or quality of water bodies or courses (Garrod and Willis 1992; Luttik 2000; Bourassa et al. 2004; Gupta et al. 2009) or agricultural land (Fleischer and Tsur 2009). On the face of it, this components approach in HPM is desirable, offering the building blocks from which can be constructed the monetary worth of a given change in the physical condition of the landscape.

Choice Experiments (CEs) and components

In the past 10 years, Choice Experiments (Hanley et al. 1998) have emerged from obscurity to challenge CVM's position as the pre-eminent method of environmental evaluation. They are very like CVM in that they seek a stated willingness to pay, but also like HPM in that they endeavour to assign portions of the willingness to pay across defined attributes of the product that is offered. Like CVM and HPM, CEs have a provenance in earlier, non-monetary assessment techniques: in this case democratic choice between bundles of social goods (Hoinville 1971). A major advantage of Choice Experiments is that they avoid the headlining of a particular issue, a defect that is inherent in a traditional CVM (about a particular habitat's conservation, for example, or degradation of a particular landscape, or enhancement of a particular recreation opportunity). By referring to a particular, polarised case, the very act of applying a CVM questionnaire may inculcate a biased view about its importance, such that respondents will offer to pay for something they do not know about, even something whose existence they doubt (Price 2000). It has been argued that giving detailed information about the issue is necessary to achieve an "informed" decision (Bishop and Welsh 1993). But doing so makes the respondent exactly atypical of that stakeholder population by which the mean valuation from the sample will be multiplied (Price 1999). The questionnaire compiler's subjective choice of what objective facts are presented to the respondent must also affect the value given, as do the words chosen to convey the information.

Choice Experiments do not, or should not, offer a not-environmental baseline. Because they have the capacity to present increments of different kinds of environmental quality or action, rather than a single hook on which to hang environmental concern, they have potential to solve the part-whole bias problem: they are explicitly concerned with what *a little more* of a specified environmental quality is worth, and so steer the respondents' attention away from making big statements about the worth, or primacy, or inviolability, of the environment generally. Choice of information and words will still be influential, but it will be less clear that there is a "correct, pro-environmental" answer to the question. Nonetheless, it is evident that sometimes an environmental attribute might be chosen at its lowest offered level, merely as a token of environmental concern. For example, Nielsen et al. (2007) found a preference for the lowest offered number of dead trees in a forest, as being symbolic of "natural forest" values.

Similarly, by offering a choice between packages containing different amounts or qualities of a number of attributes – of which money is only one – Choice Experiments help to decouple the trading of a single environmental issue against money, which is one aspect of CVM that objectors find odious.

Choice Experiments may vary different kinds of entity between packages. In some applications, elements in the package have been a number of leisure activities, as different as evenings at the theatre and recreational visits to forests (Sievänen et al. 1992). An alternative might be to vary the composition of projects constituting an environmental improvement programme. Other applications have varied the levels of attribute composing a particular kind of experience (Nielsen et al. 2007). The approach does not avoid the problems of unfamiliar, hypothetical choices inherent in CVM. But it does offer the potential of emulating HPM's teasing-out of the components of value, in a manner less conscious (and therefore less prone to deliberate biasing) than would be achieved by post-CVM discussion.

Decomposing values and subjectivity

To recap, the advantage of decomposing the value of landscape into its components, by HPM or by CEs, is that it provides an apparently objective scheme for estimating, *ex ante*, the benefit or cost of making a change in the landscape elsewhere. Also, by filtering subjective judgements of value through statistical analysis, and distancing them by transportation to another location and situation, it *seems* to put a shell of objectivity around the soft human centre in which value judgements were formed and nurtured. But even ignoring such glossing-over of the subjective source of the value, there are obstacles to decomposing and recomposing values in this way.

Choosing the components

The components to be included in a model of landscape quality are not self-evident. They may be chosen through subjective experience of aesthetic experts or consulted user groups. Or they may be selected according to which ones perform best against statistical criteria. But even in the latter case, someone must choose which components are to be tested. The four components chosen for our field exercises – namely, land form, land-uses, water and detractors (see Table 3.2) – had been considered important by experts, yet they performed rather poorly in any one field exercise, and very inconsistently between exercises.

Beauty in landscape is a subtle quality, dependent on many small things as well as on a few major components. If there are missing variables, not only will the model be incomplete, it is possible that a surrogate variable which *is* present in the model may be loaded with false significance. For example, the surprising result, that the presence of a lead mine increases the value ascribed to a landscape might arise, because most UK lead mines are located in national parks and Areas of Outstanding Natural Beauty (ONB), and because those designations or some other related variable has been omitted from the model.

Measuring the components

An objectively existing physical component has several measurable properties, some combination of which must be selected. For example, is the most appropriate quantitative predictor of the impact of an electricity transmission line on a view

(i) the number of pylons visible?
(ii) the apparent length of the line in relation to the linear dimensions of the field of view?
(iii) the apparent area of the envelope enclosing the outer limits of the structures?
(iv) other?

Defining the category boundaries

Where the representing variable is a categorical one (like “wild land”), how and by whom are boundaries of the category defined? How is it done with physically measurable terms like size category or with degree of psychological impression like “fairly suitable . . .” and “very suitable . . . to setting”?

Linearity

In order that arithmetic can be performed, both in statistical analysis of data and in compilation of a value from components, a relationship must be in place between the quantitative measure of the components, and the quantitative expression of value. Typically, a linear relationship is used, each increment in the measure of a particular component being implicitly assumed to cause an equal increment in value.

There is no general *a priori* reason to expect linear relationships, and plenty of reasons and results to favour non-linear ones. For example Whiteman (1995) found a public preference for 20–30 per cent of tree cover within recreational areas. Schroeder (1986) found a similar result for preferred density of tree cover in public parks. Starting from open land, increments of tree cover would at first increase the ascribed value, but successive increments would increase it by less, and eventually reduce it.

Nothing in the theory of hedonic pricing or Choice Experiments requires a linear relationship. But *some* functional relationship must be chosen, either on grounds of *a priori* plausibility, or by inspection of the statistical performance of alternative functions. In the latter case, the number of theoretically possible functions is boundless, for the component variable may be transformed in any number of ways, and its transformed states may be combined in any number of ways, as for example in a polynomial equation. This brings about statistical problems alluded to below.

Additive separability

Most frequently, HPMs have further assumed the additive separability of components, in their influence upon value. That is, the value ascribed to an increment in one component is independent of the current level of any other component. The overall relationship can therefore be represented by the sum of the terms which represent each component.

The typical CE has an efficient factorial design which also implicitly assumes separability. Not all possible combinations of levels of all attributes are offered for evaluation, but only sufficient to establish the value interval between levels for each attribute considered independently. This does not presuppose any particular form of functional relationship between level of attribute and value. However, such a design can only produce valid results if each difference in level is evaluated with all other attributes held constant.

Again, neither theory nor practice supports such assumptions. For example, experience has shown that the judged effect of an intrusion such as an electricity transmission line on a landscape is sensitively dependent on the topographical configuration of that landscape (Price 2008). The effect of land-use mixtures (for example, those containing geometrical commercial plantations or "improved" pasture) is also conditioned by the way in which the topography presents them visually. Hence the aesthetic objection to such land-use changes in the UK has focused on mountain areas.

Again, it is possible to construct interactive terms containing two or more components in a non-additive relationship. But this proliferates the number of possible equations, by unbounded orders of magnitude.

Composition

Additive separability is about the configuration of variables within the explanatory equation. But the effect of components in the landscape depends also on their spatial configuration on the ground, not just on their presence or quantitative level. This was what made the "picture post-card view" the overwhelming favourite of our landscape evaluation exercises. This, time and time again, is revealed by anecdotal evidence to be the key to determining famous viewpoints, within landscapes that are generally mountainous and generally endowed with diversity of land-use, presence of water, and absence of detractors. Historically, it lay behind the framing of "picturesque" views in landscaped parks (U. Price 1810). The finesse of composition also affects the impact of detractors. 'In some landscape types pylons seem hardly relevant. In others, particularly landscapes that are *delicately composed*, they may be horribly disruptive' (Price 1993: 56) [emphasis added]. Also '... the joint effect on landscape of road and pylon lines was often greater than the sum of their separate effects, because their *interaction* could be strangely distracting' (p. 61) [emphasis in original].

It could be said that composition is the key missing variable that accounts for much of the residual variation in scores given to views, when all the obvious components have been taken into account. It is of interest that in an early study of attributes of photographs, it was the length of the boundary between elements of the landscape – a *relationship* between its components – that proved the most significant variable (Shafer and Metz 1970).

Portfolios of experience

For HPMs based on travel cost, there is the additional problem of allocating cost over the multiple sites visited on a journey, in addition to the multiple attributes of each site – and the many other problems of the method.

The nub of the problem

Landscape does not lend itself to the straightforward defaults that underlie statistical regression analysis. It is subtle, complex and profoundly interactive. The possibilities for unravelling the complex relationships between components and value are, either to impose on the data a form derived from a subjective impression of the modeller, or to allow the data to impose a form in a way that covertly undermines the statistical validity of the analysis, as follows.

> It is a matter of mathematics that *n* sets of data can always be represented exactly by an expression in *n* terms (for example three points through which

> the curve $y = ax^2 + bx + c$ can always be made to pass). Such an expression implies no causal relationship. It simply describes those data, which might be entirely random: it cannot yield reliable predictions.
>
> (Price 1976: 830)

If subject to statistical analysis, such an expression will yield no significance, because there are no remaining degrees of freedom. However, many different forms of relationship containing fewer than n terms may be separately tested, one after another, without reducing formal degrees of freedom. *Sooner or later* one form of relationship will be tested, which seems to offer a high statistical significance, which may then be invested with a predictive capability which it does not in fact have. After all, if 100 data sets consisting entirely of random numbers are subject to statistical analysis, one in 100 will, on average, appear to show statistical significance purely by chance. Applying 100 widely different transformations (of functional form and of interacting variables) to a given data set would by the same token as often as not produce one relationship that was "statistically significant".

Flexible curve fitting has been commended in the literature of HPM (Price 1994). But this can be adopted only with sacrifice of real degrees of freedom. The whole tendency of the work reported at the beginning of the paper corroborates this sceptical appraisal, suggesting that components-based approaches yield unstable models, even within a given landscape context: the variables selected as significant differed substantially even for this one set of landscapes, according to the valuations of groups of evaluators drawn from a rather homogeneous population. Putting a monetary step into the process would add a dimension which is necessary to economic decision making, but would not be expected to make relationships more valid, and runs the risk of obscuring the underlying problem.

Degree of fit of the model

If the only objective was to show which components of landscape had a significant effect on existing landscape value, the statistical techniques by which HPM and CE models are unravelled, while not fully satisfactory, might still be adequate. But practical applications of landscape evaluation are mostly concerned with change. Achieving an R^2 of 0.6 for the overall model does not mean that the ascribed value of a small change in one component will be "60 per cent right". The model's fit was determined by large differences in major variables such as land form. But the developments for which a landscape assessment is often required do not, on the whole, intend to change land form categories (though in Wales and elsewhere the extractive industries have tried hard). Similarly, the creation of reservoirs, another contentious land-use issue, often only entails displacing, or changing the form of, an existing "significant" water-body or -course.

Nor is it enough to show that a given modification of a view will probably and in general change its value, and the willingness to pay for it, in a particular direction. Real economic developments and aesthetic enhancements affect specific landscapes, seen from many viewpoints and often across considerable tracts of land. The views will be of different quality, will be differently sensitive to change, and the change will affect different numbers of people. Some summation needs to be made of these multiple effects. And the balance between aesthetic and other values depends on the magnitude, not just the direction, of effect. For practical, trading-off decisions, it needs to be known *by how much* and *in what context* (see below) modification of a view affects willingness to pay for it.

Benefit transfer

Even if a stable and accurate model was found in one context, there is no guarantee that the model would apply in a different context: yet such a guarantee is needed for benefit transfer approaches to environmental evaluation. How likely is it, for example, that a model including impact of electricity transmission lines, developed for the gentle, agricultural terrain of the English Midlands (Coventry–Solihull–Warwickshire 1971), would give an accurate description of their effect in the mountainous, semi-natural physiography of the Scottish Highlands?

Diverse constituents

Finally, none of the components, in whatever mathematical representation, can consistently account for peculiarity of taste, the desire for diversity of experience, or the familiarity that people acquire, with whatever landscape it is that they habitually experience.

In seeking to break down the value of landscape into objectively described components of the *views themselves*, economists have neglected to investigate the inner constituents of value that individuals are seeking from aesthetic experience. From the viewers' perspective, a view is a view, of which the components no more require conscious evaluation than does the composition. Nor, on the whole, need viewers reflect on what inner wants are satisfied by observing the view, unless they have been encouraged to do so by experimental psychologists. And yet Lancaster's (1966) characteristics of consumer goods approach is at least as applicable to deconstructing those inner wants as it is to decomposing the physical components of landscape.

That landscape is desired for its objective quality is implied both by the perceptions of evaluators themselves, and by the strong statistical differences between different views. That cultural norms are important is hinted at by the differences between evaluators from different geographical origins. That the stability of a familiar landscape is treasured is routinely demonstrated by heartfelt expressions of opposition to *the process of change* of landscape, as well as dissatisfaction with *the changed state*. All these are influences which, by one name or another, are routine for perceptual psychologists, but have largely escaped the attention of economists. It is doubtful whether HPM and CE are capable of disentangling them. Economists may, and do, argue that it is not for them to investigate or query motivations for consumers' preferences. And yet these constituents of value change differently over time. Because landscape often can only change slowly, with the growth of vegetation, with natural amelioration of intrusions, it cannot respond instantly to perceived shift in demand: hence it is important to know what part of discovered value is transient, and what part may be sustained through generations. HPM and CE can incorporate future value only in so far as it is both *predicted* and *valued* by those who make decisions. The valuation of future landscapes adds another problem of interpretation (Price 2007a).

These manifold difficulties in composing a plausible value for an intended change, based on models transferred from elsewhere, suggest that there is, after all, relative merit in using an holistic approach. CVM's problems are particularly likely to be severe in relation to proposed changes, because of part-whole distinctions, and because of the hypothetical nature of both the visual nature of the change, and the means of "paying" for it. And yet CVM, for all its faults, essentially poses the question to which an answer is desired: what is the value to you, an affected stakeholder, of a particular change, made in a particular place? Those who seek to decrease CVM's subjectivity and bias should be careful that they do not thereby reduce its relevance.

Controlled and creative subjectivity

All quantitative approaches to landscape *evaluation* and to landscape *valuation* are fraught with intrinsic and empirical difficulties. None can escape the need to make judgements, explicitly or implicitly, if relevance to human well being is to be attained. The approach that rejects quantitative or economic measurement does not avoid judgement: far from it, it acknowledges and thrives on judgement. And it entrusts judgement to an intuition that may well be unaccountable to the affected parties.

The purpose of landscape valuation is to discover people's subjectively felt aesthetic desires, and to trade those off against other desires and against the value of the resources needed to meet those desires, in a reasonable and consistent manner. *That* indeed is the chief business of landscape economics. If the landscape preferred turns out to have evolutionary functionality for humans and to aid sustainability for the biosphere, that is a bonus. If scenic preference is contrary to ecological sustainability, that too requires trade-off (unless ecological sustainability trumps everything, everywhere, to whatever degree). To quantify is helpful in defining the terms in which such discussions can take place (Price 2004).

The important thing about subjectivity is not to try to avoid it (it would be futile to attempt to do so anyway), but to recognise its limitations and its possibilities – indeed, to recognise it for what it is, in all its wonderful and life-affirming diversity. The fundamental judgement to be made, is the overall aesthetic quality of a view, and how that quality would alter with any specified development or enhancement of the landscape. For this purpose it is desirable to engage "aesthetic expertise", widely familiar with and able to envisage the aesthetic effect of landscape change. The task is to judge how the specified change in components and in their composition would affect landscape quality. By making this judgement through explicit human mediation, all the problems of measurement and statistical representation that arise from a components approach are bypassed.

But that aesthetic expertise needs to be calibrated with reference to the judgements of the affected human population, which may have different opinions on what it is that constitutes quality. The calibration can be relatively formal, as with the repeated landscape evaluation exercises.[1] Alternatively, the calibration can be informal, by means of reading what is written on the subject by many other experts, or listening to the discourse of the population generally. This requires a degree of humility that has not, perhaps, been historically associated with aesthetes. If necessary a quantitative scale of landscape quality can be transformed into a cardinal scale by means of well-constructed, context-relevant Choice Experiments (Price 2007b).

The expertise needs also to be monetised, if it is to be incorporated in trade-offs such as are embodied in cost–benefit analysis. Judgements of quality can be translated into monetary terms by the objective observation of subjective preferences revealed by actual decisions to purchase different levels of landscape quality, via the proxies of house purchase and travel decisions. The judgements of value thus created will never be "accurate" in the sense understood in the physical sciences, but in this case it is more serviceable to be approximately right than precisely irrelevant. Whatever else might be said, this mechanism of appraisal constitutes an improvement on the present normal protocol, in which the judgement is made without quantification; and so is the trade-off with other values.

Such was the protocol applied many years ago in practical decision contexts (Price 1975; 2008; Cobham Resource Consultants and Price 1991; Price and Thomas 2001; Price 2003). As far as I know, such studies have not been replicated. But neither has it been shown that their results were less valid or less useful than those of the myriad CVM, HPM and CE studies undertaken before and since.

Careful calibration of landscape experts' judgements, against the judgements of the user group, makes those judgements more representative of its cultural norms and distribution of peculiarity of taste. To deal with subtler matters such as familiarity with the state of a landscape or desire for diversity of experience needs subtler forms of evaluation, based on analysis of the portfolio of decisions made, and profile of the group making the decisions. For example, do people appear to travel further than visiting landscape of a particular quality would require, if the visited landscape is of a very different type from that which is more accessible? Do those who have moved away from the area where they grew up appear to travel further to visit the landscape of that area, than do the generality of landscape visitors? At no stage in the described process is there any attempt to avoid or deny subjectivity in the valuation process, but on the contrary to give due regard to the subjective preferences of individuals for landscape, and for all the other good things that they desire.

Acknowledgements

The data on subjective and objective values are derived from the work of numerous student classes, to whom I am grateful for their collaboration. I acknowledge also the contribution of the late Don Harding to refining the evaluation scales used.

Note

1 As a result of these, I became better able to make my judgements representative of those of the evaluator group. The statistical results showed, in all 29 cases, that my judgement of landscape quality, calibrated to reduce idiosyncratic elements, gave a more reliable account of a population's own evaluation, than would an evaluation based on objectively measurable components of views.

References

Abelson, P. (1979) *Cost Benefit Analysis and Environmental Problems.* Farnborough, Hants, UK: Saxon House.

Anderson, L.M. and Cordell, H.K. (1988) 'Influence of trees on residential property values in Athens, Georgia (U.S.A.): A survey based on actual sales prices', *Landscape and Urban Planning*, 15(1–2): 153–164.

Anthon, S., Thorsen, B.J. and Helles, F. (2005) 'Urban-fringe afforestation projects and taxable hedonic values', *Urban Forestry and Urban Greening*, 3(2): 79–91.

Aoki, Y. (1999) 'Trends in the study of the psychological evaluation of landscape', *Landscape Research*, 24(1): 85–94.

Appleton, J. (1975) 'Landscape evaluation: the theoretical vacuum', *Transactions of the Institute of British Geographers*, 66: 120–123.

Bergin, J. and Price, C. (1994) 'The travel cost method and landscape quality', *Landscape Research*, 19(1): 21–23.

Bishop, R.C. and Welsh, M.P. (1993) 'Existence values in benefit-cost analysis and damage assessment'. In: W.L. Adamowicz, W. White and W.E. Phillips (eds), *Forestry and the Environment: Economic Perspectives*. pp. 135–154, Wallingford: CAB International.

Bourassa, S.C., Hoesli, M. and Sun, J. (2004) 'What's in a view?' *Environment and Planning*, 36(8): 1427–1450.

Brookshire, D.S., Ives, B.C. and Schulze, W.D. (1976) 'The valuation of aesthetic preferences', *Journal of Environmental Economics and Management*, 3(4): 325–346.

Brown, G. Jr and Mendelsohn, R. (1984) 'The hedonic travel cost method', *Review of Economics and Statistics*, 66(3): 427–33.

Carlson, A.A. (1977) 'On the possibility of quantifying scenic beauty', *Landscape Planning*, 4: 131–172.

Choumert, J., Travers, M., Delaître, C. and Beaujouan, V. (2009) 'Capitalization of green spaces into housing values in the city of Angers: A hedonic approach', Paper presented for the *First International Conference on Landscape Economics*, Vienna.

Clamp, P. (1976) 'Evaluating English landscapes—some recent developments', *Environment and Planning A*, 8(1): 79–92.

Clamp, P. (1981) 'The landscape evaluation controversy', *Landscape Research*, 6(2): 13–15.

Clark, J., Burgess, J. and Harrison, C.M. (2000) '"I struggled with this money business": respondents' perspectives on contingent valuation', *Ecological Economics*, 33(1): 45–62.

Cobham Resource Consultants and Price, C. (1991) *The Benefits of Amenity Trees*. London: Report to Department of the Environment.

Council of Tree and Landscape Appraisers (frequent revisions) *Guide for Establishing Values of Trees and Other Plants*. International Society of Arboriculture, Urbana, Il.

Coventry–Solihull–Warwickshire (1971) *A Strategy for the Sub-region, Supplementary Report 5: Countryside*. Coventry–Solihull–Warwickshire Sub-regional Study Group.

Davis, R. (1963) 'Recreation planning as an economic problem', *Natural Resources Journal*, 3: 239–249.

Dearden, P. (1981) 'Landscape evaluation: the case for a multidimensional approach', *Journal of Environmental Management*, 13: 95–105.

Fines, K.D. (1968) 'Landscape evaluation: a research project in East Sussex', *Regional Studies*, 2(1): 41–55.

Fleischer, A. and Tsur, Y. (2009) 'The amenity value of agricultural landscape and rural-urban land allocation', *Journal of Agricultural Economics*, 60(1): 132–153.

Garrod, G. (2002) *Social and Environmental Benefits of Forestry Phase 2: Landscape Benefits*. Edinburgh: Report to Forestry Commission. http://www.forestry.gov.uk/website/pdf.nsf/pdf/fclscaperep.pdf/$FILE/fclscaperep.pdf (accessed 11 May 2012).

Garrod, G.D. and Willis, K.G. (1992) 'Valuing goods' characteristics: an application of the hedonic pricing method to environmental attributes', *Journal of Environmental Management*, 34: 59–76.

Gilg, A.W. (1975) 'The objectivity of Linton type methods of assessing scenery as a natural resource', *Regional Studies*, 9(2): 181–191.

Gupta, V., Mythili, G. and Hegde, D.S. (2009) 'Deriving implicit prices for urban environmental amenities from Mumbai housing prices: a revealed preference approach', Paper presented for the *First International Conference on Landscape Economics*, Vienna.

Hanley, N., MacMillan, D., Wright, R.E., Bullock, C., Simpson, I., Parsisson, D. and Crabtree, B. (1998) 'Contingent valuation versus choice experiments: estimating the benefits of Environmentally Sensitive Areas in Scotland', *Journal of Agricultural Economics*, 49(1): 1–15.

Hanley, N.D. and Ruffell, R. (1993) 'The contingent valuation of forest characteristics', *Journal of Agricultural Economics*, 44(2): 218–229.

Helliwell, D.R. (1967) 'The amenity value of trees and woodlands', *Arboricultural Journal*, 1: 128–131.

Henry, M.S. (1994) 'The contribution of landscaping to the price of single family houses: a study of home sales in Greenville, South Carolina', *Journal of Environmental Horticulture*, 12(2): 65–70.

Henry, M.S. (1999) 'Landscape quality and the price of single family houses: Further evidence from home sales in Greenville, South Carolina', *Journal of Environmental Horticulture*, 17(1): 25–30.

Hoinville, G. (1971) 'Evaluating community preferences', *Environment and Planning A*, 3(1): 33–50.

Jacques, D.L. (1980) 'Landscape appraisal: the case for a subjective theory', *Journal of Environmental Management*, 10: 107–113.

Kim, Y-F. and Johnson, J.L. (2002) 'The impact of forests and forest management on neighboring property values', *Society and Natural Resources*, 15(10): 887–901.

Lancaster, K.J. (1966) 'Change and innovation in the technology of consumption', *American Economic Review, Papers and Proceedings*, 56(1/2): 14–23.
Lange, E. and Schaeffer, P.V. (2001) 'A comment on the market value of a room with a view', *Landscape and Urban Planning*, 55(2): 113–120.
Linton, D.L. (1968) 'The assessment of scenery as a natural resource', *Scottish Geographical Magazine*, 84(3): 218–238.
Luttik, J. (2000) 'The value of trees, water and open space as reflected by house prices in the Netherlands', *Landscape and Urban Planning*, 48(3–4): 161–167.
MacMillan, D.C. and Duff, E.I. (1998) 'Estimating the non-market costs and benefits of native woodland restoration using the Contingent Valuation Method', *Forestry*, 71(3): 247–259.
Mitchell, R.C. and Carson, R.T. (1989) *Using Surveys to Value Goods: the Contingent Valuation Method.* Washington: Resources for the Future.
Morales, D.J., Boyce, B.N. and Favretti, R.J. (1976) 'The contribution of trees to residential property value: Manchester, Connecticut', *Valuation*, 23(2): 26–43.
Navrud, S., Ready, R., Magnussen, K. and Bergland, O. (2008) 'Valuing the social benefits of avoiding landscape destruction from overhead power tranmission lines – do cables pass the benefit-cost test?', *Landscape Research*, 33(3): 1–16.
Nielsen, A.B., Olsen, S.B. and Lundhede, T. (2007) 'An economic valuation of the recreational benefits associated with nature-based forest management practices', *Landscape and Urban Planning*, 80(1–2): 63–71.
Payne, B.R. and Strom, S. (1973) 'The contribution of trees to the appraised value of unimproved residential land', *Valuation*, 22(2): 36–45.
Penning-Rowsell, E.C. and Hardy, D.I. (1973) 'Landscape evaluation and planning policy: A comparative study in the Wye Valley Area of Outstanding Natural Beauty', *Regional Studies*, 7(2): 152–160.
Price, C. (1970) *Social Benefit from Forestry in the UK.* Oxford University: Department of Forestry.
Price, C. (1975) *Increasing the Electricity Supply to Skye.* Report to the Public Enquiry held in Broadford, Highland Region.
Price, C. (1976) 'Subjectivity and objectivity in landscape evaluation', *Environment and Planning A*, 8(7): 829–838.
Price, C. (1978) *Landscape Economics.* London: Macmillan.
Price, C. (1991) *Landscape Valuation and Public Decision Making.* Cheltenham: Report to the Countryside Commission.
Price, C. (1993) 'Applied landscape economics: a personal journey of discovery', *Journal of Environmental Planning and Management*, 36(1): 51–63.
Price, C. (1994) 'Literature review [of landscape valuation]', *Landscape Research*, 19(1): 38–55.
Price, C. (1995) 'The pros and cons of alternative valuation methods'. In: K.G. Willis and J.T. Corkindale (eds), *Environmental Valuation: New Perspectives.* pp. 160–177, Wallingford, UK: CAB International.
Price, C. (1999) 'Contingent valuation and retrograde information bias'. In: A. Park and C. Stewart Roper (eds), *The Living Forest.* pp. 37–44, London, HMSO: Proceedings of the International Symposium on the Non-market Benefits of Forestry, Edinburgh, June 1996.
Price, C. (2000) 'Valuation of unpriced products: contingent valuation, cost–benefit analysis and participatory democracy', *Land Use Policy*, 17(3): 187–196.
Price, C. (2001) 'Exact values and vague products? Contingent valuation and passive use value'. In: T. Sievanen, C.C. Konijnendijk, L. Langner. and K. Nilsson K. (eds), *Forest*

and Social Services – the Role of Research. pp. 205–217, Vantaa, Finland: Finnish Forest Research Institute, Research Paper 815.

Price, C. (2003) 'Quantifying the aesthetic benefits of urban forestry', *Urban Forestry and Urban Greening*, 1(3): 123–133.

Price, C. (2004) 'Forest aesthetics, forest economics and ecological sustainability'. In: R. Döring and M. Rühs (eds), *Ökonomische Rationalität und praktische Vernunft*. pp. 111–128, Würzburg: Königshausen & Neumann.

Price, C. (2007a) 'The landscape of sustainable economics'. In: J.F. Benson and M. Roe (eds), *Landscape and Sustainability* (2nd ed.). pp. 37–57, Cheltenham: E & F Spon.

Price, C. (2007b) 'Putting a value on trees: an economist's perspective', *Arboricultural Journal*, 30(1): 7–19.

Price, C. (2008) 'Landscape economics at dawn: an eye-witness account', *Landscape Research*, 33(3): 263–280.

Price, C. and Thomas, A.Ll. (2001) 'Evaluating the impact of farm woodland on the landscape: a case of blending perspectives'. In: T. Sievanen, C.C. Konijnendijk, L. Langner and K. Nilsson (eds), *Forest and Social Services – the Role of Research*. pp. 191–203, Vantaa, Finland: Finnish Forest Research Institute, Research Paper 815.

Price, U. (1810) *Essays on the Picturesque, as compared with the Sublime and the Beautiful; and, on the Use of Studying Pictures, for the purpose of Improving Real Landscape*. London: Mawman.

Randall, A., Grunewald, O., Johnson, S., Ausness, R. and Pagoulatos, A. (1978) 'Reclaiming coal surface mines in central Appalachia: a case study of the benefits and costs', *Land Economics*, 54(4): 472–489.

Sagoff, M. (1988) *The Economy of the Earth*. Cambridge: Cambridge University Press.

Schroeder, H.W. (1986) 'Estimating park tree densities to maximize landscape esthetics', *Journal of Environmental Management*, 23: 325–333.

Shafer, E.L. Jr and Metz, J. (1970) *It Seems Possible to Quantify Scenic Beauty in Photographs*. USDA Forest Service Paper NE-162.

Sheppard, S.R.J. and Harshaw, H.W. (eds) (2001) *Forests and Landscapes: Linking Ecology, Sustainability and Aesthetics*. Wallingford, Berks: CABI Publishing.

Sievänen, T., Pouta, E. and Ovaskainen, V. (1992) 'Problems of measuring recreational value', *Scandinavian Forest Economics*, 33: 231–243.

Tyrväinen, L. (1997) 'The amenity value of the urban forest: an application of the hedonic pricing method', *Landscape and Urban Planning*, 37(3–4): 211–222.

Tyrväinen, L. and Miettinen, A. (2000) 'Property prices and urban forest amenities', *Journal of Environmental Economics and Management*, 39(2): 205–223.

Tyrväinen, L. and Väänänen, H. (1998) 'The economic value of urban forest amenities: An application of the contingent valuation method', *Landscape and Urban Planning*, 43(1–3): 105–118.

Whiteman, A. (1995) *The Supply and Demand for Timber, Recreation and Community Forest Outputs from Forests in Great Britain*. Edinburgh University: PhD thesis.

Willis, K.G. and Garrod, G.D. (1993) 'Valuing landscape: a contingent value approach', *Journal of Environmental Management*, 37(1): 1–22.

4 From private values to social value of landscapes

Dafna M. DiSegni

Introduction

Economic literature has devoted much attention to the individual valuation of landscapes, providing direct services that are consumed over time. As with many other ecosystems, landscapes provide a variety of benefits to people, including provisioning, regulating, cultural and supporting services. Provisioning services are the products people obtain from the landscape, such as food, bio-fuels, fiber and timber, etc. Regulating services are the benefits people obtain from the regulation of ecosystem processes, including air quality maintenance, climate regulation, control of erosion and runoff, and water purification, for example. Cultural services are the nonmaterial benefits people obtain from ecosystems through spiritual enrichment, cognitive development, recreation, and aesthetic experiences. Supporting services are those that are necessary for the production of all other ecosystem services, such as primary production, production of oxygen, and soil formation.[1]

Although the transition from private values – based on individual preferences – to the aggregate social value has gained only marginal attention, it plays a central role in determining optimal landscape use and defining policy measures. This chapter couples advances in economic literature on aggregation theory with valuation theory to clearly design the transition path from private values of landscape to the social value of landscape. The social value of landscape is a function of aggregate utilities from goods and services provided by the landscape and consumed by individuals now or in the future, as well as the benefits derived from goods and services not necessarily considered by the individual, either as a result of poor information on the nature of their flow or as a result of limited income and institutional constraints. In this chapter, we position the individual benefits within the larger framework of social benefits from a given landscape and focus our attention on two questions. First: What are the aggregate benefits from a determined landscape? Hints are provided by the Aggregate Demand theory, pending aggregation on individuals' preferences and the existence of income effects. The second question posed here is: Do the aggregate benefits represent full social benefits from landscape services, or do we have to account for additional benefits not captured by the individuals?

Landscape services are here defined as the benefits that people obtain from the landscape.

The benefits that individuals derive from services provided by a landscape at large are valued using a range of economic and statistical tools. Economic literature related to valuation approaches and techniques is often classified into two categories: valuation based on stated preferences and valuation based on revealed preferences. The individual's preferences revealed by action in a market are often estimated by a direct analysis of the demand for the specific goods or services. It is, however, common to find situations in which the specific service is not traded in any market (like the view of a landscape). In such cases, demand and expenditures' analyses of closely related markets serve to estimate the value of the good under focus. Among the common methods of valuation that are based on related market transaction we list the Travel Cost Method, where values of recreational experience are inferred from the costs individuals incur to travel to the recreation area under consideration (Ward and Beal 2000; Parsons 2003, and see Chapter 10 of this book), and the Hedonic Method, in which the value of a resource is inferred from the sensitivity of other markets (changes in prices) to changes in the properties of the resource under focus. A classical example of the Hedonic Method is to relate the value of a playground to variations in housing prices: the researcher distills the impact of vicinity of housing to the playground and associates the value of playground to the difference in housing prices (Griliches 1971; Goodman 1978; Stover 1996, and see Chapters 12 and 13 of this book).

The benefits from landscapes are not restricted, however, to the benefits revealed via market transactions. It is often the case that landscapes provide benefits beyond the benefits captured by markets, either because there is no formal market capture of these services or because markets fail to appropriate the full benefit from some goods, and in particular the benefits that are generated by public goods.[2]

Valuation based on stated preferences has been increasingly used by economists to identify the value of landscape services, or ecosystem services at large, for which no markets exist.[3] The most commonly used stated preference method is the Contingent Valuation Method. This method draws upon economic theory and uses survey questions to elicit directly people's preferences for a specific good by finding out what they would be willing to pay for a specified improvement of that good, or its supply. It circumvents the absence of markets for services provided by public goods by presenting consumers with hypothetical markets in which they have the opportunity to buy the good in question. Because the elicited willingness to pay values are contingent (i.e. dependent) upon the particular hypothetical market described to the respondent, this method is called contingent valuation (Brookshire and Coursey 1987; Portney 1994; Venkatachalam 2004; Cameron and Carson 2005). Other stated preference approaches include the Attribute-Based methods (Mackenzie 1993; Gan and Luzar 1993; Holmes and Adamowicz 1993) and the method of Paired Comparison (Brown and Peterson 2003; Holmes and Adamowicz 2003).

We begin with a sketch of the major theoretical developments in aggregation theory and outline their relevance and implications for the aggregation of individual values, based on revealed preferences. Attention is restricted to the valuation of privately-owned landscapes.[4] A move to public ownership would naturally introduce an additional complexity driven by the failure of collective action on common property, which is beyond our current scope. We then follow up with the complementary social value of landscapes, which is not captured by aggregated private values. The section includes a discussion and examples on how to determine this complementary economic value and the key difficulties in estimating it.

Consistent aggregation of revealed preferences

The Demand Aggregation theory is restricted to the aggregation of revealed preferences that fulfill the following two properties. The first is non-satiation, or the "more is better" property. This means that a bundle of goods with a large quantity of an element is preferred to a bundle with a smaller quantity of that element. The second property is that preference ordering is characterized by substitutability. This means that if the quantity of goods or services provided to the individual decreases, it is possible to increase the quantity of other goods or services provided, making the individual sufficiently indifferent to the two bundles. The substitutability reveals the value that people place on those goods. If one of the goods has a monetary price, the revealed value is also monetary in nature. The price of a marketed good is a special case of a trade-off ratio because the money spent on purchasing one unit of one element of the bundle acts as a proxy to the quantities of one or more of the other elements in the bundle that had to be reduced in order to make the purchase.

If the individual's preferred orders have the properties described above, they can be represented by ordinal preference functions or utility functions that assign a number to each bundle as a function of the quantities of each element of the bundle. This function increases in all its arguments and creates a monotonic transformation.

Let the preference ordering be represented by a utility function defined over goods. For our purposes, $X = [x_1, x_2, \ldots, x_n]$ denotes a list or vector of all the n marketed goods that the individual chooses. The k non-marketed goods are similarly listed as $Q = [q_1, q_2, \ldots, q_n]$. The utility function assigns a single number, $U(X, Q)$, for each bundle of goods $B_i = (x_i, q_i)$. For any two bundles, the respective number assigned by the utility function are such that $u(x_a, q_a) > u(x_b, q_b)$ only if bundle B_a is preferred over bundle B_b, or $u(x_a, q_a) \geq u(x_b, q_b)$ only if bundle B_a is weakly preferred to bundle B_b. The aggregate of individual valuations is given by the horizontal summation of demand, $X = \sum_j u_j^{-1}(p)$, where X is the aggregate demand of individuals to private goods and u_j^{-1} is the individual's demand for the goods, as a function of prices. Statistical estimations of the private value of landscape are often undertaken using one of the two following approaches: use the micro data level of an individual's demands and compute the aggregate demand to determine the social value; or, use aggregated data on consumers' behavior to directly compute the aggregated demand and social benefits. The distinction between the two approaches is important as, in general, aggregated data does not necessarily obey the same rules as the micro data of any individual even when, as econometricians usually assume, everyone has the same tastes.

This fundamental property of preference aggregation underlines Arrow's Impossibility Theorem (Arrow 1951).[5] Arrow rigorously investigated the general problem of finding a rule to discern social preferences from individual preferences that meet the following properties:

(i) Social preferences should be complete. Given a choice between alternatives B_a and B_b it should say whether B_a is preferred to B_b, or B_b is preferred to B_a, or that society is indifferent between B_a and B_b.
(ii) Social preferences should be transitive. That is, if B_a is preferred to B_b and B_b is preferred to B_c then B_a is also preferred to B_c.
(iii) If every individual prefers B_a to B_b then socially B_a should be preferred to B_b.
(iv) Social preferences should not depend only upon the preferences of one individual (i.e. the dictator).
(v) Social preferences should be independent of irrelevant alternatives (i.e. the social preference of B_a compared to B_b should be independent of preferences for other alternatives).

Arrow's theorem essentially tells us that under the pair wise independence condition (or the condition of independence of irrelevant alternatives), there is no social welfare function that satisfies a minimal form of symmetry among agents and a minimal form of positive representativeness, such that aggregate preferences are consistent with each individual's preferences. Following Arrow's Impossibility Theorem, an extensive body of literature has subsequently addressed the modifications of the Arrow's axiom system, in a relatively restricted domain. Along the same line, Harsanyi (1955) restricted attention to cases in which all individuals in society are von Neumann and Morgenstern (vNM) Expected Utility Maximizers (von Neumann and Morgenstern 1944). Under these conditions, Harsanyi's Theorem (1955) states that the social utility function is a linear combination of the utility functions of individuals (i.e. $u = \sum_i \alpha_i u_i + \beta$). In this case, the utility function of society is a convex combination of the utility function of individuals in society, and thus, preserves monotonicity.

Muellbauer (1975) established necessary and sufficient conditions for the aggregated data to be consistent with micro level data. These are the Generalized Linearity (GL) conditions.[6] Furthermore, it has been assessed that market demand equations are consistent with micro demand equations, which correspond to some level of income that does not vary according to price variation, unless Price Independent Generalized Linearity (PIGL) holds.[7] As a matter of fact, testing for PILG conditions on goods and services generated from landscapes of different types is complex. The conditions are easily violated if (i) the consumption of goods is income dependent (this would often be the case if the individual uses its landscape both as a source of production (grazing, for example) and as a source of consumption (tourism or leisure, for example)); and (ii) information is missing. If individuals do not understand the contribution that a landscape makes to their well being, then their observed behavior or responses to questions will reflect their ignorance rather than the true value of the landscape.

Using the Bergson–Samuelson social welfare function, we can demonstrate the way in which changes in values affect social welfare. Given that social welfare is defined as the sum of all individuals' utilities (at equal weights),

$$W(u_1, u_2, \ldots, u_I) = \sum_i u_i(x). \tag{4.1}$$

Changes in the values (prices) of goods demanded by society and income are expected to directly affect social welfare. The impact is weighted by the relative sensitivity of the individual's utility to income and prices, and is mathematically described by the following equation:

$$dW(u_1, u_2, \ldots, u_I) = \sum{}_i \frac{dW}{dv_i} \lambda_i \left(-x_i dv_{ip} + dv_{iy}\right), \tag{4.2}$$

where $\frac{dW}{dv_i}$ is the change in welfare due to a change in the indirect utility function of individual i, λ_i represents the marginal utility of income of individual i, x represents the vector of goods demanded by i, dv_{ip} is the change in the indirect utility given the change in prices, and dv_{iy} is the change in indirect utility given the changes in income for individual i.

Recent developments in microeconomic theory provide additional criteria for mapping individuals' preferences into social preferences. Of particular relevance is the contribution of Gilboa et al. (2004) who have extended the basic framework presented by Harsanyi (1955) and his followers, by considering not only heterogeneity in individual preferences, represented by a vNM utility function, but also beliefs.

The research presented above implicitly assumes that individuals differ only in their preferences but not in their beliefs. This, however, is rather restrictive. While divergence of opinions is sometimes the result of different values assigned by different individuals to outcomes, much of it is due to differences in beliefs (i.e. different probability distributions over possible states of the landscape). Everyone wants landscape to provide high returns subject to its use (as agricultural landscape, forested landscape), but people have very different views about the way to achieve this common objective (high returns from landscape use), or different views/beliefs about that climatic conditions that affects the individual's optimal strategy to achieve a given objective. Gilboa et al. (2004) formulate a Pareto condition that implies that if individuals differ only in their taste and not in their beliefs then

Harsanyi's theorem holds, and the probability measure of society is an affine combination of those individuals.

Consistent aggregation of stated preference

The approach of stated preferences has been increasingly used by economists to value non-marketed goods, which by nature, often has the character of public goods. Examples include roadways, national parks, public forests, beaches, etc. The aggregate benefit of public goods is the aggregate value of all individuals' consumption of one unit of goods. When aggregating stated preferences, the researcher/regulator has to pay particular attention to various aspects. The first is boundaries of information. There might well be values of landscapes that were not addressed or valued by the interviewer or mentioned by the respondents. Incomplete information provided by either the interviewer or the respondent would result in a biased aggregation of individual values.

The second aspect is the boundaries of aggregation over agents and space. How broad is the set of individuals benefiting from the public goods that needs to be taken into consideration? In this regard, the issue of defining the "borders" of the public goods essentially concerns the question of how broadly individuals' marginal benefits should be vertically summed. Whether this should be confined to those living in close vicinity of the supplied goods (located at a specific place), or extended across the region, country, or even further afield, has implications for the associated social value. Evidence supporting the existence and importance of this issue is discussed by Smith (1993), Loomis (2000) and Bateman et al. (2006). The answer to this question is hard as by nature, agents under-reveal their preferences for public goods, or under-provide the public goods as a result of free-riding behavior. Consequently, it is hard to determine the population and the geographic boundaries of beneficiaries and their respective values.

Whitten and Bennett (2001), for example, estimated the benefits to hunters from the wetlands in South Australia and the Murrumbidgee River floodplain (although other private values are also involved), as well as the wider community values of people living away from the wetlands. These values are an estimate of the value of this particular landscape (wetlands in Australia); however, they do not compose the social value of the landscape, which should include an aggregate of all individual benefits (at least from hunting) in addition to the community values of distant neighbors. The wetlands are valued in this case through the benefit they generate to hunters. Nonetheless, the wetlands provide additional services of value, which are not captured by the hunters' surveys. A move from individual valuation to community valuation would require, at least, an aggregation of the values of all individuals who belong to the community.

The individual's value of an improvement of certain environmental goods (public goods) or services provided by landscapes will vary according to numerous variable factors, such as an individual's income, personal taste and characteristics, and beliefs. As emphasized by Bateman et al. (2000), these characteristics also vary spatially (for example, those with higher values and/or higher incomes may live closer to a given set of sites, etc.). Consideration of space is therefore vital to the accurate aggregation of individual benefit values. The spatial analytic capability of a Geographic Information System (GIS) provides a useful medium for harmonizing the diverse data necessary to undertaking such an aggregation exercise. In particular, GIS readily allows the researcher to specify a valuation function, which varies across space according to a variety of factors, including: (i) the distribution of rivers, lakes, estuaries, etc.; (ii) the change in quality to those resources, with improvements tending to convert former non-users into users in a spatially non-random manner; (iii) the accessibility of complementary and substitute assets; and (iv) the distribution and socio-economic and demographic characteristics of the population. The inclusion of such factors allows the analyst to observe any "distance decay" in values as we progressively consider households which are more remote from a given improvement of a landscape (or on a landscape). Furthermore, once such a valuation function is estimated, by applying it within GIS to data detailing all explanatory variables for all locations, we can define the appropriate "economic jurisdiction" (i.e. the area within which values are non-zero) for calculating total benefit values. This avoids common aggregation problems associated with artificial "political jurisdictions" typically defined by convenient administrative areas.

To conclude, similar to possible estimation biases in the analysis of aggregated data, biased estimates may also come up in the analysis of vertically aggregated values, where a mean value from some subset area is used as the basis for estimating aggregate values for the entire economic jurisdiction. An example of such a situation is detailed in the study by Bateman et al. (2006) of the River Kennet case study. There, biases in estimation are explained by two sources of errors: (i) heterogeneity with respect to individuals' valuation associated with their distance from the sources, whereby values are likely to decay with increasing distance to the source; and (ii) the common reliance upon political rather than economic jurisdictions. Under such a classification, the use of samples means values within the aggregation process are liable to lead to significant errors in the computed aggregate values.

Benefits beyond individuals' aggregate benefits

Social benefits beyond individual benefits are often the result of the existence of public goods that were ignored by the individual as a result of at least one of the following: (i) the individual may have poor information about the nature of ecosystem services and how they affect their well being, especially with respect to regulating and supporting services that are not demanded on the market but indirectly consumed, or (ii) existence of free-riding behavior. The individual is aware of some common (public) benefits from goods/services provided by the landscape (for example, the benefits from maintaining a forested landscape) but has not stated his preferences, for avoiding having to pay for their consumption or having to contribute to the supply of the common goods. It is common that people's willingness to pay for public goods is suboptimal because they reject the idea of having to pay for something someone else could provide and equally benefit from it.

The landscape often provides us with services that are classified as public goods, and thus, externalities may rise out of individual behavior. This implies that deducing the value of landscape from market signals (demand/supply levels) would result in a biased estimation of individual values, and consequently, also in a biased estimation of the aggregate of individual benefits. Moreover, lack of complete understanding and awareness of services provided by a given landscape omits Arrow's (1951) fundamental properties (i)–(v) necessary for computing a social welfare function consistent with individuals' preferences.

As already mentioned above, a similar bias in the estimation of social benefit is expected when computing the aggregate benefits across individuals based on stated preferences: rational individuals would state a lower willingness to pay for the provision of a public good or a service, due to an incomplete appropriation of the benefits that this good or service provides to society. Incomplete appropriation is driven by free-riding incentives or by poor information about the nature of ecosystem services provided by landscapes.

In light of these potential failures, Disegni et al. (2010) have proposed to associate the social value of resources that provide supporting services – and which benefits are often not fully appropriated by individuals – with the impact of these resources on the optimal path of national growth. Following this approach, social value of landscapes should be determined in a macroeconomic framework, and computed as an implicit value. The implicit value is equal to the marginal benefit of the overall benefits from landscape on social welfare; i.e. the value of a specific service is equal to the increase in social welfare as a result of an increase in an additional unit of the service supplied.

Concluding remarks

The computation of the social value of landscape is often based on the aggregate of private values of landscape. The function (or concept) of aggregation selected in the computation process is important for proper results in an aggregate value that indeed represents individuals' preferences. The extended body of literature on the private valuation of natural resources and ecosystem services enables us to identify and assess the existence of social benefits from landscape. However, reported private values and results from specific studies need to be coupled with the aggregation theory in order to correctly deduce the social value of landscapes. Simple aggregation weighted by the relative income or distance factor is not sufficient to avoid biased estimates.

Valuation should clearly distinguish between private goods, public goods, and unobserved goods, where the latter often provides supporting services. Each needs to be valued by its individual approach. The literature hardly gives separate and distinctive attention to the existence of these three types of goods/services. Nevertheless, at the stage of social value computation one should consider the integration of these three categories of goods and services with the process of aggregation: (i) the benefits of goods revealed through market signals; (ii) goods that have no market signals, but are known to possess an individual value; and (iii) goods which are not considered by individuals because of an unawareness of the existence of the goods/services or due to strategic ignorance of their existence. A reasonable aggregation criterion should be selected to consider the interaction between categories, which is ethical and acceptable to most individuals. Therefore, this criterion should reflect social norms and ethical considerations supported by society (for example, an equal weight to individual welfare).

Notes

1 See Millennium Ecosystem Assessment (2005) for details on the classification of ecosystem services at large.

2 Public goods, as the name suggests, are commodities that have an inherent public character, in that consumption of a unit of goods by one agent does not preclude its consumption by another and agents often cannot be excludable from consuming the good. This characterization is independent of ownership. Public goods could be privately owned or publicly owned.

3 Brown and Peterson (2003) provide an extensive review on valuation of non-marketed goods and services, using different economic approaches.

4 A landscape that is privately owned still possesses the characteristics of a public good. That is, although it is privately owned, it could well be that it generated services to the public, as for example, an aesthetic service, supporting services of maintaining natural unique species of flora, regulatory services of climate control, etc.

5 Arrow (1951) imposed no constraints on the individuals' utility function, but restricted social welfare function, π, to satisfy the following properties: continuity, unanimity and anonymity.

6 Generalized Linearity conditions are often expressed in three alternative but equivalent restrictions:

(1) Value shares are of the form

$$w_i(y,p) = v(y,p)A_i(p) + B_i(p) \text{ with } w_i(y,p) = v(y,p)A_i(p) + B_i(p)\sum A_i = 0, \sum B_i = 0;$$

(2) Relative "marginal shares" are independent of income or utility, i.e.

$$\frac{\partial}{\partial y}\left(\frac{\partial wi/\partial y}{\partial wj/\partial y}\right) = 0 \text{ for all } i,j; \text{ and}$$

(3) The expenditure or the cost of utility function $y = m(u,p)$ is of the form $y = G(u, H(p))B(p)$, where y is total expenditure (income), p is the ($nX1$) price vector, q_i is the purchase of the good i, u is the level of utility, and $w_i = (p_i q_i)/y$ is the value share of good i.

7 PILG is defined by two equivalent restrictions:

(1) Budget shares must be of the form
either $w_i(y,p) = y^{\varepsilon} A_i(p) + B_i(p)$
or $w_i(y,p) = Log(y)A_i(p) + B_i(p)$; and

(2) The expenditure function $y = m(u, p)$ has the form
either $y = \left(a(p)^{-\varepsilon} + ub(p)^{-\varepsilon}\right)^{-(1/\varepsilon)}$
or $y = H(p)^{u} B(p)$.

References

Arrow, K.J. (1951) *Social Choice and Individual Values*. Columbia: PhD Thesis.

Bateman, I., Langford, I., Nishikawa, N. and Lake, I. (2000) 'The Oxford debate revisited: A case study illustrating different approaches to the aggregation of benefits data', *Journal of Environmental Planning and Management*, 43(2): 291–302.

Bateman, I.J., Day, B.H., Georgiou, S. and Lake, I. (2006) 'The aggregation of environmental benefit values: welfare measures, distance decay and total WTP', *Ecological Economics*, 60: 450–460.

Brookshire, D.S. and Coursey, D.L. (1987) 'Measuring the value of a public good: An empirical comparison of elicitation procedures', *American Economic Review*, 77: 554–66.

Brown, T.C. and Peterson, G.L. (2003) 'Multiple good valuation: With focus on the method of Paired Comparisons'. In: P.A. Champ, K.J. Boyle and T.C. Brown (eds), *A Premier on Nonmarket Valuation*. pp. 221–258, Dordrecht, Netherlands: Kluwer Academic Publishers.

Cameron, R. and Carson, R.T. (2005) *Using Survey to Value Public Goods: The Contingent Valuation Method*. Washington D.C.: Resources for the Future.

DiSegni, D.M., Shechter, M. and Sternberg, M. (2010) 'Estimating the economic value of climate change effects on ecosystem services via fluctuations in Saving Index of Sustainability', *Proceeding of the BIOECON Annual Conference 'From the Wealth of Nations to the Wealth of Nature: Rethinking Economic Growth*', Venice, Italy.

Gan, C. and Luzar, E. (1993) 'A contingent analysis of waterfowl hunting in Louisiana', *Journal of Agricultural and Applied Economics*, 25: 36–45.

Gilboa, I., Samet, D. and Schmeidler, D. (2004) 'Utilitarian aggregation of beliefs and tastes', *Journal of Political Economy*, 112(4): 932–938.

Goodman, A.C. (1978) 'Hedonic prices, price indices and housing markets', *Journal of Urban Economics*, 5: 471–484.

Griliches, Z. (1971) *Price Indexes and Quality Change: Studies in New Methods of Measurement*. Cambridge: Harvard University Press.

Harsanyi, J.C. (1955) 'Cardinal welfare, individualistic ethics, and interpersonal comparisons of utility', *Journal of Political Economics*, 63: 309–321.

Holmes, T.P. and Adamowicz, W.L. (2003) 'Attribute-based methods'. In: P.A. Champ, K.J. Boyle and T.C. Brown (eds), *A Premier on Nonmarket Valuation*. pp. 171–219, Dordrecht, Netherlands: Kluwer Academic Publishers.

Loomis, J. (2000) 'Vertically summing public good demand curves: An empirical comparison of economic versus political jurisdictions', *Land Economics*, 76(2): 312–321.

Mackenzie, J. (1993) 'A comparison of contingent preference models', *American Journal of Agricultural Economics*, 75: 593–603.

Millennium Ecosystem Assessment (2005) *Ecosystems and Human Well-Being: Our Human Planet*. Washington D.C.: Island Press.

Muellbauer, J. (1975) 'Aggregation, income distribution and consumer demand', *The Review of Economic Studies*, 42(4): 525–543.

Parsons, G. (2003) 'The travel cost model'. In: P.A. Champ, K.J. Boyle and T.C. Brown (eds), *A Premier on Nonmarket Valuation*. pp. 269–329, Dordrecht, Netherlands: Kluwer Academic Publishers.

Portney, P.R. (1994) 'The Contingent Valuation debate: Why economists should care', *The Journal of Economic Perspectives*, 8(4): 3–17.

Smith, V.K. (1993) 'Nonmarket valuation of environmental resources: An interpretive appraisal', *Land Economics*, 69(1): 1–26.

Stover, M.E. (1996) 'A hedonic price index for premium cigars', *Journal of Economics*, 22(2): 63–73.

Venkatachalam, L. (2004) 'The Contingent Valuation Method: A review'. *Environmental Impact Assessment Review*, 24(1): 89–124.

von Neumann, J. and Morgenstern, O. (1944) *Theory of Games and Economic Behavior*. Princeton, NJ: Princeton University Press.

Ward, F. and Beal, D. (2000) *Valuing Nature with Travel Cost Methods: A Manual*. Cheltenham, UK: Edward Elgar.

Whitten, S.M. and Bennett, J.W. (2001) *A Bio-economic Analysis of Potential Murrumbidgee River Floodplain Wetland Management Strategies (Wagga Wagga to Hay)*. Canberra: University of New South Wales, Private and Social Values of Wetlands Research Report, No. 10.

5 The governance of rural landscapes

Property, complexity and policy

Ian Hodge

Introduction

Landscapes represent major assets that are primarily privately owned but with values that are predominantly in the public realm. Governance thus involves a mix of private property and public policy operated through a variety of institutions. In this context the approach to governance follows Jessop (1998: 95) who defines governance as 'the complex art of steering multiple agencies, institutions and systems … through various forms of reciprocal interdependence'. At the same time, landscapes, especially with regard to their aspect as a major element of an ecological system, represent a complex good, one that is characterised by many elements that interact in ill-defined and often unpredictable ways. Policy thus needs to represent the public interests in landscape management decisions but in a context where there are strong elements of uncertainty and ignorance (Wätzold 2000).

This chapter first sets out the key characteristics of the benefits provided by landscapes. We then consider the institutional contexts within which landscapes are owned and managed. The problem of missing markets and public interest in landscape values indicates a major role for policy and we review the challenges facing economic valuation. Given the limitations facing a central planner in identifying the appropriate objectives for landscape, some type of institutional approach is required that represents stakeholder interests. This may take the role of a social residual claimant that will promote adaptive co-management.

Landscape values

Landscapes may be seen as a complex asset generating a variety of benefit flows both directly to consumers and indirectly as an intermediate service within ecosystems or into other production processes. The benefit flows may take a wide variety of forms, such as aesthetic qualities, space within which other activities may take place, habitat for wildlife, nutrient cycling or water collection and storage. Individual areas provide multiple benefits streams and often the benefits streams are joint products that arise from particular combinations of ecosystem functions. Landscape thus represents far more than just the physical appearance of an area of land and the concept is intimately bound together with the ecosystems with which it is associated. These benefits are now often characterised as ecosystem services (Turner and Daily 2008), although Termorshuizen and Opdam (2009) favour the term "landscape services" over "ecosystem services" which they argue highlights the importance of spatial pattern, is recognized across a broader range of disciplines, and is more often applied on the local landscape scale.

While landscapes in most contexts are substantially modified and manipulated by human activities, they are not simply man-made. Rather they represent the outcome of complex mixes of human activities within a context of natural landforms and environmental processes generating what is often referred to as a semi-natural system. This is the source of multiple streams of benefits representing both environmental service flows, such as nutrient cycling or pollution abatement, but also aesthetic and cultural benefits enjoyed directly by human populations. The value of a landscape may to some extent be characterised in terms of the values of its component parts but its total value will rarely be seen as simply their sum. The total value will depend on the interrelationships between the parts, its composition, cultural associations, history, authenticity and ecological interrelationships. The value of one area of landscape will depend on the context within which it is set; its quality will be affected by land-uses and characteristics at adjoining or nearby locations. It is possible too that some elements of landscape may have negative values such as perhaps unattractive views or perhaps negative historical associations and such negative values may similarly detract from the values of adjoining areas.

These values are associated with both demand and supply conditions. Demand will be primarily influenced by the presence, characteristics, preferences and values of a human population, although proximity may not be critical. Areas of global significance may attract visitors from great distances. And certain types of value, such as existence vales, may not depend on physical proximity at all. On the supply side, values will depend on the physical and biological qualities of the area and its human associations. Scarce and attractive landforms and or landscapes supporting rare or threatened ecosystems will be valued over common or boring landscapes.

Interdependencies

The characteristics of landscapes are fundamentally determined by land-use and management, set within an essentially given, or at least exogenously determined, topographic, biophysical and environmental context. Land-uses and management, in turn, are fundamentally determined by the decisions made by landowners and occupiers. Over most of an area, land is owned privately and so landscapes are substantially determined by the decisions of large numbers of typically small landholders, subject to the limits set by government regulation. The benefits of landscapes are thus primarily enjoyed by the private owners and users through both their value as inputs to productive activities and in direct consumption by their shaping the local environment within which such private individuals work and live. However, the term "landscape" is generally taken to represent the public aspect of landform and land cover, so that the benefits of landscapes may more often be conceptualised in terms of the values attributed to landscapes by the wider public. And it is these benefits that commonly have public good characteristics. However, we may note that Schmid (2004) argues that "public good" is not a useful term in that its definition fails to distinguish between the different combinations of exclusion cost and non-rivalry. We thus need to recognise the variety of circumstances within which these characteristics may arise and the different implications for policy.

Landscape values are generally available in a variety of ways to the general public. Because the costs of excluding the public from the enjoyment of these values are generally high relative to the level of the benefits enjoyed, the benefits are often regarded as non-excludable. In some contexts the value relative to the exclusion cost may be sufficient in order for a landowner to have an incentive to exclude people and then charge individuals for entry into an enclosed landscape, such as a park (Langholz et al. 2000). There can thus be some private markets for landscape. This is illustrated by an interesting initiative in the Netherlands under which elements of the landscape have been publicly auctioned.[1] Hedges, ponds and a nature trail were "sold" to local companies or citizens who bid most. Payments for the maintenance of the landscape elements over a ten year period are made to a regional trust fund which has a contract with the landholder for the landscape maintenance to be undertaken. The fund also monitors compliance. Launched in 2007, three subsequent auctions raised € 240,000.

Landscapes may also be non-rival in that the presence of one person enjoying the landscape may not detract from the enjoyment gained by others enjoying the view at the same time. However, the extent to which this applies to any particular landscape or to any particular person depends on the quality and sensitivity of both the landscape and the individual concerned. Some attractive landscapes, especially historical or architectural attractions in urban areas, may be robust to large numbers of visitors without substantially diminishing the enjoyment of any particular individual. In other quiet or remote locations, the presence of any other individuals may destroy the essential sense of isolation or peacefulness that is being sought. Clearly some people make considerable efforts to find isolated locations while others appear undaunted by dense crowds, as illustrated by some people's willingness to visit crowded beaches. But in probably all cases, congestion sets in at some point and reduces the value of the landscape experience to all of those who seek it.

Rights and duties towards landscapes

The rights held by an individual landowner are never complete. The state retains the right of compulsory purchase and individual rights, such as for access or to shoot game, are often partitioned from the main bundle held by the landowner. Thus, landowner decisions will be constrained by the presence of these other rights. The introduction of regulations over landscape change identifies a reference level for property rights and an associated environmental standard. This sets the baseline against which policy interventions are made. We return to this issue later. But of course, not all decisions are made simply according to the formal rights position. In practice land managers will respond to what they perceive as being the expectations of a wider social grouping, or indeed to what they judge to be "right" behaviour. Thus, landowners may feel that they have a moral duty even if not a legal duty of stewardship or some other behaviour (Colman 1994).

Landscape value is most often seen as an aesthetic pleasure derived from the view of land, or probably less significantly enjoyed vicariously through knowledge of its quality such as through pictures or film. Thus, the quality of landscape is primarily important to many who hold no rights of access over the land itself and remains important even when they are not present on that particular site. Beneficiaries of landscape will often also hold no right to the stream of benefits that the landscape can represent. In fact, it will often be uncertain as to whether the enjoyment of a landscape is a right or whether it is a privilege (Bromley 1991) that is enjoyed without any formal right. This may in fact not be clear in the absence of a legal test case. In the presence of transactions costs, it is not efficient to seek to define all potential property rights, rather, difficult or obscure cases will only be formally established by the courts when the need arises (Barzel 1989). For example, someone might object to an application for planning permission by a neighbour to erect a building that would obstruct her view. In this case, the planning system would determine whether or not she held a right to that view through its decision on the application. Caution is required when a right is asserted. More generally, the definition of what constitutes a right is not always consistent and approaches often differ as between law and economics (Cole and Grossman 2002).

However, it might be argued that society does hold certain rights to landscape collectively where the state has introduced regulations that protect certain landscapes from being harmed by landholders' decisions (Ostrom 1990). This could be said, for example, to happen within Areas of Outstanding Natural Beauty or Conservation Areas in the United Kingdom where changes to landscape are regulated by government. But even here it could be argued that this regulation still does not deliver the right to any particular individual, rather that the right is held by the state and the individual still only holds a privilege (Bromley 1991). There is certainly not an individual right in the sense that an individual can then allow a landowner manager to alter the landscape; that can only be granted by government. The difference with the regulation in place is that the privilege is at the discretion of the state rather than at the discretion of a private owner.

Markets and intermediaries

In principle, we might envisage a market for landscape under which those valuing landscape buy and sell the landed property from which the landscape value derives. However, to those other than landowners and occupiers, the landscape benefit flow enjoyed from any particular area of land is likely to be relatively small, an exception perhaps being the landscapes surrounding an individual's residence.

Thus individuals generally do not gain access to preferred landscape benefits through the acquisition of freehold rights in land.

In fact, most people would not want to acquire a full freehold in order simply to enjoy the landscape benefits, they might rather contract with a landowner with regard to the landscape separately from the many other benefit streams associated with the occupation and use of the land. However, even here an individual arrangement is unlikely to be effective unless securing the stream of landscape benefits involves only a very modest adjustment to the landowner's preferred actions given that the benefits obtained by the landowner are likely to be large relative to the benefits potentially enjoyed by one single individual from the landscape. Thus to have any real impact on land-use decisions, it would almost certainly be necessary for a group of those favouring certain landscapes to act collectively and enter into a contract with the landowner. And this of course will incur substantial transactions costs. Different individuals will have different preferences and values towards alternative landscapes and there will be strong incentives to not reveal their willingness to pay for the benefits and to free ride wherever possible. A further complication is that a contract with one landowner could become void should the landowner sell the property onto another owner. A legal covenant would be necessary for an agreement to run with the land. For these reasons, conventional market solutions to the issue of landscape quality are most unlikely to arise. Rather landscapes, where "provided" at all, are provided on a collective basis predominantly by government, but occasionally by collective demand organisations on behalf of a larger population.

The policy problem

Thus, conventional private markets are not likely to deliver the standard of landscape that is demanded generally by the public. A property rights perspective might argue that the outcome is already optimal in the absence of state intervention on the grounds that the absence of a market provides evidence that the transactions costs of establishing one exceeds the potential benefits to be gained from its operation. But this would be an extreme position and most would argue that state involvement can improve resource allocation in that superior outcomes can be identified and implemented at lower transactions costs through government involvement.

But of course, if the state is to be engaged in influencing land-use and management, some target outcome needs to be identified as an objective for collective action. What means might be adopted in seeking to identify this potential state of affairs? One immediate answer may be to adopt economic valuation techniques and to apply cost–benefit analysis. This raises a variety of challenging issues for landscape economists. Some of these issues are common to economic valuation techniques in general and will not be pursued here. However, some more fundamental issues are particularly associated with the characterisation of landscape in terms of its public good attributes, and indeed of the adoption of the metaphor of a public good at all.

Policies towards landscapes face an immediate problem in the impossibility of establishing a simple quantitative measure of the state of any particular landscape. We may sympathise with Lord Kelvin's view that 'when you cannot measure [what you are speaking about], when you cannot express it in numbers, your knowledge is of a meagre and unsatisfactory kind'. But landscape policy has to operate within this constraint.

One approach towards resolving this issue is to apply economic valuation, but in this context it is unlikely to be straightforward for a variety of reasons. There are of course the practical questions of valuation methods which may well be particularly acute with respect to landscape values, such as whether individuals may have sufficient experience so as to be able to place monetary values on public goods reliably (Bruni and Sugden 2007) or whether respondents to valuation surveys may be excessively attached to the status quo or likely to change their preferences as their experiences with different landscapes change (Kahneman et al. 1991). There is then a somewhat more fundamental question as to whether respondents to valuation surveys have sufficient knowledge of a landscape so as to be able to offer useful valuations of the asset in a broad sense. Respondents will invariably have limited knowledge of the relationship between landscapes and other ecosystem services. Notwithstanding the progress that has been made in developing techniques for the economic valuation of landscapes (Garrod and Willis 1999; Colombo and Hanley 2008; Campbell et al. 2009), presentation of information on alternative landscapes in valuation exercises tends, inevitably, to be rather superficial; in practice those asked to value it do not understand the full implications of alternative landscape scenarios and therefore cannot be expected to provide reliable valuations.

But it may be argued further that no one can fully predict the longer term implications of alternative landscapes because of the inherent complexity of the systems. There is a view in ecology (e.g. Holling et al. 1995) based on arguments of complexity and ignorance that suggests approaches that focus on the resilience of ecosystems rather than their optimality. Thus the aim of management becomes to prevent gradual degradation or unexpected radical system change rather than to move towards a static optimum. In this respect the return to the investment in the management of the system comes in the form of avoiding an essentially unknown outcome, which clearly cannot be the subject of a simple valuation exercise. This challenges the principle of a policy objective that seeks to provide an optimal level of a public good; the prescription for management changes radically from a search for an optimum towards regular adaptive management and adjustment. This may involve the maintenance of elements of a system, perhaps certain species or land-uses, that may appear to be redundant under present circumstances but which could be critical for regenerating the system following some type of disturbance (Folke 2006). In this respect, efficiency may also be an illusive goal. In the context of poorly understood ecosystems, resilience may be a more appropriate objective for policy (Hodge 2007).

In this context, the resource allocation solution is not one that can be derived by a technical economic planner but rather has to be one that emerges from a more complex (and messy) institutional infrastructure, what Sagoff (2004) refers to as civic engagement in environmental problem solving. And it is not based on a single static objective, but rather one that responds to environmental change and new information over time. This may be regretted by economists seeking to apply their tools and models directly into a policy formula, but it cannot be escaped. This is not to say that the economic models and tools have no value or function. Economic valuation studies can provide useful information about some aspects of the policy problem, but they are only one input into a wider decision-making system. Furthermore, the multiple and diverse variety of benefit streams means that it will be necessary to develop multiple and diverse policy instruments. Policy will require the development of a variety of institutional approaches towards landscape management.

Policy instruments for landscape values

The policy problems identified above present a series of challenges to policy which is likely to involve, amongst other approaches, some combinations of (i) regulation, (ii) environmental contracts, and (iii) policy by intermediary, all underpinned by some form of co-management.

(i) Regulation defines the rights and duties faced by landholders, separating out and determining separate property rights, and effectively determining the nature of property itself. It sets legal limits, especially reflecting a social judgment as to the reference level of property rights, the line below which standards of management would be regarded as causing external costs and above which would be regarded as providing external benefits (Hodge 1989). Following from this, policy towards external costs, especially in regard of pollution, would adopt the polluter pays principle. On the other hand, the promotion of land management activities which are regarded as providing external benefits would operate through the provider gets principle, i.e. positive payments for higher standards of environmental management (OECD 1999). While regulation on its own may be effective in setting well defined bounds within which actions are required to take place, it can only, in principle, be set in terms of variables that are subject to measurement and monitoring and it is unlikely to be effective in requiring positive actions to be taken by land managers.

(ii) The use of environmental contracts has come to be widely adopted in agri-environment schemes. They offer the potential to promote detailed changes in land management in order to generate quite specific changes to landscapes, although they suffer from various disadvantages such as relatively high transactions costs, asymmetric information and end of contract risks. There have also been doubts as to whether they have in practice delivered the environmental gains that have been planned. Kleijn et al. (2001) famously concluded that agri-environment schemes do not protect biodiversity in the Netherlands, and in fact in some cases reduced biodiversity by destroying important habitats. In a subsequent study Kleijn et al. (2006) carried out a series of statistically controlled comparisons of over 200 pairs of agri-environment schemes and controls in five countries, concluding that all schemes had some positive effects on biodiversity that ranged from moderate to marginal. Recently, other studies (e.g. Taylor and Morecroft 2009) have also reported evidence of their success.

(iii) An alternative approach towards policy operates where government acts indirectly through an intermediary organisation. As noted above, collective organisations can represent demand across a wider public interest. Thus conservation trusts can acquire property and manage land directly or else can enter into agreements with landholders to influence the way in which the land is managed (Dwyer and Hodge 1996). The objectives of such non-profit, non-governmental organisations are likely to be closer to public policy objectives than those of the typical private owner and public funding can extend trusts' activities as well as influence their priorities. But in practice, such organisations are substantially reliant on public funding given the inevitable problem of the free-rider in raising funds for the generation of public benefits, hence it is more realistic to view them as policy by intermediary rather than as free market environmentalism. On the supply side, state support can promote collective action by landholders so as to co-ordinate and redirect their activities, as illustrated for instance by support for environmental co-operatives in the Netherlands (Franks and McGloin 2007). Groups of landholders can co-ordinate their actions to deliver benefits at a landscape scale and can have good information about each others' costs and opportunities so as to improve monitoring and enforcement.

The potential for voluntary transactions in support of landscape improvement and conservation can also be facilitated by the unbundling and rebundling of property rights. Thus, for instance, the creation of a legal framework for conservation covenants or easements can enable private bodies to enter into long term legal arrangements with land holders that can support landscape conservation. Alternatively, or in combination, there may be scope for assurance arrangements with product certification and independent verification to create markets for environmentally linked commodities. This may be seen as creating a private good where the benefits were previously a public good (e.g. Bougherara and Combris 2009). This suggests the potential and requirement for a variety of policy mechanisms and approaches as illustrated in Table 5.1. Each is likely to have strengths and limitations and will be best suited to different circumstances; but none will be perfect.

The disintegration of interests in landscapes

The partitioning of property is a common characteristic of modern management systems. Privatisation and contracting out from both public and private organisations can potentially promote an increase in cost effectiveness. But it runs the

Table 5.1 Goods characteristics and governance arrangements

Goods characteristics	*Governance arrangement*
Private goods	Private market
Potential private goods	Market creation
Collective goods	Commons management and clubs
Charismatic public goods	Non-profit: Conservation Trusts
Public bads and uncharismatic public goods (e.g. insects)	Government regulation and incentives (agri-environment)
Residual: Unexpected costs and benefits	Co-management

risk of introducing agency problems of asymmetric information and externality. Outputs need to be defined precisely in formal terms, alongside careful monitoring and enforcement to avoid shirking and excess cost. Within rural land management there is often an increasing array of contracting and licensing arrangements under which agricultural land is occupied and managed. This disintegration of ownership of agricultural land brings many different interests onto a single site and may open possibilities for costs to be passed on to other agents when in the past they would have been internalised within a single farm organisation. In particular, it is the environmental interests of land management that can suffer as those occupying and using the land have a less immediate or shorter term concern for its environmental quality and little concern for wider landscape or biodiversity values.

There is a similar element in policy making whereby, perhaps following Tinbergen, specific policy issues are addressed by separate policy instruments. Typically separate government agencies will develop policies in respect of particular flows of costs or benefits, such as in the regulation of nitrates or pesticides, the conservation of biodiversity, support for afforestation, flood control and so on. But there will be interactions amongst these policies and there will be unexpected events and changes, such as associated with new pollutants, unpredicted floods or droughts, the arrival of invasive species or the development of increased human demands for recreation or agricultural uses. These changes will be specific to a particular locality and have a variety of local implications. However, we need to recognise that effective conservation of landscapes and ecosystem services relies on consistent long term decision-making, in securing resources and management over the long term. But management can also face unexpected shocks which require unplanned and unanticipated changes in management. This implies that there is some optimal level of insecurity in environmental management. Not all social benefits and costs are captured by the various policy processes that are likely to be introduced, especially in the context of managing dynamic and unpredictable ecosystem processes. There is imperfect information, imperfect specification of the targets for particular policies and imperfect monitoring and enforcement of outcomes. Agents acting within local areas will seek to push costs and risks onto other parties. The issue arises as to whom or which organisation takes an oversight over the operation of the local landscape or ecosystem as a whole?

Social residual claimant and adaptive co-management

A holistic or ecosystems approach towards landscapes recognises the full range of environmental processes and the known and potential interactions at various scales and over long time periods. The partitioned arrangements for the private control of resources and the policy instruments for separate policy objectives are set out in various contractual forms but they will primarily deal with measurable and predicted quantities. This suggests that the partitioning of ownership and policy instruments into separate elements may increase the potential for costs and risks to be passed on from one to another group of stakeholders. And this implies that there will be some optimal degree of partitioning of property rights. Especially in the context of a dynamic and uncertain system, some mechanism is required that takes oversight of the management of the system as a whole and represents the interests of those who are not present to represent themselves, especially those not present within the local area and future generations. The position here has some parallels with that relating to the management of commons, but this tends to focus on resource users with similar interests and in the rules that can promote cooperative behaviour of commoners (Ostrom 1990) rather than with the wider issues of ecosystem or landscape management.

If all costs and benefits were enjoyed by a single owner then the costs and benefits would be internalised, but this can never be the case in practice. The agent or organisation taking such oversight on behalf of the wider society might be seen as the "social residual claimant" of the overall system, effectively the "owner" of the system. The residual claimant receives what is left after all the payments have been made and receipts received and has the incentive to maximise the value of the asset (North 1990). When events are unexpected, the residual will fall to the residual claimant because they cannot have been foreseen when the terms and conditions of the contractual arrangements were drawn up and so are unlikely to be covered by them. Further, private individuals will have relatively short time horizons and high discount rates and hence an incentive to push costs onto future occupiers and users. In this context we may think of the "owner" of an ecosystem as being the agent that receives the residual flows of costs and benefits that arise through landscape and ecosystem functions within a particular area of land.

But the management of the system needs to take account of both the ecological and the social aspects of governance, in what Folke et al. (2005) term 'adaptive governance of socio-ecological systems'. This represents some form of co-management (e.g. Baland and Platteau 1996), a participatory process amongst all stakeholders which allows exchange of information and reveals values in a process of decision-making. This needs to allow an element of trial and error, building information and experience over time and also building social capital so that the governance of the system can accumulate and exchange information and adapt to new information and changes in the environmental context or in preferences and priorities. The co-management organisation (Hodge 2007) needs to engage with both public, private and non-profit organisations, to promote co-ordinated actions amongst property owners, liaise with higher level organisations acting as a bridge across scales and facilitating flows of information and resources (Adger et al. 2005).

The organisation needs to be able to act with some authority, derived from a higher level of government, such as would be available to a legal owner. In many respects, this model is equivalent to an idealised traditional village where the community, governed by some internal authority, sharing the full ownership of the land and then allocating use rights to land and other assets to individual households subject to specific constraints and time limits. But of course, this is not straightforward and this sort of organisation involves significant transactions costs. It has to take account of the interests and preferences of all stakeholders including those both physically and temporally distant from the landscape being managed in a way that may not be possible in a democratic organisation that only reflects local interests. Also, such an organisation may be subject to institutional capture, whereby a narrow sectional interest group could gain control over the operation of the organisation as a whole, and this could distort its decisions away from those more generally favoured. It will need to be subject to external scrutiny and accountability but will be very difficult to evaluate in formal terms.

Given the presence of transactions costs, the level of investment in governance arrangements will depend on the potential gains to be achieved from better ecosystem management. The case for this type of organisation is apparent when asset values are relatively high and there is a substantial degree of uncertainty. We can identify some types of existing organisation, such as National Park Authorities, as emerging along these lines, but there is much to do to work out how such organisations should be operationalised in practice.

Conclusion

Landscape is a complex good that presents distinct challenges to governance. Property is concerned with the relationships among individual agents relating to the benefit flows associated with natural and man-made assets. Many of the benefit and cost streams between stakeholders within landscape systems are not financial and institutional arrangements need to incentivise stakeholders to take them into account in their decision-making. Private property arrangements and policy instruments are tending increasingly to separate out and partition rights. This focuses decision-makers' attention on these particular flows of costs and benefits, but runs the risk that it also increases potential agency problems amongst stakeholders and the potential for externalities. The problem is exacerbated in the context of uncertainty, ignorance and change such that some overall management of the system is required. The solution is likely to lie in some form of adaptive co-management that provides governance for both the natural and social systems and more needs to be done to determine when such arrangements are most important and how they may be operationalised.

Note

1 See http://www.cbd.int/doc/newsletters/news-biz-2008-02/?articleid=35 (accessed 25 March 2010).

References

Adger, W.N., Brown, K. and Tomkins, E. (2005) 'The political economy of cross-scale networks in resource co-management', *Ecology and Society*, 10(2): 9 [online] http://www.ecologyandsociety.org/vol10/iss2/art9/.

Baland, J.-M. and Platteau, J.-P. (1996) *Halting Degradation of Natural Resources. Is there a Role for Rural Communities?* Oxford: Oxford University Press.

Barzel, Y. (1989) *Economic Analysis of Property Rights*. Cambridge: Cambridge University Press.

Bougherara, D. and Combris, P. (2009) 'Eco-labeled food products: what are consumers paying for?' *European Review of Agricultural Economics*, 36(3): 321–341.

Bromley, D.W. (1991) *Environment and Economy: Property Rights and Public Policy*. Oxford: Blackwell.

Bruni, L. and Sugden, R. (2007) 'The road not taken: How psychology was removed from economics, and how it might be brought back', *Economic Journal*, 117: 1383–1393.

Campbell, D., Hutchinson, G. and Scarpa, R. (2009) 'Using choice experiments to explore spatial distribution of willingness to pay for rural landscape improvements', *Environment and Planning A*, 41: 97–111.

Cole, D.H. and Grossman, P.Z. (2002) 'The meaning of property rights', *Land Economics*, 78(3): 317–330.

Colman, D. (1994) 'Ethics and externalities: Agricultural stewardship and other behaviours', *Journal of Agricultural Economics*, 45: 299–311.

Colombo, S. and Hanley, N. (2008) 'How can we reduce errors from benefit transfer?' *Land Economics*, 84: 128–147.

Dwyer, J. and Hodge, I. (1996) *Countryside in Trust, Land Management by Conservation, Amenity and Recreation Organisations*. Chichester: John Wiley and Sons.

Folke, C. (2006) 'Resilience: The emergence of a perspective for social-ecological systems analyses', *Global Environmental Change*, 16: 253–267.

Folke, C., Hahn, T., Olsson, P. and Norberg, J. (2005) 'Adaptive governance of socio-ecological systems', *Annual Review of Environment and Resources*, 30: 441–473.

Franks, J. and McGloin, A. (2007) 'Environmental co-operatives as instruments for delivering across-farm environmental and rural policy objectives: Lessons for the UK', *Journal of Rural Studies*, 23(4): 472–489.

Garrod, G. and Willis, K. (1999) *Economic Valuation of the Environment: Methods and Case Studies*. Cheltenham: Edward Elgar.

Hodge, I. (1989) 'Compensation for nature conservation', *Environment and Planning A*, 21: 1027–1036.

Hodge, I. (2007) 'The governance of rural land in a liberalised world', *Journal of Agricultural Economics*, 58(3): 409–432.

Holling, C.S., Schindler, D.W., Walker, B.W. and Roughgarden, J. (1995) 'Biodiversity in the functioning of ecosystems: An ecological synthesis'. In: C. Perrings, K.-G. Mäler, C. Folke, C.S. Holling and B.-O. Jansson (eds), *Biodiversity Loss: Economics and Ecological Issues*. pp. 44–83, Cambridge: Cambridge University Press.

Jessop, B. (1998) 'The governance of complexity and the complexity of governance: preliminary remarks on some problems and limits of economic guidance'. In A. Amin and J. Hausner (eds) *Beyond Market and Hierarchy: Interactive Governance and Social Complexity*. pp. 95–128, Cheltenham: Edward Elgar.

Kahneman, D., Knetsch, J. and Thaler, R. (1991) 'Anomalies: The endowment effect, loss aversion and status quo bias', *Journal of Economic Perspectives*, 5(1): 193–206.

Kleijn, D., Berendse, F., Smit, R. and Gilissen, N. (2001) 'Agri-environment schemes do not effectively protect biodiversity in Dutch agricultural landscapes', *Nature*, 413: 723–725.

Kleijn, D., Baquero, R.A., Clough, Y., Dıaz, M., De Esteban, J., Fernández, F., Gabriel, D., Herzog, F., Holzschuh, A., Jöhl, R., Knop, E., Kruess, A., Marshall, E.J.P., Steffan-Dewenter, I., Tscharntke, T., Verhulst, J., West, T.M. and Yela, J.L. (2006) 'Mixed biodiversity benefits of agri-environment schemes in five European countries', *Ecology Letters*, 9(3): 243–254.

Langholz, J.A., Lassoie, J.P., Lee, D. and Chapman, D. (2000) 'Economic considerations of privately owned parks', *Ecological Economics*, 33(2): 173–183.

North, D. (1990) *Institutions, Institutional Change and Economic Performance*. Cambridge: Cambridge University Press.

OECD (1999) *Cultivating Rural Amenities: An Economic Development Perspective*. Paris: Organisation for Economic Co-operation and Development.

Ostrom, E. (1990) *Governing the Commons*. Cambridge: Cambridge University Press.

Sagoff, M. (2004) *Price, Principle and the Environment*. Cambridge: Cambridge University Press.

Schmid, A. (2004) *Conflict and Co-operation. Institutional and Behavioral Economics*. Oxford: Blackwell Publishing.

Taylor, M.E. and Morecroft, M.D. (2009) 'Effects of agri-environment schemes in a long term ecological time series', *Agriculture, Ecosystems and Environment*, 130(1–2): 9–15.

Termorshuizen, J. and Opdam, P. (2009) 'Landscape services as a bridge between landscape ecology and sustainable development', *Landscape Ecology*, 24(8): 1037–1052.

Turner, R.K. and Daily, G.C. (2008) 'The ecosystem services framework and natural capital conservation', *Environmental and Resource Economics*, 39(1): 25–35.

Wätzold, F. (2000) 'Efficiency and applicability of economic concepts dealing with environmental risk and ignorance', *Ecological Economics*, 33(2): 299–311.

6 Governance of protected landscapes and its implications for economic evaluation

Nora Mehnen, Ingo Mose and Dirk Strijker

Introduction

The goal of this chapter is to provide an introduction to the governance of protected landscapes and to shed some new light on how the increasing complexity of this topic requires a multifaceted and nuanced economic evaluation approach. Because the term "governance" – and especially its relation to "protected areas" – is crucial to this chapter, we will in this introductory section first elaborate on these two items, before we briefly outline the structure of this chapter.

Governance – a general description

The last decade has seen rapid advances in the field of nature policy and regional development, especially in the direction of non-hierarchical structures and bottom-up decision making. As such, the term "governance" has become crucial in the ongoing scientific and policy discussions about the development of regions and landscapes.[1]

Governance primarily describes the social decisions which develop from (in principal) equal and voluntary cooperation between state, private industries and civic actors. It concerns the interactions and relationships of actors through networks and the combination of different co-ordination mechanisms such as markets and negotiations. Due to this, governance implies that policy decisions are not under direct political control alone, but negotiated between the different actors and political levels. Due to their involvement in governance arrangements, economic and civil actors have the ability to influence political processes. This contrasts with the traditional role of the state as a hierarchical political decision maker. The potential of governance for environmental policy is currently broadly discussed (Böcher et al. 2008).

A variety of governance terms have emerged (see Figure 6.1). What is relevant here is that some of these terms refer to different spatial scales and levels, such as local governance, regional governance, urban governance, rural governance or landscape governance (see the bold elements in Figure 6.1). They almost always share equal cooperation of all actors and have a bottom-up approach in common, but it is questionable whether this really is the case. Only the geographic scale differs with implications for the context.

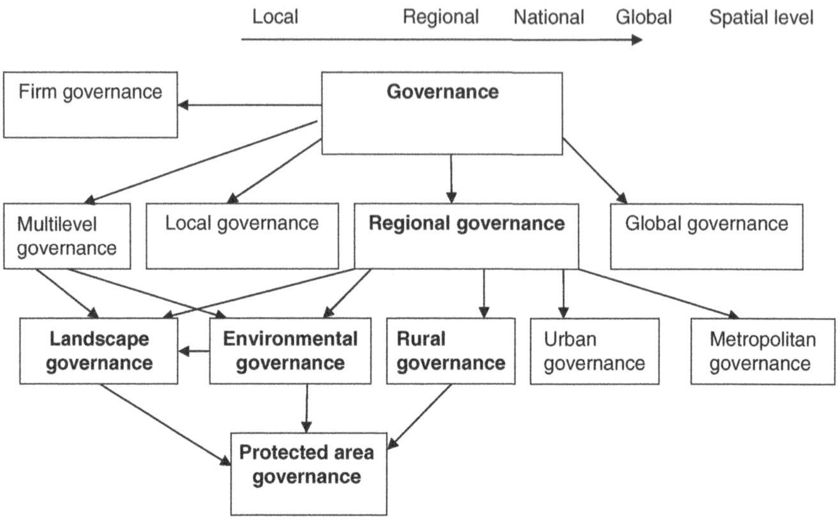

Figure 6.1 Different terms of governance.

Regional governance, for example, refers first and foremost to the regional level (Fürst 2003; Pollermann 2006), while the term "rural governance" applies particularly to peripheral, rural areas. In addition, the concept of environmental governance has received considerable attention over the last few years (Hempel 1996; Weale et al. 2003). Davidson and Frickel (2004: 471) define environmental governance as 'attempts by governing bodies or combinations thereof to alleviate recognized environmental dilemmas.' A vast volume of literature on environmental governance focuses on the top-down approach. According to most of these treatments, an improvement in governance is to be sought by reforming government and government-related institutions, assuming that the reason for environmental problems is an institutional weakness.

Landscape governance is a component of environmental governance. It is neither exclusively nor primarily involved in the maintenance of untouched, natural landscapes, nor is it necessarily in concordance with the maintenance of cultural landscapes. In the next sub-section, we will describe in further detail the governance of protected areas and how it relates to the preservation of landscapes. But before doing so, we first quote Görg, who argues:

> What seems to make the landscape concept useful as a link between governance processes in multi-level-politics and natural-spatial conditions is precisely its hybrid character, that is, that societal and "natural" factors are intrinsically linked to one another. Cultural, aesthetic, economic and social dimensions are as much involved as ecological functioning or abiotic conditions. Therefore, using the concept of landscape instead of the concept

> of region (cp. to regional governance: Pütz, 2004) is a more appropriate way of incorporating these dimensions. Herein, landscape provides the rescaling of politics with a material foundation, without returning to presumptions regarding ontologically prescribed spaces.
>
> (Görg 2007: 961)

However, the concept of regional governance is more valuable when it comes to concrete regions with distinctive (administrative) boundaries and management systems, which is the case in protected landscapes.

Governance of protected areas

For several years, the term governance has been applied to protected areas which has resulted in the coining of the term "protected area governance" (Borrini-Feyerabend 2004; Borrini-Feyerabend et al. 2006; Dearden et al. 2005; Fürst et al. 2006; Stoll-Kleemann et al. 2006). Protected area governance is useful because resources such as biodiversity and heritage create conflicts over their use and preservation. But also a kind of tension emerges, as resource protection is usually regulated top-down (through state administration) and governance is bottom-up (for example, through local initiatives).

Protected areas are a cornerstone of modern conservation policy – a legacy of the world's most valuable assets and places. By 2008, there were over 120,000 protected areas covering a total of about 21 million square kilometers. The terrestrial protected areas listed in the World Database on Protected Areas cover 12.2 per cent of the Earth's land area, while marine protected areas currently cover 5.9 per cent of the Earth's territorial seas and only 0.5 per cent of the extraterritorial seas. Among nations there is a great deal of variation in protection: only 45 per cent of the 236 countries and territories assessed had more than 10 per cent of their terrestrial area protected, and only 14 per cent had more than 10 per cent of their marine area protected (UNEP-WCMC 2010). The roots of the protected area idea go back thousands of years – long before governments created national parks – to the conservation regimes that human societies have been devising for millennia, among which are community-conserved areas (Borrini-Feyerabend 2002).

The modern foundation for protected areas was established in the late nineteenth century in the United States, with the designation of Yosemite as a State Park in 1864 and followed shortly by Yellowstone as a National Park in 1872. The "Yellowstone model" is seen as representing the preservation of large and wild areas by governments, where people are allowed to visit for recreation and pleasure but not reside. While in many places the public image of protected areas is still rooted in this national park model, in reality the protected area idea has evolved, moving beyond a single model to include many different kinds of protected areas.

Today the world's protected areas vary in almost every respect, including the purposes for which they are managed, their size, the resources they protect, and the management body responsible for the area (Phillips 2002). For this reason, the International Union for Conservation of Nature (IUCN) has created a classification system which identifies six categories of protected areas according to different management objectives. Within this system, the protected areas of Category V ("Protected Landscapes/Seascapes") are based on the interaction of *people and nature*, which makes the concept of governance a cornerstone in the maintenance of this category.

The importance of the governance concept is further substantiated by the paradigmatic shift or extension from a static-preservation to a dynamic-development oriented approach of areas, which occurred in recent years (Mose 2007). Among a number of relevant aspects, this shift includes a reallocation of responsibility: formerly protected areas were primarily managed by the central government while now several actors or stakeholders are involved. In general, however, protection remains the formal responsibility of the government.

Protected landscapes are a recent development and today play a growing role in national systems of protected areas, and in regional and global conservation strategies. Significant progress has been made over the last 25 years, running parallel to broader trends in conservation and in new approaches to protected areas in general. Protected landscapes are the predominant category, especially in Europe. The most recent United Nations List of Protected Areas 2003 declares that 46.1 per cent of the total protected area in Europe is classified as protected landscape (IUCN category V) (Chape et al. 2003). According to this classification, which we shall return to later in this chapter, protected landscapes take into account multiple interests and (because of this) are less strict and less contentious than other protected areas, like for example national parks which are mostly created for the preservation of flora and fauna and for the benefit and enjoyment of the public. Because the term "national park" existed long time before the IUCN classification system, some countries have categorized their national parks under other IUCN categories (e.g. English national parks are described as lived-in landscapes and classified under IUCN category V (Dudley 2008)). Due to this characteristic, protected landscapes have become important for regional development. Some regions apply for designation as a protected landscape to increase the value of the area in order to obtain social, cultural and economic benefits. A good example of this is the Black Forest Nature Park Central/North in Germany, where local stakeholders and the local population applied together for the status of protected landscape area (Verband Deutscher Naturparke 2008). The same applies for the UNESCO Biosphere Reserve Entlebuch in Switzerland (Hammer 2007) and the UNESCO Biosphere Park[2] Großes Walsertal in Austria (Coy and Weixlbaumer 2009).

All in all, governance structures in protected landscapes have become more complex and the goals are less concrete. This, of course, has consequences for the evaluation of protected landscapes: what is their economic and societal impact, and are the costs of protecting these landscapes higher or lower than their benefits?

Structure of the chapter

The remainder of this chapter is organized as follows. In the following section we further elaborate on the concepts of protected landscapes. Here, we also conceptualize the governance of these landscapes, before we conclude with a few considerations about the state-of-the-art in different European countries. Then, attention will be paid to the existence of multiple functions and various values of protected landscapes, including the production of (agricultural) goods and biodiversity. In the last part, we will trace the objectives of the different actors, and conclude with the consequences of multiple and possibly conflicting objectives for the economic evaluation of protected landscapes, and introduce a step-by-step-plan for the economic evaluation (of governance) of protected landscapes.

Protected landscapes – different approaches, categories and emphases

Protected landscapes are categorized by the interaction of man and nature, and thus the question of governance emerges. Apart from the already mentioned IUCN classification, there is also a classification by the United Nations Educational, Scientific and Cultural Organization (UNESCO), which we will discuss in the following sub-section. Table 6.1 summarizes the various types of protected landscapes.

In the table, the term "multi-stakeholder processes" is used to indicate that these types of protected landscapes require and allow for new forms of communication and decision-finding and possibly decision-making. Decision-finding or decision aid relates to the finding of all different possible arguments and 'keys which might enable the actors to go forward' (Bana e Costa and Pirlot 1997: 565). Decision-making, as the term itself already suggests, relates to the making of concrete decisions (based on weights). The geographical scope of the application in most types of protected landscapes is global, only the areas classified under IUCN category V and category VI (see hereafter) are more of national or sub-national (and hence local or regional) importance.

UNESCO designations

UNESCO designates places of cultural and scientific significance. The most important types are UNESCO World Natural Heritage, UNESCO World Cultural Heritage, UNESCO Cultural Landscapes and the UNESCO Biosphere Reserves (UNESCO 2008) (see Table 6.1). One of the UNESCO types is "cultural landscapes", described as having outstanding universal value and developed through dynamic and evolving human relationships and interactions with the environment ("living landscapes").[3] There are three categories of cultural landscapes.

Table 6.1 Comparison of the different protected landscape initiatives

Initiative	*Character of affected landscape*	*Geographical scope of application*	*Areas covered by initiative*	*Main aims*	*Governance*
UNESCO World Natural Heritage	Outstanding natural value, from the point of view of science, conservation or natural beauty	Global	Natural features	Protect natural values	No specific governance objectives
UNESCO World Cultural Heritage	Outstanding cultural value, from the point of view of history, art or science	Global	Monuments, group of buildings, sites	Protect cultural values	No specific governance objectives
UNESCO Cultural Landscape	Outstanding universal value	Global	Any appropriate area	Protect heritage values	Multi-stakeholder approach
UNESCO Biosphere Reserve	Outstanding universal value	Global	Areas with high biodiversity values	Conservation of biodiversity and biological resources	Multi-stakeholder approach
IUCN Category V	Landscape/Seascape that deserves protection	National/Sub-national	Landscapes that typically have been modified extensively by people over time	Integrate activities and enhance natural and cultural values	Multi-stakeholder approach with different protected area governance types
IUCN Category VI	Areas in a natural condition	National/Sub-national	Unmodified natural systems	Sustainable use of natural resources	Management by public bodies or through local custom

(i) Landscape designed and created intentionally by man: this embraces garden and parkland landscapes constructed for aesthetic reasons which are often (but not always) associated with religious or other monumental buildings and ensembles.
(ii) Organically evolved landscape: this results from an initial social, economic, administrative, and/or religious imperative and has developed its present form by association with and in response to its natural environment. They fall into two sub-categories: a relict (or fossil) landscape (where the evolutionary process came to an end at some time in the past) or a continuing landscape (where the evolutionary process is still in progress, and which retains an active social role in contemporary society closely associated with the traditional way of life).
(iii) Associative cultural landscape: the inscription of such landscapes on the World Heritage List is justifiable by virtue of the powerful religious, artistic or cultural associations of the natural element rather than material cultural evidence, which may be insignificant or even absent (UNESCO 2008).

UNESCO Biosphere reserves are maybe the most prominent example of protected landscapes; they are an instrument of UNESCO's Man and the Biosphere (MAB) programme, dedicated to sustainable development and the conservation of biodiversity, as well as the support of environmental education, research, and the monitoring of the most important natural areas of the world. One key element in this program is the involvement of local actors and especially the participation of local inhabitants. Because these biosphere reserves are places that seek to reconcile economic development, social development and environmental protection through partnerships between people and nature, they are ideal to test and demonstrate approaches to sustainable development at a regional scale (Hammer 2007).

IUCN categories

The International Union for Conservation of Nature (IUCN) designates two categories which concern protected landscapes: IUCN category V and IUCN category VI. IUCN category V is defined as "Protected Landscape/Seascape" and

> represents a protected landscape or seascape where the interaction of people and nature over time has produced an area of distinct character with significant ecological, biological, cultural and scenic value. Safeguarding the integrity of this interaction is vital to protecting and sustaining the area and its associated nature conservation and other values.
>
> (Dudley 2008: 20)

An area that is categorized as "Protected Landscape/Seascape" may be owned by a public authority, but is more likely to comprise a mosaic of private and public ownerships operating under a variety of management regimes, such as associations or limited companies. These regimes should be subject to a degree of planning or other control and are supported, where appropriate, by public funding and other incentives, to ensure that the quality of the land- or seascape and the relevant local customs and beliefs are maintained in the long term.

The other category of protected landscapes is IUCN category VI: "Managed Resource Protected Areas", which shares with Category V an emphasis on sustainable use of natural resources. However, Category VI differs from Category V in that "Protected Landscape/Seascape" areas involve landscapes that typically have been modified extensively by people over time. They are based on the interaction of people and nature. Category VI, on the other hand, includes areas with predominantly unmodified natural systems that are managed to ensure long term protection and maintenance of nature and biodiversity. These areas must be managed in such a way that at least two-thirds of their natural system remains unmodified (Phillips 2002). This means that management should be undertaken with an unambiguous remit for conservation, and carried out in partnership with the local community. In practical terms, management of Category VI areas may be provided through local customs supported and advised by governmental or non-governmental agencies. Ownership is in the hands of the national (or other levels of) government, the community, private individuals, or a combination of these.

Emphases of protected landscapes in Europe

Protected landscapes in Europe have different emphases and goals (Mose 2007), such as traditional forms of nature protection (Italy, Spain, Slovenia, Croatia, East Germany), recreation and tourism (West Germany, Great Britain) and motor for rural development (France, Luxembourg, Belgium, Austria, Germany (from 2002), Switzerland (from 2006)). In the Netherlands, the so-called National Landscapes (*Nationale Landschappen*) are characteristic for the Dutch countryside, emphasizing the landscape scale of history and the connectivity between people, places and heritage items. They integrate natural and cultural heritage conservation by examining them at a landscape level. The Dutch National Landscapes are not classified within the IUCN classification system or within any other global initiative yet.

Most, if not all protected landscapes in Europe aim to integrate protection and land-use and, as such, are instruments of regional development. Hence, protected landscapes contribute to the regional economy as well as to the preservation of biodiversity and the services provided by nature. In the future, this integration will be of even greater importance, due to increasing impacts of settlements and other forms of land-use. In this context protected landscapes have a central role, because they open the perspective of multi-purpose land-use explicitly (Mose 2007).

Multi-functionality and values of protected landscapes

Protected landscapes are not only characterized by their classification and their complex governance structures but also by their multiple functions (e.g. regulatory, habitat, support, development and information). These functions have been described extensively in the literature (e.g. Mose 2007) and will not be discussed further here. Rather, we wish only to state that until recently most (rural) landscapes were treated as "mono-functional" with emphasis almost exclusively on (agricultural) productive functions. While for many parts of Europe such functions are still important to some extent, other (explicitly consumptive) functions are gaining in significance, such as leisure, residence, and ecological and environmental functions. As a result of this, the primary production function has decreased further and further in importance.

Transitions between functions, and the changing relationships that these transitions imply are encapsulated in the concept of landscape. In fact, landscapes and especially protected landscapes are multifunctional by definition: they are more than just shapes and morphological features of the surface of the earth, more than habitats, more than images, more than elements of culture. They are the interfaces where the social and the natural interact and a space where the global and the local meet. Multifunctionality refers to the fact that an economic activity may have multiple outputs and may contribute to several societal objectives at once (e.g. Brouwer and Van der Heide 2009). It is thus an activity-oriented concept related to specific properties of the production process and its multiple outputs. Interestingly, multifunctionality represents a shift from a rather traditional production-centric view to one that incorporates other outputs that are often beyond the private domain.

While multifunctionality and multi-purpose land management is becoming increasingly important, land-users and policy makers must nevertheless make choices between different, and usually competing, land-uses. In view of the various functions of the protected landscape and the multiple stakeholders and land-users involved (e.g. farmers, environmentalists, tourists and local inhabitants), land-use allocation problems are spatial planning problems. The interaction among different stakeholders on a common stage highlights the complex interaction between ecological processes and economic activities. It is therefore not surprising that problems of competing claims can be severe in protected landscapes.

The various functions of protected landscapes reflect a variety of values (which can be tangible but also intangible). They indicate that protected landscapes serve not only as a conservation tool but also as economic engines for tourism and outdoor recreation, food and fiber production, and (ecological) education and research. Very important values are environmental and natural, cultural, educational, scientific, recreational, and spiritual. There is a broad range of literature focusing on these values (Getzner et al. 2005; Amend et al. 2008; Mallarach 2008). For several years (and reflected by the concept of "ecosystem services"), the economic or utilitarian value of protected areas has become more and more important. Most studies differentiate between use values and non-use values of protected areas and concentrate on regional value added through protected area tourism (e.g. Task Force on Economic Benefits of Protected Areas 1998; Küpfer 2000; Job et al. 2005, 2009, and various other contributions to this book).

Many stakeholders with differing interests

One key issue for protected landscapes is the engagement of a diverse set of stakeholders on equal terms in order to facilitate government. A legal framework can shape the form of governance and can provide recognition of traditional management systems and customary law. Strict principles of (voluntary) participation, decentralization, transparency, and a search for consensus combined with administrative flexibility, a lack of bureaucracy and a process that involves simultaneous and equal participation of all levels of government and non-government bodies (NGOs; members of the scientific community, the private sector and local population) characterize contemporary governance.

These new governance structures have a different influence on economic evaluation – the process of scoring or rating the quality of landscapes. Strict rules and objectives defined by the state are easier to evaluate than fuzzy rules from bottom-up. Governance could increase or decrease the different functions of protected landscapes depending on which parties are most successful in navigating the governance process. For example, if civil society has a stronger influence on the process than private business interests then sustainable development could become a prominent concern while private business would become secondary.

Is local economic development compatible with environmental concerns such as the protection of biodiversity or landscapes? After numerous debates in the scientific world and in society, there is a growing consensus on this question. In many protected areas, institutional arrangements try to reconcile the environmental objectives and economic interests of local populations. This multi-purpose land-use is not entirely restricted to protected landscapes, but is rather typical for an increasing number of non-protected rural landscapes as well. In most of the protected landscapes, settlements (often established before the official designation of a protected area) are included and even the extension of built-up areas is allowed, while the stimulation of "heavy" industrial activities in protected landscapes is forbidden. As such, the multi-purpose land-use is restricted to a limited number of less polluting and less disturbing (economic) activities (e.g. agriculture, tourism, crafts). As a result the integration of environmental objectives and economic interests leads to many different types of governance structures, in which local communities and actors regularly participate in decision-making processes. Many protected landscapes provide good examples of governance and especially of adaptive co-management of these areas (see also Chapter 5 in this book; Elbakidze et al. 2010).

In the United Kingdom, the practice of government regarding protected landscapes has changed recently as a result of processes of devolution and regionalization. The experience of regional governance in England has been that state and public agencies dominate new governance arrangements. Regional institutions have become recognized as significant players with whom the National Park Authorities and Area of Outstanding Natural Beauty (AONB) partnerships increasingly interact especially with respect to the regional delivery of European funding. However, the national government is still important in terms of annual funding allocations and policy steer (Thompson 2003).

In Germany, the role of governance in protected landscapes is not so well researched. There are only a few studies focusing on governance in Biosphere Reserves (Stoll-Kleemann and Bertzky 2004; Fürst et al. 2006) and in Nature Parks (Gailing and Keim 2006). One general remark based on that research is that the conservation success of protected landscapes strongly depends on the appropriateness of their governance and management systems with regard to the local context (e.g. organizational structure, adaptive planning tools), but also on broader economic and governance issues.

In Finland, the management model for National Parks is rather centralized from a European perspective (Grönholm 2009). There is, however, a tendency to put more emphasis on stakeholder governance through public participation within the participatory planning processes of management plans for National Parks. Nevertheless, only very few of the affiliated local stakeholders are involved in institutional participation processes. Furthermore (and despite sincere efforts and good intentions) a majority of the local stakeholders share the feeling that there is no possibility of participation. The main reason for this is a general lack of information and communication between the administrators and the local stakeholders. Because the lack of certain social values, such as social responsibility, an active local democracy and tolerance and civility, is at the root of many environmental conflicts, Grönholm (2009) suggests a more extensive local stakeholder involvement in the management of protected areas.

To perform a thorough evaluation, it is essential to identify and include all actors, interpret their interests and assess their costs and benefits of how to use the protected landscapes (see Figure 6.2).

Implications for evaluation

There are many actors actively involved in new and modern ways of governance of protected landscapes, and these actors have different resources, interests, and relations (Stoll-Kleemann and Welp 2006; Stokman and Vieth 2004; Stokman 2004). Some actors, such as the park management, will have broad interests in the ecological and recreational development of the protected area. Others will have more limited or partial interests. The owner of a rent-a-bike for instance is primarily interested in the accessibility of the park, the quality of the infrastructure, and probably the routing of the visitors, but not in protection of water resources, or the working routines of the park management. Local farmers will judge a plan primarily in terms of land-use, water levels and building restrictions, but not in terms of visitors or the quality of the biodiversity.

In such circumstances, the emphasis in the assessment of the impact of a certain project on a protected landscape shifts from a discussion between specialists with a common language (discourse) to a kind of communication tool with different actors with unequal interests, knowledge and language. These interests cannot be denied, even when specialists regard them as less important: the articulated

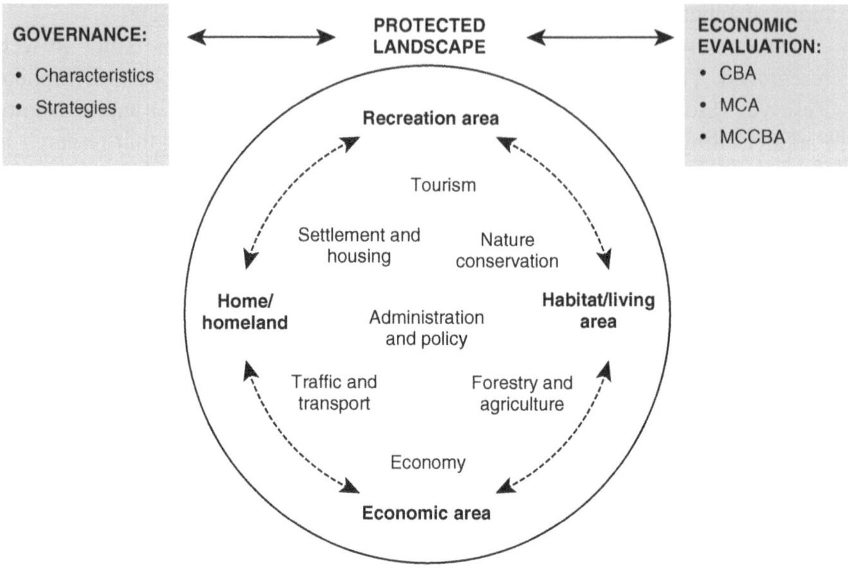

Figure 6.2 Governance, protected landscapes and economic evaluation.

interests of the impacted actors should always be covered in the assessment. Their concerns should be taken into account in order to connect them to the project (Pannell et al. 2006; Stolp et al. 2002; Craig and Vanclay 2005). These concerns will be manifold for almost all projects, because usually not only economic, but also all kinds of social or environmental aspects play a role.

The traditional cost–benefit analysis (CBA) evaluation technique, which typically quantifies the costs and benefits of a project in monetary terms, is not very well suited to the challenge of considering all interests and concerns, because it does not take into account the distribution of the impacts. Recently, Sijtsma et al. (2011) have elaborated on this issue in the light of sustainability. Their conclusion is that a social benefit in monetary terms is only an increase in welfare, "when, and only when" society conceives it as such. That implies that all individual actors should conceive the project as yielding a social benefit. Besides the distributional shortfall of CBA, there is often fundamental disagreement between actors about the relevance of monetary judgment, or even its ethical desirability.

The complement of monetary evaluation techniques are non-monetary evaluation techniques, of which multi criteria analysis (MCA) is the most well known. As such, MCA is able to take into account all possible criteria of a project. The technique allows for the recognition and active participation of stakeholders, which fits very well with modern governance structures. However, in the case of many actors with differing interests and positions, which is often the case in protected areas, there are practical limits to the application of MCA. One problem is how the preferences of stakeholders with partial interests should be judged against those with a broader interest. Moreover, long lists of criteria are difficult to analyze, and lead to serious risks of double counting (Strijker et al. 2000b). In the case of multi-criteria decision making (Strijker et al. 2000a), long lists of criteria will result in very low weights for the individual criteria (Sijtsma et al. 2011). And, very practically, MCA with many different criteria is costly to perform – not only in terms of money, but also in time.

An approach advocated in recent years is to combine both techniques – CBA and MCA – in a mixed approach: multi criteria cost–benefit analysis (MCCBA) (Sijtsma et al. 2011). Their idea dates back to the 1970s (Nijkamp and Van Delft 1977). Criteria which can be easily summarized in monetary terms, and on which the different actors can agree (investment costs, maintenance, benefits of the production of market goods) are treated in a traditional CBA-framework, leading to some kind of net present value (NPV). This NPV is then brought as one specific criterion in a MCA-framework. Non-market goods and elements on which the actors do not (potentially) agree, or which are relevant for only a limited number of actors, are measured in their own dimension or unit and are also inserted in the MCA. The risk of long lists of criteria as mentioned above is diminished, although not completely solved.

A great advantage of this approach in relation to modern governance structures is that different actors, also the less important ones with partial and specific interests, can participate in the process of effect assessment, which will lead to a greater support for the ultimate outcome of the project.

Our suggestion for a step-by-step-plan for the economic evaluation (of governance) of protected landscapes is based on the MCCBA approach of Sijtsma et al. (2011) (see Table 6.2).

In the first step the function and the scale of the evaluation should be identified, but also possible alternatives need to be considered. Negotiation is crucial at this point. The second step is especially important for the evaluation of protected landscapes. (Representatives of) all actors from state, economy and civil society should be involved in the evaluation process. The whole evaluation process, its function, objective, alternatives and scale should be identified and negotiated with all actors involved in step 2. In the third step, when the clear judgment criteria on governance of the protected landscape are organized and negotiated, it is important that all interests and opinions should be taken seriously, also if specialists do not agree; otherwise the evaluation will not work. In step 4, the criteria should be quantified physically with the knowledge, experience and resources of all actors involved. Steps 5 and 6 deal with the aggregation of monetary scores and non-monetary scores. All actors must reach a consensus and have an equal and effective voice. Maybe it is necessary to use weights for actors who are more affected by a project (e.g. because they live in the area) than actors who are less affected. In step 7 trade-offs should be interpreted. Here, the consensus based aggregation of the performance matrix should be clarified and annotated, for example with a ratio-analysis (Sijtsma et al. 2011). The ratio-analysis is similar to a cost-effectiveness analysis and sets one criterion alongside another criterion (Stewart and Losa 2003; Sijtsma et al. 2011). A crucial step is the last one, when a sensitivity analysis

Table 6.2 Step-by-step-plan for economic evaluation of (governance of) protected landscapes

Blocks	*MCCBA*	*Economic evaluation of (governance of) protected landscapes*
I: Providing the basic evaluation structure	Stage 1: Identify function, project alternatives and scale of the evaluation	Step 1: Identify function, alternatives and scale of the evaluation, should be negotiated afterwards with all actors
	Stage 2: Involve a broad group of stakeholders	Step 2: Involve actors from all sectors (state, economy and civil society)
	Stage 3: Organize judgment criteria on sustainability impacts	Step 3: Organize clear judgment criteria on governance of the protected landscape: value all meanings and opinions
II: Fact finding on physical impacts	Stage 4: Quantify impacts physically	Step 4: Quantify criteria physically (with knowledge of all actors)
III: Aggregation of impacts to a compact format	Stage 5: Aggregate monetary scores consensus based	Step 5: Aggregate monetary scores consensus based – all actors should have an equal voice
	Stage 6: Aggregate non-monetary scores consensus based	Step 6: Aggregate non-monetary scores consensus based – all actors should have an equal voice
IV: Communication of problem understanding	Stage 7: Interpret trade-offs	Step 7: Interpret trade-offs
	Stage 8: Perform sensitivity analysis and reconsider project alternatives	Step 8: Perform sensitivity analysis and reconsider project alternatives, evaluate the process

should be performed and project alternatives should be reconsidered. The whole economic evaluation process should be evaluated at the end.

Conclusion

Compared to National Parks, protected landscapes have come relatively late upon the protected area scene. Nevertheless, they play today a major role in national systems of protected areas and in regional and global conservation strategies. And although they are classified within different classification systems (UNESCO or IUCN), protected landscapes always have a combination of economic and/or societal uses and biodiversity conservation in common. Governance structures of such areas can be very diverse, with many stakeholders and land-users being involved (each with different interests and objectives) and this requires a multifaceted and nuanced economic evaluation approach. Economic (valuation) knowledge and information can be very useful for the governance process of protected areas. For these very purposes, it is crucial that actually all actors are getting involved and that all their interests are being covered by the evaluation.

Clearly, it is unlikely that actors will adopt economic valuation results (e.g. CBA or MCCBA) wholeheartedly, but these results can nevertheless play an important role in negotiating and in the outcome of the process. Of course, there will always be actors who are likely to use the results as an argument (e.g. entrepreneurs), and others who are less likely to use them (e.g. nature conservationists). However, if every position and every stake is taken into account, it is more likely that the whole process will end with a well-grounded result, which all actors can accept and support. As we have explained above, the development of suitable governance arrangements is exactly about gaining sufficient acceptance and gets as many actors as possible actively involved in "their" protected landscape.

Notes

1 According to the European Landscape Convention, "Landscape" means: 'an area, as perceived by people, whose character is the result of the action and interaction of natural and/or human factors' (Council of Europe 2006; see also Chapter 2 of this book). The definition applies to the whole territory of states including all urban and periurban landscapes, towns, villages and rural areas, the coast and inland areas. It applies to ordinary or even degraded landscape as well as those areas that are outstanding or protected. The European Landscape Convention argues that the protection, management and planning of all landscapes in Europe is a task not just for governments but for all sectors of civil society, entailing "rights and responsibilities for everyone" (Governance). The concept of landscape as understood by all three official languages (English, French and Spanish) of IUCN and in the European Landscape Convention embodies both the natural world and the human impact on it. We follow Antrop et al. (Chapter 2 of this book), who distinguish between landscape and land (a piece of terrain, bounded in space and bordered, and very often owned by someone or some institution. Land refers to (private) property that can be used more or less freely by its owner). Land-use is the result of indirect and direct human impact.

2 Biosphere parks are identical to biosphere reserves. The Austrian Man and the Biosphere (MAB) National Committee decided to use the term "biosphere park" for reasons of acceptability (http://www.grosseswalsertal.at).
3 Cultural landscapes are often conceived as a 'conceptual bridge between culture and nature, between tangible and intangible heritage, and across space and time.' (Brown 2008: 5).

References

Amend, T., Brown, J., Kothari, A., Phillips, A. and Stolton, S. (eds) (2008) *Protected Landscapes and Agrobiodiversity Values*. Volume 1, Protected Landscapes and Seascapes, IUCN & GTZ, Heidelberg: Kasparek Verlag.

Bana e Costa, C.A. and Pirlot, M. (1997) 'Thoughts on the future of the multicriteria field: Basic convictions and outline of a general methodology'. In J. Clímaco (ed.) *Multicriteria Analysis*. pp. 562–568, Berlin: Springer Verlag.

Böcher, M., Giessen, L. and Kleinschmit, D. (eds) (2008) *Environmental and Forest Governance – The Role of Discourses and Expertise*, Proceedings of the International Conference, Göttingen 2007: Universitätsverlag Göttingen.

Borrini-Feyerabend, G. (2002) 'Indigenous and local communities and protected areas: rethinking the relationship', *Parks*, 12(2): 5–15.

Borrini-Feyerabend, G. (2004) *Governance of Protected Areas – Innovation in the Air...* http://www.earthlore.ca/clients/WPC/English/grfx/sessions/PDFs/session_1/Borrini_Feyerabend.pdf (accessed 10 January 2011).

Borrini-Feyerabend, G., Johnston, J. and Pansky, D. (2006) 'Governance of protected areas'. In M. Lockwood, G.L. Worboys and A. Kothari (eds) *Managing Protected Areas: A Global Guide*. pp. 116–145, London, UK and Sterling, VA, USA: Earthscan.

Brouwer, F. and Heide, C.M. van der (eds) (2009) *Multifunctional Rural Land Management Economics and Policies*. London: Earthscan Publications Ltd.

Brown, S. (2008) *Cultural Landscapes and Park Management: A Literature Snapshot*. Sydney: Department of Environment and Climate Change. http://www.environment.nsw.gov.au/resources/cultureheritage/07137cultlandresearch.pdf (accessed 10 January 2011).

Chape, S., Blyth, S., Fish, L., Fox, P. and Spalding, M. (2003) *United Nations List of Protected Areas*. Cambridge, UK/Gland, Switzerland: UNEP – World Conservation Monitoring Centre/IUCN – The World Conservation Union.

Council of Europe (2006) *Landscape and sustainable development: challenges of the European Landscape Convention*. Strasbourg: Council of Europe Publishing. http://www.coe.int/t/dg4/cultureheritage/heritage/landscape/Publications/PaysageDeveloppement%20_en.pdf (accessed 10 January 2011).

Coy, M. and Weixlbaumer, N. (Hrsg.) (2009) 'Der Biosphärenpark als regionales Leitinstrument', *Das Große Walsertal im Spiegel der Nutzer. Alpine Space – Man & Environment*, Vol. 10. Innsbruck: University Press (in German).

Craig, A. and Vanclay, F. (2005) 'Questioning the potential of deliberativeness to achieve acceptable natural resource management decisions'. In: J. Martin and R. Eversole (eds), *Participation and Governance in Regional Development*. pp. 155–172, Aldershot: Ashgate.

Davidson, D.J. and Frickel, S. (2004) 'Understanding environmental governance: a critical review', *Organization & Environment*, 17(4): 471–492.

Dearden, P., Bennett, M. and Johnston, J. (2005) 'Trends in global protected area governance 1992–2002', *Environmental Management*, 36(1): 89–100.

Dudley, N. (ed.) (2008) *Guidelines for Applying Protected Area Management Categories*. Cambridge, UK/Gland Switzerland: IUCN – The World Conservation Union.

Elbakidze, M., Angelstam, P.K., Sandström, C. and Axelsson, R. (2010) 'Multi-stakeholder collaboration in Russian and Swedish Model Forest initiatives: adaptive governance toward sustainable forest management?' *Ecology and Society*, 15(2): 14. http://www.ecologyandsociety.org/vol15/iss2/art14/ES-2009-3334.pdf (accessed 10 January 2011).

Fürst, D. (2003) *Steuerung auf regionaler Ebene versus Regional Governance. Informationen zur Raumentwicklung* 8/9.2003: 441–450 (in German).

Fürst, D., Lahner, M. and Pollermann, K. (2006) *Entstehung und Funktionsweise von Regional Governance bei dem Gemeinschaftsgut Natur und Landschaft – Analysen von Place-making- und Governance-Prozessen in Biosphärenreservaten in Deutschland und Großbritannien*, Hannover: Beiträge zur räumlichen Planung, H. 82 (in German).

Gailing, L. and Keim, K.D. (2006) *Analyse von informellen und dezentralen Institutionen und Public Governance mit kulturlandschaftlichem Hintergrund in der beispielregion Barnim*. Berlin: Berlin-Brandenburgische Akademnie der Wissenscgaften. Materialien 6 (in German).

Getzner, M., Spash, C.L. and Stagl, S. (eds) (2005) *Alternatives for Environmental Valuation*. London and New York: Routledge.

Görg, C. (2007) 'Landscape governance: the "politics of scale" and the "natural" conditions of places', *Geoforum*, 38(5): 954–966.

Grönholm, S. (2009) 'Governing national parks in Finland: the illusion of public involvement', *Local Environment*, 14(3): 233–243.

Hammer, T. (2007) 'Biosphere Reserves: An instrument for sustainable regional development? The case of Entlebuch, Switzerland'. In: I. Mose (ed.), *Protected areas and Regional Development in Europe : Towards a New Model for the 21st Century*. pp. 39–54, Aldershot: Ashgate, Studies in Environmental Policy and Practice.

Hempel, L.C. (1996) *Environmental Governance: The Global Challenge*. Washington, DC and Covelo, California: Island Press.

Job, H., Harrer, B., Metzler, D. and Hajizadeh-Alamdary, D. (2005) *Ökonomische Effekte von Großschutzgebieten*. BfN-Skripten 135. Bonn-Bad Godesberg. (in German).

Job, H., Woltering, M. and Harrer, B. (2009) *Regionalökonomische Effekte des Tourismus in deutschen Nationalparken*. Naturschutz und Biologische Vielfalt, 76, Bonn-Bad Godesberg, Bundesamt für Naturschutz (in German).

Küpfer, I. (2000) *Die regionalwirtschaftliche Bedeutung des Nationalparktourismus – Untersuchung am Beispiel des Schweizerischen Nationalparks*, Zürich: Geographisches Institut der Universität Zürich-Irchel (in German).

Mallarach, J.-M. (ed.) (2008) *Protected Landscapes and Cultural and Spiritual Values*. Heidelberg, Kasparek Verlag, Volume 2, Values of Protected Landscapes and Seascapes, IUCN, GTZ and Obra Social de Caixa Catalunya.

Mose, I. (ed.) (2007) *Protected Areas and Regional Development in Europe: Towards a New Model for the 21st Century*. Aldershot: Ashgate, Studies in Environmental Policy and Practice.

Nijkamp, P. and Delft, A. van (1977) *Multicriteria Analysis and Regional Decision-making*. Leiden: Nijhoff.

Pannell, D.J., Marshall, G., Barr, N., Curtis, A., Vanclay, F. and Wilkinson, R. (2006) 'Understanding and promoting adoption of conservation practices by rural landholders', *Australian Journal of Experimental Agriculture*, 46(11): 1407–1424.

Phillips, A. (2002) *Management Guidelines for IUCN Category V Protected Areas: Protected Landscapes/Seascapes*. Cambridge, UK/Gland, Switzerland: IUCN – The World Conservation Union.

Pollermann, K. (2006) *Regional Governance and Biosphere Reserves – Empiric results from case-studies in Germany and Great Britain. http://www.ttef.org.uk/Braunton_* Burrows_File/TTEF%20-%20Biosphere%20Reserve%20-%20Regional%20Governance%20-%20English%20Summary.pdf (accessed 10 January 2011).

Sijtsma, F.J., Heide, C.M. van der and Hinsberg, A. van (2011) 'Biodiversity and decision-support: integrating CBA and MCA'. In: A. Hull, A. Khakee, J. Woltjer and E. Alexander (eds), *Methodological Innovation in Planning Evaluation*. London: Routledge.

Stewart, T.J. and Losa, F.B. (2003) 'Towards reconciling outranking and value measurement practice', *European Journal of Operational Research*, 145(3): 645–659.

Stokman, F.N. (2004) 'Frame dependent modeling of influence processes'. In: A. Diekmann and T. Voss (eds), *Rational-Choice-Theorie in den Sozialwissen-schaften. Anwendungen und Probleme*. pp. 113–127, München: Oldenbourg Verlag.

Stokman, F.N. and Vieth, M. (2004) *Was verbindet uns wann mit wem? Inhalt und Struktur in der Analyse sozialer Netzwerke*, Kölner Zeitschrift für Soziologie. Sonderheft 44, pp. 274–302 (in German).

Stoll-Kleemann, S. and Bertzky, M. (2004) *Linking governance and management perspectives with conservation success in Biosphere Reserves – The GoBi-Project.* http://www.mnf.uni-greifswald.de/fileadmin/Geowissenschaften/geographie/angew_geo/Publikationen/Stoll-Kleemann-Publikationen/Stoll-Kleemann_Bertzky_Artikel_BRIM_2004_final_korr.pdf (accessed 10 January 2011).

Stoll-Kleemann, S., Bender, S., Berghöfer, A., Bertzky, M., Fritz-Vietta, N., Schliep, R. and Thierfelder, B. (2006) *Linking Governance and Management Perspectives with Conservation Success in Protected Areas and Biosphere Reserves*, Berlin: Humboldt-Universität, Discussion Paper 1. http://www.gis.uni-greifswald.de/agnw/shared-data/Files/Publications/GoBi_discussionpaper01.pdf (accessed 10 January 2011).

Stoll-Kleemann, S.and Welp, M. (eds) (2006) *Stakeholder Dialogues in Natural Resources Management. Theory and Practice*. Berlin, Heidelberg: Springer.

Stolp, A., Groen, W., Vliet, J. van and Vanclay, F. (2002) 'Citizen values assessment: incorporating citizens' value judgements in environmental impact assessment', *Impact Assessment & Project Appraisal*, 20(1): 11–23.

Strijker, D., Sijtsma, F.J. and Bettels, K. (2000a) 'Evaluating nature conservation: the case of meadow birds in the Netherlands', *Agricultural Economics Review*, 1(2): 57–70.

Strijker, D., Wiersma, D. and Sijtsma, F.J. (2000b) 'Evaluation of nature conservation', *Environmental & Resource Economics*, 16(4): 363–378.

Task Force on Economic Benefits of Protected Areas of the World Commission on Protected Areas (WCPA) of IUCN, in collaboration with the Economics Service Unit of IUCN (1998) *Economic Values of Protected Areas: Guidelines for Protected Area Managers.* Cambridge, UK / Gland, Switzerland: IUCN – The World Conservation Union.

Thompson, N. (2003) *Governing National Parks in a Devolving UK.* Leeds: University of Leeds, School of Geography, PhD Dissertation.

UNEP-WCMC (2010) *2010 Biodiversity Indicators Partnership – Factsheet Coverage of Protected Areas.* http://www.unep-wcmc.org/wdpa/statistics/2010BIP_Factsheet_Coverage_of_Protected_Areas.pdf (accessed 20 February 2011).

UNESCO (2008) *Convention Concerning the Protection of the World Cultural and Natural Heritage*. http://whc.unesco.org/archive/convention-en.pdf (accessed 10 January 2011).

Verband Deutscher Naturparke (2008) Ergebnisbericht "Optimierte Umsetzung von Naturparkplänen". Ausführliche Ergebnisse des Forschungs- und Entwicklungsvorhabens, bearbeitet von BTE Tourismusmanagement, Regionalentwicklung in Kooperation mit Institut für Umweltplanung der Leibniz, Universität Hannover (Prof. Dr. Christina von Haaren) (in German). http://www.naturparke.de/downloads/Perspektive_fuer_laendliche_Raeume_in_Europa.pdf (accessed 10 January 2011).

Weale, A., Pridham, G., Cini, M., Konstadakopoulos, D., Porter, M. and Flynn, B. (2003) *Environmental Governance in Europe: an ever closer ecological Union?* Oxford: Oxford University Press.

Part II

Reorientations and reflections – building blocks of landscape economics

7 A procedure for determining an optimal landscape and its monetary value

Wim J.M. Heijman and Pierre v. Mouche

Introduction

Goods in general and landscapes in particular are composed of characteristics. For the (representative) consumer an optimal landscape is a combination of possible landscape characteristics that maximises his net benefits. An optimal landscape provides useful information for decision makers. A specifically important topic is the net benefits of an optimal landscape, also referred to as the monetary value of the landscape. Only if the monetary value is positive is it worthwhile to invest.

In Heijman and Goossen (2009) a procedure, based on expressed consumer preferences (Pruckner 1995), was introduced to determine, in the case of two characteristics and net-benefit functions of Cobb–Douglas form, an optimal landscape and its monetary value. These preferences concerned the revelation of the optimal combination of characteristics. The aim of our chapter is nothing more than to correct, generalise, and formalise this procedure but now for a class of C(onstant) E(lasticity of) S(ubstitution) net-benefit functions which are more flexible. In addition to this an additive separable class of net-benefit functions will be dealt with and an attempt will be made to give a welfare theoretic foundation for this class.

The setting in Heijman and Goossen (2009) is as follows: the landscape has area A and the possible landscapes are characterized by all possible combinations (x_1, x_2) of two characteristics 1 and 2 such that $x_1 + x_2 = A$. For example, x_1 represents the number of hectares under grass and x_2 represents the number of hectares under forest. The price of characteristic $i = 1, 2$ is $w_i > 0$. The consumer is supposed to have a benefit function of Cobb–Douglas form

$$kx_1^{\gamma_1} x_2^{\gamma_2}$$

where k, γ_1, γ_2 are positive numbers and $\gamma_1 + \gamma_2 = 1$. An optimal landscape is a combination of characteristics (x_1, x_2) that maximizes the net-benefit function

$$kx_1^{\gamma_1} x_2^{\gamma_2} - w_1 x_1 - w_2 x_2.$$

The procedure is the following: the consumer is asked to give a for him optimal combination of landscape characteristics, not only in case of the prices

$\mathbf{w} = (w_1, w_2)$ but also for the case of (well-chosen) other prices (w'_1, w'_2) of the characteristics. Out of these two answers k and γ_1 are derived that are compatible with these choices. Having k and γ_1, the benefit function of the consumer is identified. Finally, the monetary value of the landscape is the value (may be negative) of the net-benefit function for prices $\mathbf{w}$. So the monetary value has been determined not by asking for it directly, but rather indirectly via asking for optimal landscapes at different price levels of the characteristics. Normally speaking it is easier for a consumer to express his preferences for a certain landscape in actual characteristics than attaching monetary values to that preferences.

The value added by this chapter to the emerging discipline of landscape economics concerns the following four contributions to the article of Heijman and Goossen (2009). First, the class of benefit functions is enlarged by adding two parameters $h > 0$ and $\rho \in [0, 1[$ with $\rho \leq h \leq 1$:

$$k(\gamma_1 x_1^{\rho} + \gamma_2 x_2^{\rho})^{h/\rho}. \tag{7.1}$$

So this class concerns CES-functions. h and ρ are fixed and k and γ_1 have to be determined.[1] Second, the following class of benefit functions will also be dealt with:

$$k\left(\frac{x_1^{\beta_1}}{\beta_1} + \frac{x_1^{\beta_2}}{\beta_2}\right), \tag{7.2}$$

where k, β_1, β_2 are positive numbers and $\beta_1 + \beta_2 = 1$. Here k and β_1 have to be determined. Third, the formula in Heijman and Goossen (2009) to derive k and γ_1 may give k and γ_1 that are not compatible with a net-benefit function from the Cobb–Douglas class; the problem is that this formula is only correct for specific answers. We will correct the procedure for this. Fourth, the procedure is formalised and its correctness is mathematically proved.

The new procedure is as follows: the consumer is asked to give a for him optimal combination of landscape characteristics for two well-chosen prices of the characteristics, one of them being $\mathbf{w}$. Not all answers are compatible with the assumption that the consumer has a benefit function of the desired form (7.1) or (7.2). Sufficient and necessary conditions for this will be provided. Moreover, in case they are compatible it will be shown that k and γ_1, respectively k and β_1 are unique.[2]

The next section deals with the foundation of the procedure by presenting a general theoretical model for a general class of benefit functions; Appendix B deals with the economic interpretation of these functions thereby focussing explicitly on the net-benefit functions of the form (7.2). In the subsequent sections this is worked out for the classes (7.1) and (7.2). Also, we illustrate the procedure by an example with fictive numbers. We conclude the main text with final remarks. Appendix A contains proofs.

Theoretical model

Denote by C the set of continuous functions $\mathbb{R}^2_+ \to \mathbb{R}$. By a *net-benefit class* we understand a 2-tuple (P,b) where P is a non-empty (parameter) subset of $\mathbb{R}^2$ and $b: P \to C$; instead of $b(\pi)$ we write b_π. Moreover, the following has to hold: for each $\pi \in P$ and $\mathbf{q} = (q_1, q_2) \in \mathbb{R}^2_+$, the maximisation problem

$$\mathrm{MAX}_{(x_1,x_2)\in X}\ b_\pi(x_1,x_2) - q_1 x_1 - q_2 x_2 \qquad \spadesuit_{\pi;\mathbf{q}}$$

where

$$X := \{(x_1,x_2) \in \mathbb{R}^2_+ \mid x_1 + x_2 = A\}$$

with $A > 0$, has a unique maximiser and this maximiser belongs even to $\{(x_1,x_2) \in \mathbb{R}^2_{++} \mid x_1 + x_2 = A\}$.

Given a net-benefit class (P,b), the function

$$\mathbf{y} = (y_1, y_2) : P \times \mathbb{R}^2_+ \to X$$

is well-defined by: $\mathbf{y}(\pi, \mathbf{q})$ is the unique maximiser of $\spadesuit_{\pi;\mathbf{q}}$.

Fundamental Problem. Let (P,b) be a net-benefit class and $(\mathbf{w}^{(1)}, \mathbf{w}^{(2)}) \in \mathbb{R}^2_+ \times \mathbb{R}^2_+$. Give a procedure to determine whether for a given $(\mathbf{a}^{(1)}, \mathbf{a}^{(2)}) \in X \times X$ there exists $\pi \in P$ such that $(\mathbf{a}^{(1)}, \mathbf{a}^{(2)}) = (\mathbf{y}(\pi, \mathbf{w}^{(1)}), \mathbf{y}(\pi, \mathbf{w}^{(2)}))$. If such an π exists, then determine all such π.

In the following sections we show that the classes of functions (7.1) and (7.2) form net-benefit classes and we consider the Fundamental Problem for these classes for prices $\mathbf{w}^{(1)}, \mathbf{w}^{(2)}$ such that $w_1^{(1)} = w_2^{(1)}$ and $w_1^{(2)} \neq w_2^{(2)}$.

Implementation for the class (7.1)

Notation: for $k > 0$, $\rho \in [0,1[$, $\gamma_1 \in]0,1[$ and $h > 0$ with $\rho \le h \le 1$, writing $\gamma_2 := 1 - \gamma_1$, the function

$$f_{k,\rho,\gamma_1,h} : \mathbb{R}^2_+ \to \mathbb{R},$$

is defined by

$$f_{k,\rho,\gamma_1,h}(x_1,x_2) := \begin{cases} k(\gamma_1 x_1^\rho + \gamma_2 x_2^\rho)^{h/\rho} & \text{if } 0 < \rho < 1, \\ k x_1^{h\gamma_1} x_2^{h\gamma_2} & \text{if } \quad \rho = 0. \end{cases} \tag{7.3}$$

Note that for $0 < \rho = h < 1$ one has

$$f_{k,\rho,\gamma_1,\rho}(x_1,x_2) = k(\gamma_1 x_1^\rho + \gamma_2 x_2^\rho).$$

For given $\rho \in [0,1[$ and $h > 0$ with $\rho \leq h \leq 1$ let

$$P^{\rho,h} := \mathbb{R}_{++} \times]0,1[.$$

Define $b : P^{\rho,h} \to C$ by

$$b_{k,\gamma_1} := f_{k,\rho,\gamma_1,h}.$$

Proposition 1. $(P^{\rho,h}, b)$ is a net-benefit class. ◇

Proof. See Appendix A. Q.E.D.

Theorem 1. Fix $\rho \in [0,1[$ and $h > 0$ with $\rho \leq h \leq 1$,

$$\mathbf{w}^{(1)} \in \mathbb{R}^2_+ \text{ with } w_1^{(1)} = w_2^{(1)} \text{ and } \mathbf{w}^{(2)} \in \mathbb{R}^2_+ \text{ with } w_1^{(2)} \neq w_2^{(2)}.$$

Let $\mathbf{a}^{(1)}, \mathbf{a}^{(2)} \in X$. Then there exists $\pi \in P^{\rho,h}$ such that $(\mathbf{a}^{(1)}, \mathbf{a}^{(2)}) = (\mathbf{y}(\pi, \mathbf{w}^{(1)}), \mathbf{y}(\pi, \mathbf{w}^{(2)}))$ if and only if[3]

$$a_1^{(1)} > 0, \; a_2^{(1)} > 0, \tag{7.4}$$

$$\frac{w_1^{(2)} - w_2^{(2)}}{(a_2^{(1)} a_1^{(2)})^{\rho-1} - (a_1^{(1)} a_2^{(2)})^{\rho-1}} > 0. \tag{7.5}$$

If (7.4) and (7.5) hold, then such π is unique and given by $\pi = (k, \gamma_1)$ where

$$\gamma_1 = \begin{cases} \dfrac{(a_2^{(1)})^{\rho-1}}{(a_1^{(1)})^{\rho-1} + (a_2^{(1)})^{\rho-1}} & \text{if } \rho \neq 0 \\ \dfrac{a_1^{(1)}}{A} & \text{if } \rho = 0 \end{cases},$$

$$k = \begin{cases} \dfrac{w_1^{(2)} - w_2^{(2)}}{h\left(\frac{(a_2^{(1)})^{\rho-1} (a_1^{(2)})^{\rho} + (a_1^{(1)})^{\rho-1} (a_2^{(2)})^{\rho}}{(a_1^{(1)})^{\rho-1} + (a_2^{(1)})^{\rho-1}}\right)^{\frac{h}{\rho}-1} \frac{(a_2^{(1)} a_1^{(2)})^{\rho-1} - (a_1^{(1)} a_2^{(2)})^{\rho-1}}{(a_1^{(1)})^{\rho-1} + (a_2^{(1)})^{\rho-1}}} & \text{if } \rho \neq 0 \\ \dfrac{A(w_1^{(2)} - w_2^{(2)})}{h (a_1^{(2)})^{h\frac{a_1^{(1)}}{A}} (a_2^{(2)})^{h\frac{a_2^{(1)}}{A}} \left(\frac{a_1^{(1)}}{a_1^{(2)}} - \frac{a_2^{(1)}}{a_2^{(2)}}\right)} & \text{if } \rho = 0. \end{cases} \quad \diamond$$

Proof. See Appendix A. Q.E.D.

Implementation for the class (7.2)

Notation: for $k > 0,\ \beta_1 \in]0,1[$, write $\beta_2 := 1 - \beta_1$ and define the function

$$f_{k,\beta_1} : \mathbb{R}_+^2 \to \mathbb{R},$$

by

$$f_{k,\beta_1}(x_1, x_2) := k\left(\frac{x_1^{\beta_1}}{\beta_1} + \frac{x_2^{\beta_2}}{\beta_2}\right). \tag{7.6}$$

Let

$$P := \mathbb{R}_{++} \times]0,1[$$

and define $b : P \to C$ by

$$b_{k,\beta_1} := f_{k,\beta_1}.$$

Proposition 2. (P, b) is a net-benefit class. ◇

Proof. See Appendix A. Q.E.D.

Theorem 2. Fix

$$\mathbf{w}^{(1)} \in \mathbb{R}_+^2 \text{ with } w_1^{(1)} = w_2^{(1)} \text{ and } \mathbf{w}^{(2)} \in \mathbb{R}_+^2 \text{ with } w_1^{(2)} \neq w_2^{(2)}.$$

Let $\mathbf{a}^{(1)}, \mathbf{a}^{(2)} \in X$. Then there exists $\pi \in P$ such that $(\mathbf{a}^{(1)}, \mathbf{a}^{(2)}) = (\mathbf{y}(\pi, \mathbf{w}^{(1)}), \mathbf{y}(\pi, \mathbf{w}^{(2)}))$ if and only if[4]

$$a_1^{(1)} > 0,\ a_2^{(1)} > 0, \tag{7.7}$$

$$\frac{w_1^{(2)} - w_2^{(2)}}{(a_1{}^{(2)})^{-\frac{\ln(a_2^{(1)})}{\ln(a_1^{(1)}a_2^{(1)})}} - (a_2{}^{(2)})^{-\frac{\ln(a_1^{(1)})}{\ln(a_1^{(1)}a_2^{(1)})}}} > 0. \tag{7.8}$$

If (7.7) and (7.8) hold, then such π is unique and given by $\pi = (k, \beta_1)$ where

$$\beta_1 = \frac{\ln(a_2^{(1)})}{\ln(a_1^{(1)}a_2^{(1)})}$$

$$k = \frac{w_1^{(2)} - w_2^{(2)}}{(a_1{}^{(2)})^{-\frac{\ln(a_2^{(1)})}{\ln(a_1^{(1)}a_2^{(1)})}} - (a_2{}^{(2)})^{-\frac{\ln(a_1^{(1)})}{\ln(a_1^{(1)}a_2^{(1)})}}} > 0. \text{ ◇}$$

Proof. See Appendix A. Q.E.D.

Illustrations of the procedure

For the net-benefit class $(P^{\rho,h}, b)$

Consider the procedure for the net-benefit class $(P^{0,1}, b)$. Thus

$$\rho = 0,\ h = 1.$$

Suppose

$$A = 100 \text{ and } \mathbf{w} = (2, 1).$$

Take $\mathbf{w}^{(2)} = \mathbf{w} = (2, 1)$, and, for example, $\mathbf{w}^{(1)} = (0, 0)$. Suppose the answers for these prices are $\mathbf{a}^{(1)} = (50, 50)$ and $\mathbf{a}^{(2)} = (40, 60)$. Then the inequalities (7.4) and (7.5) hold and we obtain

$$\gamma_1 = \frac{50}{100} = 1/2,\ k = \frac{100(2-1)}{40^{1/2}60^{1/2}(\frac{50}{40} - \frac{50}{60})} = 2\sqrt{6} = 4,89...$$

So the benefit function is $2\sqrt{6}\sqrt{x_1}\sqrt{x_2}$ and the monetary value is

$$2\sqrt{6}\sqrt{40}\sqrt{60} - 2 \cdot 40 - 1 \cdot 60 = 100.$$

Remark: if the answers were $\mathbf{a}^{(1)} = (40, 60)$ and $\mathbf{a}^{(2)} = (50, 50)$, then the inequality (7.5) no longer holds and therefore these answers are not compatible with a net-benefit function from $P^{0,1}$.

Next let us illustrate the procedure for the net-benefit class $(P^{1/2,1/2}, b)$. Thus

$$\rho = 1/2,\ h = 1/2.$$

Again, suppose

$$A = 100 \text{ and } \mathbf{w} = (2, 1).$$

Take $\mathbf{w}^{(2)} = \mathbf{w} = (2, 1)$ and $\mathbf{w}^{(1)} = (0, 0)$. Again, suppose the answers for these prices are $\mathbf{a}^{(1)} = (50, 50)$ and $\mathbf{a}^{(2)} = (40, 60)$. Then the inequalities (7.4) and (7.5) hold and we obtain

$$\gamma_1 = \frac{50^{-1/2}}{50^{-1/2} + 50^{-1/2}} = 1/2,$$

$$k = \frac{2-1}{\frac{1}{2}\frac{(50 \cdot 40)^{-1/2} - (50 \cdot 60)^{-1/2}}{50^{-1/2} + 50^{-1/2}}} = \frac{24\sqrt{10}}{3 - \sqrt{6}} = 137.86\ ...$$

Thus the benefit function is $\frac{24\sqrt{10}}{3-\sqrt{6}}(\frac{1}{2}\sqrt{x_1} + \frac{1}{2}\sqrt{x_2})$ and the monetary value is

$$\frac{24\sqrt{10}}{3 - \sqrt{6}}\left(\frac{1}{2}\sqrt{40} + \frac{1}{2}\sqrt{60}\right) - 2 \cdot 40 - 1 \cdot 60 = 829.89\ ...$$

For the net-benefit class (P, b)

Consider the procedure for the net-benefit class (P, b). Suppose

$$A = 100 \text{ and } \mathbf{w} = (600, 500).$$

Take $\mathbf{w}^{(2)} = \mathbf{w} = (600, 500)$, and, for example, $\mathbf{w}^{(1)} = (0, 0)$. Suppose the answers for these prices are $\mathbf{a}^{(1)} = (50, 50)$ and $\mathbf{a}^{(2)} = (3664)$. Then the inequalities (7.7) and (7.8) hold and we obtain

$$\beta_1 = \frac{\ln(50)}{\ln(2500)} = 1/2, \quad k = \frac{600 - 500}{36^{-1/2} - 64^{-1/2}} = 2400.$$

So the benefit function is $2400(2\sqrt{x_1} + 2\sqrt{x_2})$ and the monetary value is

$$2400(2\sqrt{36} + 2\sqrt{64}) - 600 \cdot 36 - 500 \cdot 64 = 13600.$$

Concluding remarks

Of course, it is important to test the method in practice. This would require a carefully designed research phase in the decision procedure. Moreover, it is good to realise that in order to build optimal landscapes, a positive monetary value would be required. If this is not the case these landscapes would not contribute to social welfare. Therefore, a suggestion for further theoretical research is to investigate the sign of the monetary value for the two classes of benefit functions we deal with.

Notes

1 For $h = 1$ and $\rho = 0$ this is just the above Cobb–Douglas class (also see (7.6)).
2 Of course, one not only may apply the procedure for (7.1) with fixed ρ and h, but one can do this for various ρ and h and then determine (in some way) ρ and h that fits best to the given answers.
3 Of course, the validity of (7.5) implies that the denominator of the fraction on the left-hand side of this inequality is not zero.
4 Of course, the validity of (7.8) implies that the denominator of the fraction on the left-hand side of this inequality is not zero.

References

Pruckner, G. (1995) 'Agricultural landscape cultivation in Austria: an application of the CVM', *European Review of Agricultural Economics*, 22(2): 173–190.

Heijman, W. and Goossen, M. (2009) 'Consumer preference in landscape design', Vienna, *First International Conference on Landscape Economics*, Papers, pages 9–18.

Appendix A: Proofs

First we deal with the net-benefit class $P^{\rho,h}$. For $\pi = (k, \gamma_1) \in P^{\rho,h}$, define $\tilde{\pi} \in P^{\rho,h}$ by

$$\tilde{\pi} = (1, \gamma_1).$$

Note that $b_\pi = k b_{\tilde{\pi}}$.

Proof of Proposition 1. $P^{\rho,h}$ is non-empty and each b_π is continuous. Now consider for $\pi = (k, \gamma_1) \in P^{\rho,h}$ and $\mathbf{q} \in \mathbb{R}^2_+$ the maximisation problem $♠_{\pi;\mathbf{q}}$. In order to study it, define the function $w : \mathbb{R}^2_+ \to \mathbb{R}$ by

$$w(x_1, x_2) := b_\pi(x_1, x_2) - q_1 x_1 - q_2 x_2.$$

$♠_{\pi;\mathbf{q}}$ is in an obvious way related to the maximisation problem of the function $W : [0, A] \to \mathbb{R}$ defined by

$$W(x_1) := w(x_1, A - x_1) = b_\pi(x_1, A - x_1) - q_1 x_1 - q_2(A - x_1). \tag{7.9}$$

W is a continuous function with non-empty compact domain and therefore has according to the Weierstrass' theorem at least one maximiser. Therefore, $♠_{\pi;\mathbf{q}}$ has this too. $W \restriction]0, A[$ is smooth and (in short notations),

$$W' = D_1 b_\pi - D_2 b_\pi - q_1 + q_2, \quad W'' = D_{11} b_\pi - D_{21} b_\pi - D_{12} b_\pi + D_{22} b_\pi.$$

A straightforward calculation shows that one has for the Hesse matrix $\begin{pmatrix} D_{11} b_\pi & D_{12} b_\pi \\ D_{21} b_\pi & D_{22} b_\pi \end{pmatrix}$ of $b_\pi \restriction \mathbb{R}^2_{++}$ the matrix

$$G \begin{pmatrix} \frac{1}{\gamma_2} x_2^{1-\rho} x_1{}^{-1} (\gamma_1 (h-1) x_1^\rho + \gamma_2 (\rho - 1) x_2^\rho) & h - \rho \\ h - \rho & \frac{1}{\gamma_1} x_1^{1-\rho} x_2{}^{-1} (\gamma_2 (h-1) x_2^\rho + \gamma_1 (\rho - 1) x_1^\rho) \end{pmatrix}$$

where

$$G := k \gamma_1 \gamma_2 h f_{1,\rho,\gamma_1,h}^{1 - 2\frac{\rho}{h}} (x_1 x_2)^{\rho - 1}.$$

We see that $D_{11} b_\pi < 0$, $D_{22} b_\pi < 0, D_{12} b_\pi = D_{21} b_\pi \geq 0$. This implies that $W''(x_1) < 0$ $(0 < x_1 < A)$. Therefore $W \restriction]0, A[$ is strictly concave. Because W is continuous, it even follows that W is strictly concave. It follows that W, and therefore $♠_{\pi;\mathbf{q}}$ too, has at most one maximiser. Thus $♠_{\pi;\mathbf{q}}$ has a unique maximiser.

The following identity on $\mathbb{R}^2_{++}$ between the partial derivative $D_i b_\pi$ and $b_{\tilde{\pi}}$ holds:

$$D_i b_\pi = k h \gamma_i b_{\tilde{\pi}}^{1-\rho/h} x_i^{\rho - 1}.$$

From this identity we now obtain for $x_1 \in]0,A[$,

$$W'(x_1) = khb_{\tilde{\pi}}^{1-\rho/h}(x_1, A - x_1)(\gamma_1 x_1^{\rho-1} - \gamma_2(A-x_1)^{\rho-1}) - w_1 + w_2. \quad (7.10)$$

W is a concave function with domain the non-degenerate interval $[0,A]$. Therefore its derivative $W'(0)$ exists as element from $\mathbb{R} \cup \{+\infty\}$ and its derivative $W'(A)$ exists as element from $\mathbb{R} \cup \{-\infty\}$. Moreover,

$$W'(0) = \lim_{x_1 \downarrow 0} W'(x_1) \text{ and } W'(A) = \lim_{x_1 \uparrow A} W'(x_1).$$

Now, from (7.10),

$$W'(0) = +\infty \text{ and } W'(A) = -\infty.$$

This implies that neither 0 nor A is a maximiser of W. Thus the unique maximiser belongs to $\{(x_1,x_2) \in \mathbb{R}^2_{++} \mid x_1 + x_2 = A\}$. Q.E.D.

Lemma 1. For $\pi = (k, \gamma_1) \in P^{\rho,h}$ and $\mathbf{q} \in \mathbb{R}^2_+$ one has that $\mathbf{y}(\pi, \mathbf{q})$ is the unique solution of the equation

$$khb_{\tilde{\pi}}^{1-\rho/h}(\mathbf{x})(\gamma_1 x_1{}^{\rho-1} - \gamma_2 x_2{}^{\rho-1}) = q_1 - q_2 \quad (7.11)$$

in $\mathbf{x} \in \mathbb{R}^2_{++}$ with $x_1 + x_2 = A$. ◇

Proof. Write $\mathbf{y}$ instead of $\mathbf{y}(\pi, \mathbf{q})$. Because $y_1 \in]0,A[$, the function W defined by (7.9) is differentiable in y_1. y_1 is the (unique) maximiser of W and therefore $W'(y_1) = 0$. By (7.10) and $A - y_1 = y_2$,

$$khb_{\tilde{\pi}}^{1-\rho/h}(\mathbf{y})(\gamma_1 y_1^{\rho-1} - \gamma_2 y_2^{\rho-1}) - q_1 + q_2 = 0.$$

Thus $\mathbf{y}$ is a solution of (7.11). If also $\mathbf{x} \in \mathbb{R}^2_{++}$ with $x_1 + x_2 = A$ is a solution of (7.11), then, by (7.10) $W'(x_1) = 0$. Because W is concave, x_1 is a maximiser of W. We know that y_1 is the unique maximiser of W. Thus $x_1 = y_1$ and therefore $\mathbf{x} = \mathbf{y}$. Q.E.D.

Proof of Theorem 1. Suppose that there exists $\pi \in P^{\rho,h}$ such that $(\mathbf{a}^{(1)}, \mathbf{a}^{(2)}) = (\mathbf{y}(\pi, \mathbf{w}^{(1)}), \mathbf{y}(\pi, \mathbf{w}^{(2)}))$. Because $y_1(\pi, \mathbf{w}^{(1)}) > 0$ and $y_2(\pi, \mathbf{w}^{(1)}) > 0$, it follows that (7.4) holds. Because $y_1(\pi, \mathbf{w}^{(2)}) > 0$ and $y_2(\pi, \mathbf{w}^{(2)}) > 0$, it also follows that $a_1^{(2)} > 0$ and $a_2^{(2)} > 0$. Let $\pi = (k, \gamma_1)$. By Lemma 1, we have the identities

$$khf_{1,\rho,\gamma_1,h}^{1-\rho/h}(\mathbf{a}^{(1)})(\gamma_1 (a_1^{(1)})^{\rho-1} - \gamma_2 (a_2^{(1)})^{\rho-1}) = 0, \quad (7.12)$$

$$khf_{1,\rho,\gamma_1,h}^{1-\rho/h}(\mathbf{a}^{(2)})(\gamma_1 (a_1^{(2)})^{\rho-1} - \gamma_2 (a_2^{(2)})^{\rho-1}) = w_1^{(2)} - w_2^{(2)}. \quad (7.13)$$

Because $\mathbf{a}^{(1)}, \mathbf{a}^{(2)} \neq \mathbf{0}$, one has $f_{1,\rho,\gamma_1,h}^{1-\rho/h}(\mathbf{a}^{(i)}) \neq 0$ $(i = 1,2)$. Also note that $w_1^{(2)} \neq w_2^{(2)}$ implies that $\gamma_1 (a_1^{(1)})^{\rho-1} - \gamma_2 (a_2^{(1)})^{\rho-1} \neq 0$. Now consider two cases:

- $\rho \neq 0$. Now $f_{1,\rho,\gamma_1,h} = (\gamma_1 x_1^\rho + \gamma_2 x_2^\rho)^{h/\rho}$ and (7.12) and (7.13) are equivalent with

$$\gamma_1 = \frac{(a_2^{(1)})^{\rho-1}}{(a_1^{(1)})^{\rho-1} + (a_2^{(1)})^{\rho-1}},$$

$$k = \frac{w_1^{(2)} - w_2^{(2)}}{h(\gamma_1 (a_1^{(2)})^\rho + \gamma_2 (a_2^{(2)})^\rho)^{\frac{h}{\rho}-1} (\gamma_1 (a_1^{(2)})^{\rho-1} - \gamma_2 (a_2^{(2)})^{\rho-1})}.$$

These equations in turn are equivalent with

$$\gamma_1 = \frac{(a_2^{(1)})^{\rho-1}}{(a_1^{(1)})^{\rho-1} + (a_2^{(1)})^{\rho-1}}, \tag{7.14}$$

$$k = \frac{w_1^{(2)} - w_2^{(2)}}{h\left(\frac{(a_2^{(1)})^{\rho-1}(a_1^{(2)})^\rho + (a_1^{(1)})^{\rho-1}(a_2^{(2)})^\rho}{(a_1^{(1)})^{\rho-1} + (a_2^{(1)})^{\rho-1}}\right)^{\frac{h}{\rho-1}} \frac{(a_2^{(1)} a_1^{(2)})^{\rho-1} - (a_1^{(1)} a_2^{(2)})^{\rho-1}}{(a_1^{(1)})^{\rho-1} + (a_2^{(1)})^{\rho-1}}}. \tag{7.15}$$

(7.15) implies that (7.5) holds.
- $\rho = 0$. Now $f_{1,\rho,\gamma_1,h} = x_1^{h\gamma_1} x_2^{h\gamma_2}$ and (7.12) and (7.13) are equivalent with

$$\gamma_1 = \frac{(a_2^{(1)})^{-1}}{(a_1^{(1)})^{-1} + (a_2^{(1)})^{-1}},$$

$$k = \frac{w_1^{(2)} - w_2^{(2)}}{h(a_1^{(2)})^{h\gamma_1} (a_2^{(2)})^{h\gamma_2} (\gamma_1 (a_1^{(2)})^{-1} - \gamma_2 (a_2^{(2)})^{-1})}.$$

These equations in turn are equivalent with

$$\gamma_1 = \frac{a_1^{(1)}}{A}, \tag{7.16}$$

$$k = \frac{A(w_1^{(2)} - w_2^{(2)})}{h(a_1^{(2)})^{h\frac{a_1^{(1)}}{A}} (a_2^{(2)})^{h\frac{a_2^{(1)}}{A}} (\frac{a_1^{(1)}}{a_1^{(2)}} - \frac{a_2^{(1)}}{a_2^{(2)}})}. \tag{7.17}$$

(7.17) implies that (7.5) holds.
Now suppose (7.4) and (7.5) hold. Define $\pi = (k, \gamma_1)$ by (7.14) and (7.15). Now $\pi \in P^{\rho,h}$ and (7.12) and (7.13) hold. By Lemma 1, $(\mathbf{a}^{(1)}, \mathbf{a}^{(2)}) = (\mathbf{y}(\pi, \mathbf{w}^{(1)}), \mathbf{y}(\pi, \mathbf{w}^{(2)}))$.

Finally, assume $\pi = (k, \gamma_1) \in P^{\rho,h}$ is such that $(\mathbf{a}^{(1)}, \mathbf{a}^{(2)}) = (\mathbf{y}(\pi, \mathbf{w}^{(1)}), \mathbf{y}(\pi, \mathbf{w}^{(2)}))$. Then, as above, if $\rho \neq 0$, (7.14) and (7.15) hold. And if $\rho = 0$, then (7.16) and (7.17) hold. Thus π is unique. Q.E.D.

Now we deal with the net-benefit class (P, b).

Proof of Proposition 2. P is non-empty and each b_π is continuous. Now consider for $\pi = (k, \beta_1) \in P$ and $\mathbf{q} \in \mathbb{R}^2_+$ the maximisation problem $\spadesuit_{\pi;\mathbf{q}}$. In order to study it, define the function $w : \mathbb{R}^2_+ \to \mathbb{R}$ by

$$w(x_1, x_2) := b_\pi(x_1, x_2) - q_1 x_1 - q_2 x_2.$$

$\spadesuit_{\pi;\mathbf{q}}$ is in an obvious way related to the maximisation problem of the function $W : [0, A] \to \mathbb{R}$ defined by

$$W(x_1) := w(x_1, A - x_1) = k(\frac{x_1^{\beta_1}}{\beta_1} + \frac{(A - x_1)^{\beta_2}}{\beta_2}) - q_1 x_1 - q_2(A - x_1). \quad (7.18)$$

W is a continuous function with non-empty compact domain and therefore has according to the Weierstrass' theorem at least one maximiser. Therefore, $\spadesuit_{\pi;\mathbf{q}}$ has this too. $W \restriction]0, A[$ is smooth and (in short notations),

$$W' = k(x_1^{\beta_1 - 1} - (A - x_1)^{\beta_2 - 1}) - q_1 + q_2, \quad (7.19)$$

$$W'' = k((\beta_1 - 1)x_1^{\beta_1 - 2} + (\beta_2 - 1)(A - x_1)^{\beta_2 - 2}).$$

This implies that $W''(x_1) < 0 \, (0 < x_1 < A)$. Therefore $W \restriction]0, A[$ is strictly concave. Because W is continuous, it even follows that W is strictly concave. It follows that W, and therefore $\spadesuit_{\pi;\mathbf{q}}$ too, has at most one maximiser. Thus $\spadesuit_{\pi;\mathbf{q}}$ has a unique maximiser.

$$W'(0) = +\infty \text{ and } W'(A) = -\infty.$$

This implies that neither 0 nor A is a maximiser of W. Thus the unique maximiser belongs to $\{(x_1, x_2) \in \mathbb{R}^2_{++} \mid x_1 + x_2 = A\}$. Q.E.D.

Lemma 2. For $\pi = (k, \beta_1) \in P$ and $\mathbf{q} \in \mathbb{R}^2_+$ one has that $\mathbf{y}(\pi, \mathbf{q})$ is the unique solution of the equation

$$k({x_1}^{-\beta_2} - {x_2}^{-\beta_1}) = q_1 - q_2 \quad (7.20)$$

in $\mathbf{x} \in \mathbb{R}^2_{++}$ with $x_1 + x_2 = A$. $\diamond$

Proof. Write $\mathbf{y}$ instead of $\mathbf{y}(\pi, \mathbf{q})$. Because $y_1 \in]0, A[$, the function W defined by (7.18) is differentiable in y_1. y_1 is the (unique) maximiser of W and therefore

$W'(y_1) = 0$. By (7.19) and $A - y_1 = y_2$,

$$k(y_1^{-\beta_2} - y_2^{-\beta_1}) = q_2 - q_1.$$

Thus $\mathbf{y}$ is a solution of (7.11). If also $\mathbf{x} \in \mathbb{R}^2_{++}$ with $x_1 + x_2 = a$ is a solution of (7.20), then, by (7.19) $W'(x_1) = 0$. Because W is concave, x_1 is a maximiser of W. We know that y_1 is the unique maximiser of W. Thus $x_1 = y_1$ and therefore $\mathbf{x} = \mathbf{y}$. Q.E.D.

Proof of Theorem 2. Suppose that there exists $\pi \in P$ such that $(\mathbf{a}^{(1)}, \mathbf{a}^{(2)}) = (\mathbf{y}(\pi, \mathbf{w}^{(1)}), \mathbf{y}(\pi, \mathbf{w}^{(2)}))$. Because $y_1(\pi, \mathbf{w}^{(1)}) > 0$ and $y_2(\pi, \mathbf{w}^{(1)}) > 0$, it follows that (7.7) holds. Because $y_1(\pi, \mathbf{w}^{(2)}) > 0$ and $y_2(\pi, \mathbf{w}^{(2)}) > 0$, it also follows that $a_1^{(2)} > 0$ and $a_2^{(2)} > 0$. Let $\pi = (k, \beta_1)$. By Lemma 2, we have the identities

$$k((a_1{}^{(1)})^{-\beta_2} - (a_2{}^{(1)})^{-\beta_1}) = 0, \tag{7.21}$$

$$k((a_1{}^{2)})^{-\beta_2} - (a_2{}^{(2)})^{-\beta_1}) = w_1^{(2)} - w_2^{(2)}. \tag{7.22}$$

Note that $w_1^{(2)} \neq w_2^{(2)}$ implies that $((a_1{}^{2)})^{-\beta_2} - (a_2{}^{(2)})^{-\beta_1}) \neq 0$.
(7.21) and (7.22) are equivalent with

$$\beta_1 = \frac{\ln(a_1^{(1)})}{\ln(a_1^{(1)} a_2^{(1)})},$$

$$k = \frac{w_1^{(2)} - w_2^{(2)}}{(a_1{}^{2)})^{-\beta_2} - (a_2{}^{(2)})^{-\beta_1}}.$$

These equations in turn are equivalent with

$$\beta_1 = \frac{\ln(a_1^{(1)})}{\ln(a_1^{(1)} a_2^{(1)})}, \tag{7.23}$$

$$k = \frac{w_1^{(2)} - w_2^{(2)}}{(a_1{}^{2)})^{-\frac{\ln(a_2^{(1)})}{\ln(a_1^{(1)} a_2^{(1)})}} - (a_2{}^{(2)})^{-\frac{\ln(a_1^{(1)})}{\ln(a_1^{(1)} a_2^{(1)})}}}. \tag{7.24}$$

(7.24) implies that (7.8) holds.

Now suppose (7.7) and (7.8) hold. Define $\pi = (k, \gamma_1)$ by (7.23) and (7.24). Now $\pi \in P$ and (7.21) and (7.22) hold. By Lemma 2, $(\mathbf{a}^{(1)}, \mathbf{a}^{(2)}) = (\mathbf{y}(\pi, \mathbf{w}^{(1)}), \mathbf{y}(\pi, \mathbf{w}^{(2)}))$. Finally, assume $\pi = (k, \beta_1) \in P$ is such that $(\mathbf{a}^{(1)}, \mathbf{a}^{(2)}) = (\mathbf{y}(\pi, \mathbf{w}^{(1)}), \mathbf{y}(\pi, \mathbf{w}^{(2)}))$. Then, as above, (7.23) and (7.24) hold. Thus π is unique. Q.E.D.

Appendix B: Welfare theoretic foundation

Below a first attempt will be made to give an economic interpretation of the net-benefit functions of the form (7.2). The interpretation will be that of a total willingness to pay.

Suppose there are well defined inverse demand functions $p_1, p_2 : \mathbb{R}^2_+$ for the characteristics. For the combination of characteristics (x_1, x_2) this leads to the prices $p_1(x_1, x_2)$ and $p_2(x_1, x_2)$. Suppose the functions $\text{wtp}_1, \text{wtp}_2 : \mathbb{R}^2_+ \rightarrow \mathbb{R}$ are well-defined by the (maybe improper) Riemann integrals

$$\text{wtp}_1(x_1, x_2) := \int_0^{x_1} p_1(x, x_2)\, dx, \quad \text{wtp}_2(x_1, x_2) := \int_0^{x_2} p_2(x_1, x)\, dx.$$

The interpretation of $\text{wtp}_1(x_1, x_2)$ is the willingness to pay for an amount x_1 of characteristic 1 given an amount x_2 of characteristic 2. The interpretation of $\text{wtp}_2(x_1, x_2)$ is the willingness to pay for an amount x_2 of characteristic 2 given an amount x_1 of characteristic 1. The total willingness to pay wtp is

$$\text{wtp} := \text{wtp}_1 + \text{wtp}_2.$$

Now the benefit function in the introduction is just the function wtp. The authors are aware that this interpretation is disputable. (The second author even feels uncomfortable with it.)

Here is an example: suppose

$$p_1(x_1, x_2) = Kx_1^{-\beta_2}, \; p_2(x_1, x_2) = Kx_2^{-\beta_1}$$

where k, β_1, β_2 are positive and $\beta_1 + \beta_2 = 1$. This leads to

$$\text{wtp}_1(x_1, x_2) = \frac{k}{1-\beta_2} x_1^{1-\beta_2}, \quad \text{wtp}_2(x_1, x_2) = \frac{k}{1-\beta_1} x_2^{1-\beta_1}.$$

So

$$\text{wtp}(x_1, x_2) = k\left(\frac{x_1{}^{\beta_1}}{\beta_1} + \frac{x_2{}^{\beta_2}}{\beta_2}\right).$$

8 Evaluation of landscape impacts – enriching the economist's toolbox with the HotSpotIndex

Frans J. Sijtsma, Hans Farjon, Sandy van Tol, Peter van Kampen, Arjen Buijs and Arjen van Hinsberg

Introduction

In the Netherlands, cost–benefit analysis (CBA) has gained an increasingly important role in the evaluation of so-called "integrated" spatial projects (spatial transformation projects) that aim to simultaneously improve the economic, social and ecological qualities of an area. Such projects strive to realize, for example, a combination of infrastructure, housing, nature development, and business parks. The evaluation of these projects is thus a challenging task due to the wide range of complex impacts that intervene at different geographical scales, such as impacts of building, the development of nature areas, and landscaping, to name just three. As developers of local and regional projects request financial support from the Dutch government, a clear distinction between impacts at local, regional and national scale is more often required.

This chapter will focus on the measurement of landscape impacts. Landscape is defined in the European Landscape Convention as an area, as perceived by people, whose character is the result of the action and interaction of natural and/or human factors. The evaluation of landscape impacts should, accordingly, be based on perceptions of landscapes. Several authors have identified a number of different discourses, each having its own perception on the values of the landscape (Van den Berg 1999; Stobbelaar and Pedroli 2011).

In this chapter we focus on evaluation of landscape impacts from the perspective of the scenic beauty discourse and we do so for several reasons. First, this perspective has the most direct link to the CBA approach, as it relates directly to the significance of landscape for the wealth and wellbeing of society. Second, within the scenic beauty discourse empirical evidence from surveys is, in our view, underused. The standard CBA toolbox for capturing scenic beauty impacts is well known. The economist can select revealed preference or stated preference tools (Boardman et al. 2011). In so far as landscape issues are concerned, the most widely used revealed preference tools are hedonic pricing and travel cost approaches as set out in the comprehensive text of Herriges and Kling (2008). Of the 76 articles in the handbook, 40 focus on travel cost/recreational demand and 28 discuss hedonic models. With regard to stated preferences, contingent valuation and discrete choice models are used most often (Kapper 2004; Powe 2007;

Hanley and Barbier 2009; Boardman et al. 2011). The approach of these CBA techniques is to evaluate landscape appreciation using monetary values exclusively. For this reason we here maintain that an in-depth knowledge of the landscape preferences of citizens is a fundamental prerequisite before one can reasonably calculate landscape impacts in monetary terms. Towards the development of such new knowledge we describe the Hotspotmonitor database and the HotSpotIndex in the fourth section of this chapter.

Although it is not essential to our argument, it may help the reader to understand our position if we explain that we have argued elsewhere for a hybrid alternative to the standard CBA, multi criteria cost–benefit analysis (MCCBA), which is an appropriate methodology in complex public decision making contexts (Sijtsma 2006; Sijtsma et al. 2011). MCCBA is a carefully designed combination of CBA and Multi Criteria Analysis (MCA) (Pomerol and Barba-Romero 2000; Belton and Stewart 2002) and is able to retain the analytical rigor of CBA while allowing for more flexibility in non-monetary measurements (Sijtsma 2006). In an MCCBA it is not the monetary or non-monetary character of measurements that matters most, but rather the fact that the measurement must be on interval or ratio scales, preferably using standardized measurements (Sijtsma et al. 2011). We will argue that this shift of emphasis from monetary to standardized measurement of preferences using interval or ratio scale measurements – consistent at different spatial levels – will substantially enrich the current toolbox of the CBA analyst in evaluations of landscape impacts. In an MCCBA approach, monetary valuation is not an absolute necessity; whether or not it is possible is a matter of operational feasibility and data availability in each evaluation case. However, to explicitly adopt an MCCBA stance is also not required for the argument of this paper; the argument also holds for use in regular CBAs. In a standard CBA setting our approach would then simply provide a stronger foundation for monetary valuation.

The structure of our study is formulated as follows (see Figure 8.1). The next section discusses the actual practice of landscape valuation in Dutch CBA studies and in a review clarifies the unexpected main weaknesses in the actual use of theoretically available CBA tools. The subsequent section identifies which tools Dutch landscape researchers have to offer to the CBA analyst in order to evaluate landscape change. These methods of non-economist landscape researchers certainly have strengths which economists' tools lack, but they also have their own drawbacks when applied within CBA. Thereafter, we introduce the so-called Hotspotmonitor: a new tool used to gather landscape preferences in a standardized, systematic and spatially-precise way. Next, we discuss the key indicator derived from measuring landscape attractivity: the HotSpotIndex (HSI) and examine its ability to assess landscape impacts (within a CBA) in the upgrade of the city of Almere. The chapter concludes with a brief discussion.

What Dutch economists do: two case studies

In this section we review the Dutch practice of evaluating landscape impacts of integrated spatial projects. We analyze a representative set of high-quality Dutch

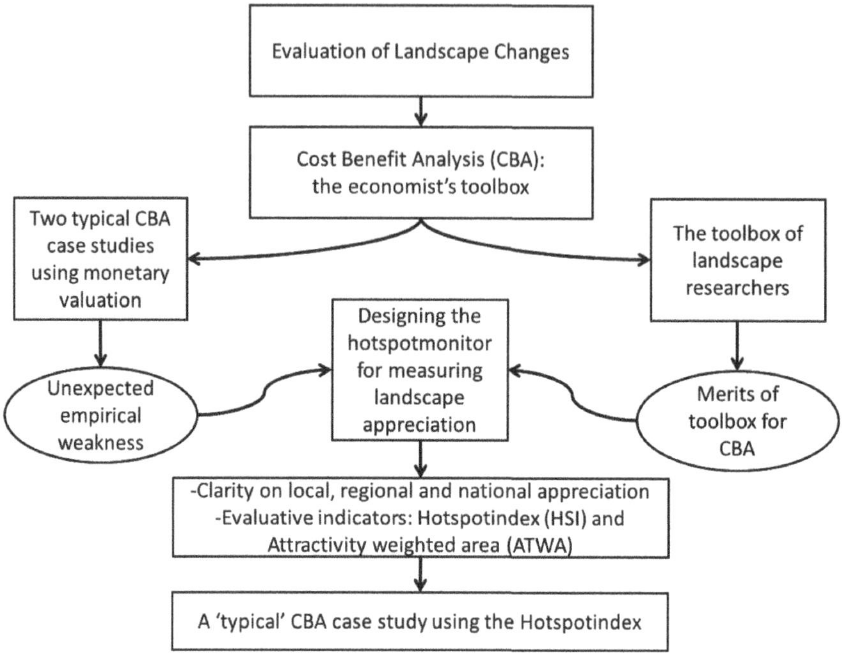

Figure 8.1 The structure of this chapter.

CBAs – carried out by professional consultancy firms and used in policy making (Sijtsma et al. 2008, unpublished). We study four aspects of these CBAs.

- Which perspectives of landscape quality are evaluated?
- How are landscape impacts priced?
- How are distinct (changes in) landscape preferences measured?
- How is the impact population defined?

Measurement of the appreciation of landscape changes in Dutch CBA follows the international state-of-the-art: monetary valuation by revealed preference or stated preference tools. We will now briefly discuss two actual CBA case studies that clearly illustrate the perhaps unexpected limitations of the Dutch practice of monetary valuation of landscape appreciation impacts.

(i) A CBA of two urbanization options for the Randstad – the Netherlands' most densely populated and urbanized region, in the west of the country.
(ii) A CBA of a new business park development at Hoeksche Waard in the southern part of the Randstad.

Case (i): Urbanization of the Randstad

This urbanization CBA focuses on the costs and benefits of building 27,000 new houses according to two different strategies: (1) urban densification by building within currently built-up areas and (2) urban sprawl in (semi-)rural landscapes. Urban densification would presumably save some 1,100 hectares of now (semi-)rural landscape. The CBA monetarily valued this saved open space between €60 and 210 million, as the landscape costs in strategy 2. But how is the price of lost rural landscape determined? The basic assumption is that the market price of land reflects whether government has imposed restrictions against building on a particular area of land. In this assumption, the difference in the market price of land with (and without) spatial planning restrictions reflects the societal preference to preserve rural landscapes. According to the authors of the CBA study, there was little time for this analysis in the research, so they used secondary sources (Ecorys 2005). The maximum estimate is based on the value of the impacted hectares if spatial policy were to allow (semi-)rural landscapes to be developed for housing projects. A GIS-based analysis values this change from open space to housing land by comparing the market values of open space – agricultural land, nature or urban green – with those of a housing site. They obtain values between €0.10 and 1.40 million per hectare. Minimum values are determined by estimating the costs of realizing a nature or recreation area on agricultural land. The estimates vary between €51,000 per hectare and 111,000 per hectare. This CBA can be described as a revealed preference analysis with scant attention being given to landscape values and, as a consequence, little differentiation among landscape preferences. The method does not evaluate which inhabitants are confronted with which costs and benefits.

The CBA of the house building project throws up a few unexpected non-theoretical observations.

- The CBA analysts have spent little time calculating impacts and price.
- The spatial precision of part of the GIS-based data on land-use is very high but the monetary valuation part has no such spatial precision. Only general valuation numbers are used (benefit transfer or averting behavior estimates); these numbers can, however, be used anywhere in the Netherlands.
- The final monetary valuations of landscape impacts in this CBA have quite a large factor (28) between the minimum and maximum estimates per hectare (50,000 versus 1.4 million) and unfortunately, the authors have little clue as to what would be "the best" valuation.

The CBA analysts of this study are discomfited by this situation. Given the wide range in the valuation estimates, they acknowledge that the issue has generated robust discussion. The authors conclude their analysis, disappointed that:

> All valuation methods have disadvantages and are little developed in the Netherlands. Even further away is a database of benefit transfer numbers

> for different landscape types or regions. Specific landscape qualities are extremely hard to value. Above that, none of the available methods incorporates the spatial configuration or the precise degree of landscape openness ….
>
> (Ecorys 2005, Appendix: 47)

Case (ii): CBA business park Hoeksche Waard

The Hoeksche Waard park is situated in an area with distinct landscape qualities: it is part of a nationally protected landscape. The 180 hectare large business park to be built on the site is geared towards businesses of the region but is mostly for businesses coming from a wider area. The plan involves a landscaping effort to mimic a dike structure with trees surrounding the industrial sites on a 23 hectare plot. The CBA calculated benefits total €118 million net present value (NPV), most of which are travel-related. Of the total benefits, €108 million correspond to reduced travel time and €7 million are travel safety gains. The remaining €3 million benefits are landscape benefits. The CBA calculated three types of benefits resultant from the plan's landscaping efforts: a more attractive housing environment (€1.8 million), existence value of the landscape pattern (€0.9 million), and recreational benefits (€0.3 million). The cost of the landscaping effort comes to €14.5 million. The calculated benefits of a more attractive housing environment may be seen as contradictory, since realizing the site ostensibly deteriorates housing attractiveness, but the additional landscaping efforts could restore some of this loss. The CBA is not transparent about this and instead only calculates the "gain" from additional landscaping: an averted loss of 5 per cent (benefit transfer from another study) of the value of an estimated 300 houses. For the non-use "existence value" of the landscape pattern, benefit transfer from a Contingent Valuing (CV) study is applied, thereby showing that 'households in a range of 10 km were willing to pay €11 per year for upkeep of [a] river landscape. Since the case of Hoeksche Waard is about reconstructing and maintaining 23 hectares (compared to the total size of the industrial site of 180 hectares), a conservative estimate of $1/8 \times 11 =$ €1.38 per household is used.' (Ecorys 2006: 215). A total number of 33,000 households were assumed to be willing to pay this amount.

And lastly in the CBA, recreational benefits were calculated with the use of a day-recreation demand model. This model estimates deficits of green area in relation to the size of the population nearby, "nearby" being operationalized as within walking or biking distance from their own home. The demand for day trips in this model is determined on the basis of the national average of the number of people who take day trips on sunny days. The supply of available green areas for this modeled demand is determined using national "capacities" of different nature types. All 10 new (waterfront) hectares are assumed to have recreation qualities. More than 10,000 day trips are calculated to take place in these areas, and are assumed, on a benefit transfer basis, to have a monetary value of €1.68 per day trip.

Once again, this CBA reveals a number of unexpected non-theoretical observations:

- Substantial effort was made in this CBA to calculate the landscape impacts but confusion remains as to whether there is a gain or a loss to the landscape. The overall monetary value of the change turned out to be quite low compared to other benefits, and seemed unable to compensate for the landscape-related extra costs. But why was it protected in the first place? Because it was appreciated?
- The spatial precision of the valuations is very low. Although the area is well-defined and the proposed spatial changes are specific, valuations depend only on vaguely-related "benefit transfer" numbers.

Interpretation

The methodological interpretation of our observations above is that, although there seem to be distinct and well-established landscape qualities which are affected in a clearly localized area, nowhere in the CBA analysis are any actual landscape preferences measured. We do not suggest that CBA or its analysts are indifferent to these issues. It is likely that either lack of time or insufficient basic data are reasonable explanations for the measurement gap. The time taken in a robust CBA to measure landscape issues, in addition to 10–20 impacts, often prevents the use of primary sources. There is also no useful spatial-specific landscape preference database on which to build secondary data valuation efforts.

Cost–benefit analysis theoretically has several tools for the capture of landscape impacts; each of these monetary valuation tools has different methodologically positive and negative qualities. What our underlying analysis of representative Dutch cases has shown and what the above two cases highlight, however, is that most tools in practice seem to share one common drawback: they do not measure real landscape preferences with any systematic precision. There is no strong relation between the physical spatial changes and the valuation of the involved impact population. The CBAs are "distant" in a negative sense; they do not seem to digest real preferences of impact populations regarding the differences between project alternatives, as we think they should.

It is important to stress that this way of working is at odds with the theoretical foundation of CBA in which individual welfare and individual preferences are central (Pearce et al. 2006; Boardman et al. 2011). From such a foundation it is unwarranted to simply assume to know how welfare functions look, without first trying to establish the empirically real preferences. This is reflected in CBA's approach to accurate monetary valuation impacts. In this case the analyst often works with revealed preferences through market changes, which are precise and systematic; markets can even distinguish between products with only slightly different characteristics. Although in practice, aggregate units may typically be involved in CBAs (prices, quantities, surpluses) the foundation is nevertheless comprised of empirically systematic and precise individual preferences.

Our case studies of the Dutch best practice in CBA for landscape preferences show something completely different. As mentioned above, the lack of a systematic and precise preference pool of a relevant impact population seems to be the main problem. The valuation tools are not inferior; the sticking point is the amount of knowledge the tools have to work with. Against this background, we will in the next section address the issue of getting a more systematic and precise grip on landscape impacts.

What Dutch non-economists do

Introduction

This section discusses three direct landscape evaluation tools developed and used in public policy evaluation in the Netherlands. Unlike the CBA approach, they focus explicitly on actual landscape preferences: appreciation of landscapes or landscape characteristics such as openness or variety, etc. The influence of nature management practices on the landscape and their effects on people's preferences have been an object of study in environmental psychology for over 40 years (for an overview see Aoki 1999). Psychological studies of landscape preferences in particular have emerged as an alternative to the widely criticized landscape assessments based on expert judgments (Dakin 2003). Although expert judgment still plays an important role in environmental management practices and CBA, empirical studies clearly show that lay people's perceptions of, and preferences for, natural landscapes differ significantly from those of experts (Daniel 2001; Buijs et al. 2011). Most empirical studies in environmental psychology are based on what is known as the perceptual approach (also called the experimental approach: Zube et al. 1982; Daniel 2001). The focus of the approach is on the evaluation of the environment through individual perceptual processes, whereby landscape is an external stimulus to which individuals respond. Their responses are typically measured by rating overall preference, scenic beauty, attractiveness, or simply "liking".

Buijs and van Kralingen (2003) identify the three methods relative to the perceptual approach which score best on their set of validity, detail and usability criteria.

- The SPEL method, using standardized questionnaires.
- A GIS-based Landscape Appreciation Model (GLAM).
- The grounded theory based on Citizen Values Assessment (CVA).

None of the methods uses monetary valuation, but in no way do we dismiss them for that reason. We briefly discuss each method in order to clarify their merits in a CBA setting.

Written questionnaire: SPEL method

The Scales for Perception and Evaluation of Landscape (SPEL) questionnaire was developed by psychologist Coeterier (2000) based on 20 years of interviewing individuals on their landscape perceptions. The questionnaire measures the appreciation of the attractiveness of the landscape and eight basic qualities of the landscape: unity, use, own use, naturalness, historical character, spaciousness, management, and sensual experience. Two to four sub-qualities are assessed for each basic quality. Technically, SPEL uses 16 questions scored on a 1–10 scale, where one is extremely bad and 10 is extremely good. Although the questionnaire has a number of open questions, its standardized format nevertheless allows for the comparison of attractiveness of different landscapes.

SPEL is applied in a large number of regional case studies in the Netherlands and to a national monitoring program of landscape qualities of the Netherlands Environmental Assessment Agency (PBL in Dutch)). Every three years the monitoring program carries out a national enquiry involving 10,000 respondents (De Boer and De Groot 2010). The current version of the SPEL questionnaire enquires into the attractiveness of whole landscape regions where respondents are asked to assess attractiveness of a region of about 10 km^2, or a region within 15 km of their home address. But it is not known whether a respondent evaluates the whole region with all its diversity or a just a specific part of the region. For this reason, the outcomes of SPEL-based inquiries are not yet adequate evidence for location decisions relating to integrated spatial projects.

Usability of the SPEL method for CBA

The SPEL method measures the landscape preferences of individuals, which is good news in a CBA approach, since its focus is on individual preferences. The method unravels overall landscape appreciation by breaking it down into different aspects. Useful insights into the perceived qualities of the landscape are thereby drawn from among the impact population. This quality of sub-values may also be judged as extremely useful since conceptually, it corresponds nicely to market segmentation and logic about markets capturing consumer surpluses that are widely used in CBA.

The application of SPEL-based inquiries in CBA does, however, pose serious limitations. Most relevant to our argument is that, as SPEL generally concerns a whole "landscape", the results have too little spatial precision, even though in principle, the scale can be adjusted to smaller or larger units. But whether a respondent evaluates the whole region with all its diversity or a just a specific part of the region is not evident. The outcomes of the SPEL-based inquiries therefore cannot offer sufficient spatial precision in location decisions in integrated spatial projects, in for instance, the choice of a more eastern or western route for an infrastructure link or housing site.

Another limitation of SPEL is that, for practical reasons, the enquiries relate only to a segment of the impact population. Enquiries carried out to generate sufficient respondents at a reasonable cost are only possible if none other than locals are asked to evaluate a landscape in direct proximity to areas with which they are familiar. Regional or national preferences are not considered in SPEL. A final drawback of SPEL is the somewhat prohibitive cost, since it would be necessary to conduct the research for each CBA under evaluation. Although current internet-based research can help to cut the costs, the method is only economically feasible for large-scale projects. Therefore, as a means of overcoming this final limitation, GLAM was developed in order to predict stated landscape appreciation based on geo-data.

The GIS-based Landscape Appreciation Model (GLAM)

GLAM (Roos-Klein Lankhorst et al. 2005; De Vries et al. 2007) predicts the mean stated landscape appreciation of a 250 × 250 m grid, based on geo-data of four physical factors of which two are positive and two are negative. These factors are drawn from extensive literature on the types of physical features that are known to inform landscape perception. The two positive factors are naturalness and historical character. Naturalness is mainly defined by the amount of forests, heathlands, wetlands, and grasslands; the presence of natural waters also adds to naturalness. Historical character is operationalized mainly by the number of historical monuments. The two negative criteria are visual disturbance by high buildings, power lines and wind turbines, and the presence of nearby urban areas. The selection of these factors was determined by the availability of geo-data and the estimated predictive power of the factors. Version 2 of GLAM was re-calibrated using data from a recent study commissioned by the Netherlands Environmental Assessment Agency (PBL), resulting in a new version of GLAM (Van der Wulp 2008).

Usability of GLAM for CBA

The experience with GLAM thus far indicates that the spatial precision is still inadequate for its application in CBAs of integrated spatial plans. Although the model grid is 250 × 250 m, it may be more realistic to use output at a lowest level of grids at 3 × 3 km, as the model is calibrated with the mean stated appreciation of regions of about 10 km^2. The main problem for a GIS-based model is the limited availability of relevant information in national GIS datasets. At present, geo-datasets can provide quick but rough characterizations of the perceived landscapes. Thus far, for important perceptual landscape characteristics such as landscape coherence, no sensible geo-dataset whatsoever can be defined. What we learn from the present GLAM model is that people appreciate landscapes that have a high density of nature areas, and they appreciate large-scale open agricultural landscapes much less. But whether they prefer one particular forest over another is not clear.

Citizen Values Assessment (CVA)

The experience value research or Citizen Values Assessment (CVA) (Stolp 2006) is developed from what is known as grounded theory (Glaser 2008). A CVA ("Belevingswaarde onderzoek" in Dutch) ideally has three essential phases, starting with an explorative experience value research in the shape of 'a preliminary qualitative study to provide in-depth understanding of citizen's connections to the area affected by the project' (Stolp 2006: 42). Semi-structured interviews (with say, 75 respondents) are conducted from all relevant affected and interested citizen groups, and the outcome comprises a list of chosen key values. Phase two consists of a testing experience value research which culminates in a so-called Citizen Values Profile: a representative quantitative survey to validate preliminary key values, determine their relative importance (using ranking or five-point scales), and ascertain how respondents currently feel about their living environment in relation to the key values. This elaborate quantitative information (based on say, 500 respondents), may be gleaned using factor analysis. The final phase three is the determination of impacts. Experts determine measureable and testable indicators using information from the Environmental Impact Assessment (EIA) for different criteria, and then assess whether the different alternatives score well or poorly on these indicators. They are generally scored, as is common in Dutch EIAs, on 5-point scales (++, +, 0, –, ––) and weighted overall scores for Citizen Values are determined for the different project alternatives.

Usability of CVA for CBA

The added value of CVA is that it takes individual preferences very seriously and is consequently flexible in its content, based on specific landscape characteristics as experienced by local people (and/or tourists). For example, safety as landscape quality is experienced by people and turns up regularly in CVA, but this aspect is neither in GLAM nor in SPEL. The CVA characteristic of taking individuals and their preferences seriously resembles CBA and its welfare theoretical base. As such, it is much closer to the social CBA tradition than to the more technical EIA tradition, in which it was developed in the Netherlands. Furthermore, the representativeness of the questionnaire also fits well with CBA, since CBA always takes a firm stance on analyzing costs and benefits "to whomever they accrue".

But the CVA has two major drawbacks: time and money. To conduct the in-depth interviews and carry out the survey not only takes a substantial amount of time but the cost is prohibitive (€50,000–100,000). The cost exceeds the total budget available for most of the CBAs we analyzed, at around €20,000 per CBA. A CVA with regard to the required workload is in fact comparable to an original stated preference Contingent Valuation research, which asks for Willingness To Pay.

Further interpretation

The non-economist toolbox seems to show strengths that CBA lacks in practice. CVA is very specific on the quality of individual preferences and spatial specificity.

GLAM is low cost with regard to time and money and yields highly comparable outcomes. The standardized SPEL questionnaire also achieves highly comparable results across projects and is able to assess many specific landscape qualities. However, our discussion does not give rise to a convincing winner; we have not yet identified one flexible tool usable with or within CBA. CVA is expensive and not standardized. GLAM is powerful, fast and cheap but rather superficial in the measurement of real preferences of a well defined impact population, and its spatial specificity is narrow. The most feasible option is probably SPEL, as this method is indeed standardized and has some spatial specificity; however, it focuses on whole landscapes despite that, in principle, the size is adjustable to areas under study. But SPEL is not inexpensive; it would therefore be difficult to address the full impact population at reasonable cost. None of the methods uses monetary valuation of preferences; this factor may not be essential as such, but in order to assess trade-offs in project evaluation, it is indeed necessary to have as many cardinal or ratio measurement scales of measurement as possible. And none of these methods is able to offer a strong cardinally-measured alternative for monetary valuation.

The Hotspotmonitor

Theoretical principles

Our approach to the problem of capturing landscape impacts in CBA is "first things first": CBA is ineffectual as an evaluation tool without a firm grasp of the preferences of individuals (Pearce et al. 2006). The main priority is to make the gathering of preferences fast and easy by using a (potentially) standardized, systematic and spatially-precise method. The "repair tool" we have developed, known as the Hotspotmonitor, is the latest addition to the economist's toolbox (see www.hotspotmonitor.eu). The aim of the Hotspotmonitor is to gather hotspots of landscape experience: places with high attractiveness in general, and attractiveness for particular experiences (quiet, biking, bird watching, etc.). The Hotspotmonitor measures "mere" current preferences; it does not measure or value change. Hotspotmonitor can be understood as a SoftGIS approach (Kyttä 2011a, 2011b), as a "value mapping" technique (Brown and Raymond 2007; Brown and Weber, 2011), and as part of the trend to integrate the potential of GIS into CBA (Bateman et al. 2005).

The following guiding design principles apply for the Hotspotmonitor. First, in line with CBA, but unlike non-economist landscape methods, the tool should be clear in its definition and use of impact population – the population whose welfare counts. For projects with only local impacts, local preferences should count. For projects with national impacts, national preferences should count as well, and so on. The tool should be able to make these distinctions. Second, following CBA and CBA thinking, the choice set of spots or places from which people choose when stating their preferences should be crystal clear. Third, the tool should not fall back on predefined "landscapes", as this unit of appreciation may be spatially vague, it may not relate to individuals' experience, and (in the Dutch context of evaluation of integrated spatial projects) it may be far too large-scale to make a difference in most spatial choices. The tool should therefore allow for far greater spatial precision. Fourth, the tool should certainly inquire into people's preferences that are not discovered through revealed preference data or otherwise. Finally, following market segmentation thinking, which is highly relevant to the interpretation of consumer surpluses, the tool should identify which qualitative preferences are at stake, and at which place or area. People may have preferences for, say, places to appreciate quiet but also other preferences, such as areas for biking and places to walk the dog; these different preferences may be meaningful in spatial decisions so it is crucial to gather these aspects.

Practical constraints

The most important prerequisite for landscape preference gathering is that techniques for assembling these data should be cheap and fast. The Hotspot-monitor is web-based and builds on the ubiquitous Google maps tool in order to develop software for respondents to mark attractive spots or places on this map. The Hotspotmonitor has different versions and is available in different languages. Our discussion of results in this chapter correspond to versions 1.2 and 1.9, in which respondents are free to mark any place, but the questions about the qualities of the marked places are largely closed questions. Versions 1.2 and 1.9 differ in the emphasis of their questions, but not in the way respondents mark places on the map.

The central question for respondents in the Hotspotmonitor is the following: 'Which places do you find very attractive, valuable or important? And why?' The answer to the first question may very well be the supermarket or the city center, which might be interesting, but for two reasons it diverges too far away from landscape impacts and the guiding principles and practical constraints. First, if the Hotspotmonitor were to allow purely urban answers, it would require more respondents in order to develop a clear picture of the "greener" landscape preferences (our focal point). And secondly, we already have plenty of information on urban preferences from other sources. Therefore, we limit the possible answers and instruct respondents as follows: places may be

> both within or outside a city or village. The only condition is that it should be places with green, or water or nature. You can think of a place in a park, by a lake, at the sea, in the forest, near cows in the meadow, in a tulip field, a place to watch birds, etc. It doesn't matter whether they are places you never visit or if they are places you visit frequently.

Many small choices had to be made in the actual software design of our SoftGIS software (Kyttä 2011), but we will only highlight the essential aspects here. The Hotspotmonitor has a location-centered design; it does not focus on opinions at the outset. Its starting point is the place (house or street) where people live, which is marked by a red flag. People are completely free to mark their own places by dragging a pin to the map; the Hotspotmonitor does not offer pre-defined places or areas.

In the Hotspotmonitor we choose to let people mark only their *most preferred* places. Choosing only positively preferred places is in line with individual marketplace preferences used elsewhere in CBA: markets merely measure that someone has, for example, bought a Toyota, but they do not measure his or her dislike, for example, of a BMW or a Ford.

The Hotspotmonitor asks respondents for the most attractive places within a given range and within a given spatial level. In our version we identify three spatial levels:

(i) a circle with a range of 2 km from home;
(ii) a circle with a range of 20 km from home; and
(iii) the whole of the Netherlands (shown twice).

The first level is a typical distance for walking the dog and other small everyday recreational behaviors around the house. The second level is, in the Netherlands, often seen as the operationalization of people's "living environment": they often commute, go to school, do their shopping and engage in more recreational activity within that range. The third level is geared towards policy, culture and evaluation: in the development of many Dutch national policy choices it is crucial to know the preferences of people within the Netherlands, for places within the Netherlands.

Respondents were shown separate maps on which to place one marker. After marking a place, four (groups of) questions were asked: whether they meant the exact spot or a broader area; what mark on a 1–10 scale they would give the place (1 as extremely unattractive and 10 as perfect); why they find the place attractive (using 14 key words derived from the SPEL questionnaire); how often they visit the place, and the activities they undertake at this place (23 activities designated by icons).

Hotspotmonitor results

Highly appreciated Dutch places

This section will illustrate the working of the Hotspotmonitor for CBA. In May 2010 approximately 3,300 respondents in the Netherlands grouped in six spatial concentrations were asked to select their most attractive places using the Hotspotmonitor (version 1.2). The respondents comprised one of the largest internet panels in the country. These large-scale results were obtained within two weeks at a cost of €16,000. Analysis by the authors within a couple of days produced a database of 6,600 nationally-appreciated places, 3,300 regional places and 3,300 local places, of which regional and local most attractive places are spatially clustered around respondents' homes.

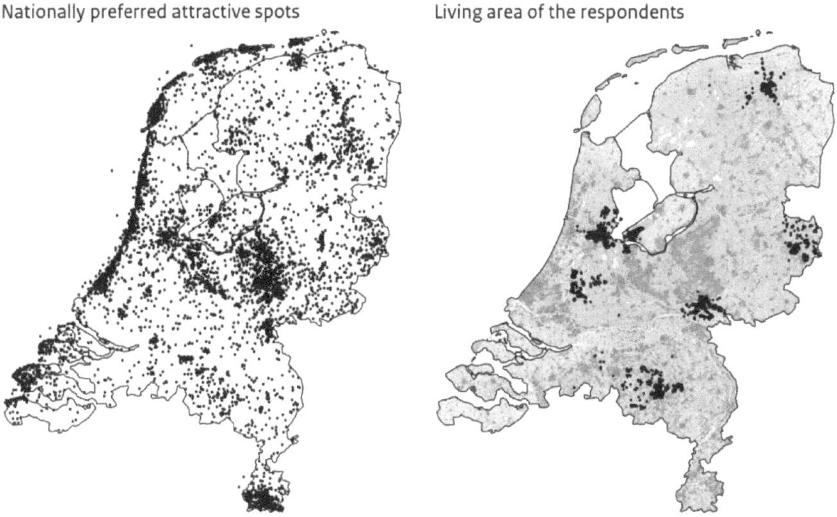

Figure 8.2 Nationally-preferred attractive spots of 3,300 respondents in the Netherlands (left), the living areas of respondents (right).

Although respondents are densely clustered, as shown on the right side of Figure 8.2, the selection of 3,300 respondents already generates a detailed spatially-precise picture of landscape preferences of this group for the whole of the Netherlands. The relative *strength* of the preferences of this group for certain areas, which is the most important prerequisite for further cardinal measurement, is provided by the relative number of hits per area. Even without having to break down the scores of different areas, one can easily recognize that the coastal areas, the Wadden isles, the Veluwe, and the southern hilly area of Limburg are hotspots with high attractiveness.

Of these data we will discuss two vital aspects for use in CBA. The first noteworthy aspect we expand on is the degree of clarity that the Hotspotmonitor provides on local, regional or national appreciation of places; such information is essential to CBA, since the impact population can be twinned with the suitability level. Secondly, we will discuss the use of the HotSpotIndex (HSI) to calculate the appreciation of specific ecosystem and land-use categories.

Highly appreciated places: locally, regionally, and nationally

We notice in Figure 8.3 the conspicuous differences between spatial levels. On the right, in the Groningen region, local markers show city parks and for instance many smaller green or water-related spots. But there are striking differences at the regional level. The strong cluster below the city of Groningen (the "Hoornse plas"

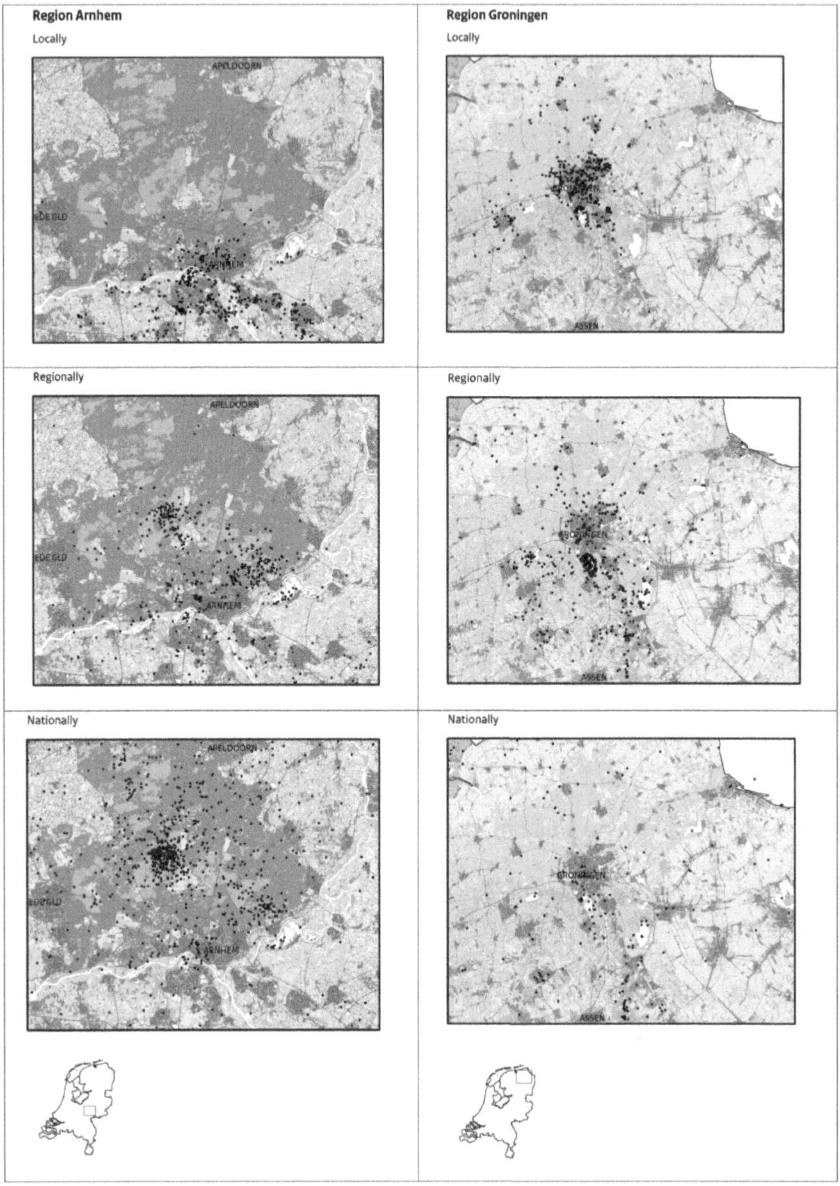

Figure 8.3 Hotspotmarkers for Groningen and Arnhem – at local, regional and national scale levels.

and the connected Paterswoldse Lake) indeed shows up as regional hotspots, but at national level this lake is not a hotspot of significance.

On the left side of Figure 8.3 we see that the city of Arnhem is situated below the Veluwe nature area. Locally, we notice the popularity of places close to the river Rhine, which runs through Arnhem. At the regional level, the Veluwe area comes within reach of more respondents (moving from the 2 km to 20 km range) and is subsequently marked as attractive by many people. Other than the Paterswoldse meer in the Groningen region, however, the Veluwe area is also found to be a highly attractive place by many respondents throughout the Netherlands.

Measuring the intensity of landscape appreciation: the HotSpotIndex (HSI) and Attractivity Weighted Area (ATWA)

In order to use the Hotspotmonitor data for evaluations of land-use changes, we develop two measurement procedures by using the intensity of preferences for different places in a landscape. To understand the measurement, let us turn to Figure 8.4 and Tables 8.1(a) and 8.1(b). Figure 8.4 shows the spatial structure of the hotspotmarkers of a hypothetical set of 27 respondents. On the left-hand side of the figure they are shown on grid cells. Table 8.1(a) shows how to calculate the HotSpotIndex for the 16 grid cells. Each grid cell has a surface of one square, which implies a surface share of six per cent. The HotSpotIndex compares the actual number of markers with the expected number of markers if the pattern were to be random (every part would be appreciated equally). Table 8.1(a) shows that areas 2 and 15 have the highest HSI and are, so to speak, the "hottest spots" on the map. This method of calculating the HSI is a very "clean" one. However, in land-use evaluation one may often work with evaluations of land-use categories: landscape types, ecosystems and land-use types. We develop a different procedure for these uses. The Hotspotmonitor data have a double function here: the pattern

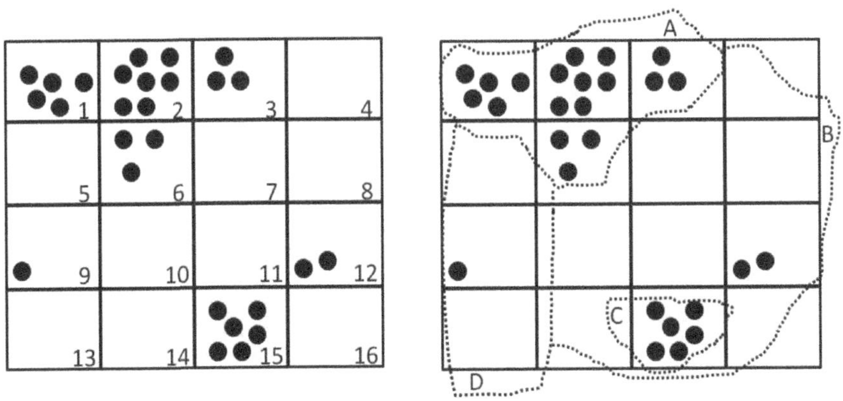

Figure 8.4 An area with its hotspotmarkers: in different grid cells on the left; in different ecosystem/land-use types on the right ($n = 27$).

Table 8.1a The HotSpotIndex calculation for the different areas in Figure 8.4 ($n = 27$)

Area number	*Area surface*	*% share surface*	*Hotspotmarkers*	*Random exp markers*	*HotSpotIndex (HSI)*
1	1	6%	5	1.7	2.96
2	1	6%	7	1.7	4.15
3	1	6%	3	1.7	1.78
4	1	6%	0	1.7	0.00
...	...	...	...	...	...
...	...	...	...	...	...
15	1	6%	6	1.7	3.56
16	1	6%	0	1.7	0.00
Subtotal*	**16**	100%	**27**	**27**	**1.00**
Ecosystem/land-use type					
A (nos 1,2,3,6)	4	25%	18	6.8	2.67
B (nos 4,7,8,10,11,12,14,16)	8	50%	2	13.5	0.15
C (no 15)	1	6%	6	1.7	3.56
D (nos 5,9,13)	3	19%	1	5.1	0.20
Subtotal*	**16**	**100%**	**27**	**27**	**1.00**

*average for HSI.

Table 8.1b The calculation of the Attractivity Weighted Area (ATWA) for two alternative land-use changes

	Land-use			*Attractivity weighted area (ATWA)*		
Ecosystem/ land-use type	*Current situation*	*Alternative 1*	*Alternative 2*	*Current situation*	*Alternative 1*	*Alternative 2*
A	4	5	4	10.7	13.3	10.7
B	8	7	9	1.2	1.0	1.3
C	1	1	0	3.6	3.6	–
D	3	3	3	0.6	0.6	0.6
Total	16	16	16	16.0	18.5	12.6
Change (in %)				–	16%	–21%

on the map can be used to define appreciation types by analyzing the "fit" of clustered markers with different types of land-use categories.

Consider the right-hand side of Figure 8.4, showing four different ecosystem/ land-use types on our grid: A, B, C, and D, which may represent woodland or a lake, etc. Figure 8.4 clearly depicts two clusters of hotspotmarkers that coincide with areas A and C, respectively. On the basis of this fit, one may define these areas as "units" of appreciation. As with the grid cell, the HSI can be calculated for these units. Table 8.1(b) shows that both A and C have high scores

(2.7 and 3.6, respectively). These unit scores can in turn be used for scoring different project alternatives by multiplying the areas and their HSI for different options, thereby yielding what we call the Attractivity Weighted Area (ATWA). Table 8.1(b) shows how this works for two alternative land-use changes, with both changing one grid cell (see Figure 8.4). The first alternative involves a land-use change of grid cell number 7 from type B to type A. The second alternative involves exchanging grid cell number 15 from type C to type B. The ATWA scores change positively from 16 to 18.5 for Alternative 1, and negatively to 12.6 for Alternative 2.

The appreciation of different ecosystem/land-use types

Figure 8.5 shows the procedure to calculate an HSI for different landscape categories with the actual Hotspotmonitor data from Figure 8.2. We took the spatial results of the Hotspotmonitor and placed two different types of ecosystem maps and land-use maps below them. We interactively combined both types of maps until we visually found the best fit. For this best-fit typology we have calculated the HSI scores shown in Figure 8.5.

Figure 8.5 shows that the "Open dunes" is by far the most attractive category. It also shows that Dutch agricultural areas have relatively few markers compared to the hits one might randomly expect. In addition, closed agricultural area is better appreciated than open agricultural area. The results also show that, for instance, dry natural forests have a much higher appreciation than wet natural forests. The results of this analysis have been used to evaluate different policy alternatives for

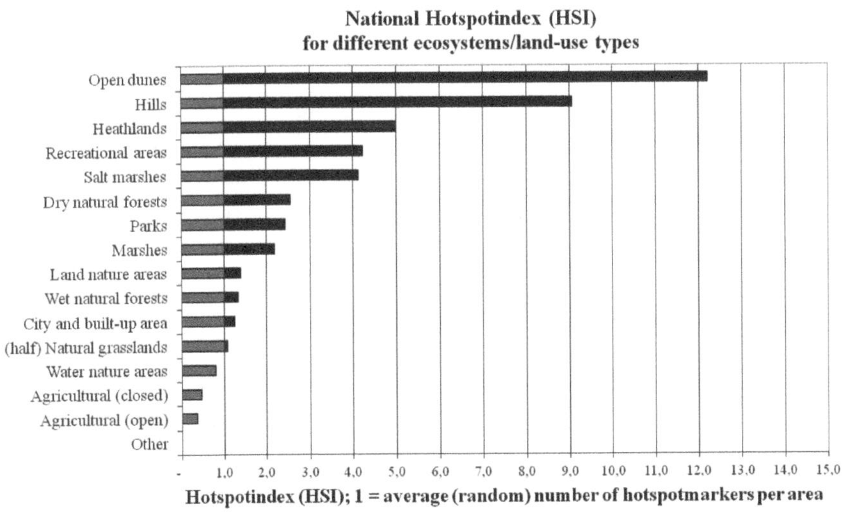

Figure 8.5 The HotSpotIndex for different ecosystem/land-use types (national data, based on 6,600 hotspotmarkers of 3,300 respondents; see Figure 8.2).

future Dutch nature policy (Sijtsma et al. 2011), for which attractivity scores were calculated within an MCCBA approach. In this chapter, however, we will examine the level of "integrated" spatial plans by discussing the use of the Hotspotmonitor in the CBA of the upgrading of the Dutch city of Almere.

Cost–benefit analysis: upgrading the city of Almere

Plan for the upgrade of Almere

Almere is the youngest city of the Netherlands with 190,000 inhabitants. The first houses were built in 1975 on new land reclaimed from the former Southern Sea. As is the case in most new towns, Almere does not have much in the way of urban allure, as it is characterized as having relatively few expensive houses and offers little employment. Moreover, most jobs are situated in the Amsterdam region, so consequently the bridges connecting new and old land are jammed daily with commuters. The government anticipates further growth of Almere to reach 350,000 inhabitants by 2030; so to handle this, the municipality of Almere has developed Plan Almere 2.0 in order to upgrade the city not only by building new houses, but also by improving accessibility and quality of life in the city. Plan Almere 2.0 includes (see Figure 8.6):

- building 60,000 new houses during the period 2010–2030 (at four alternative locations) with a high share of expensive houses. Alternative Pampus Buiten implies reclamation of about 100 hectares of Lake Markermeer.

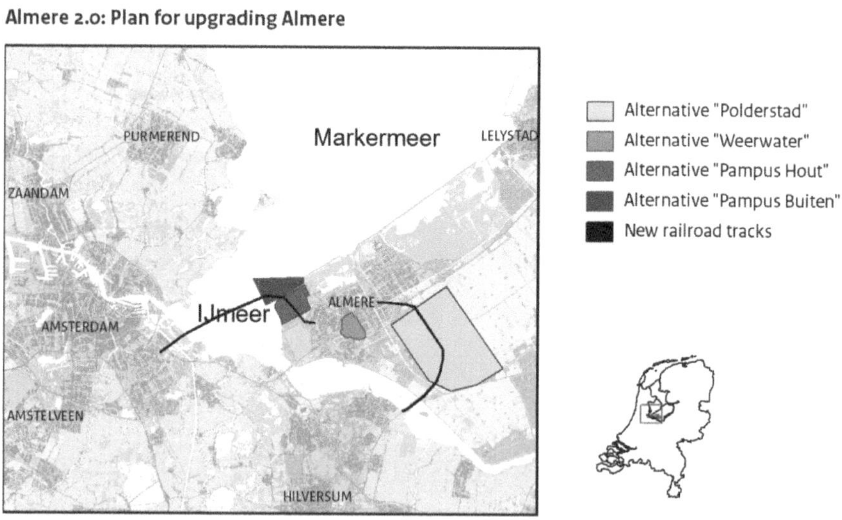

Figure 8.6 Plan Almere 2.0. Four locations for building 60,000 new houses and two new railroad tracks to connect Almere with the old land.

- building two new railway tracks and bridges to connect Almere with the old land.
- measures to improve the ecological quality of the former Southern Sea, now called IJmeer and the Markerwaard.

Plan Almere 2.0 will only be implemented by the municipality if the national government finances most of the investment in railroad tracks and takes measures to improve the ecological quality of the lakes IJmeer and Markermeer. A new railroad track crossing the IJmeer will cost €4–6.5 billion. To support the decision of this national investment, a cost–benefit analysis was made (Zwaneveld et al. 2009), but it turned out to be difficult to quantify the impacts on landscape quality. This is an omission, considering that most civil resistance against investments in new rail projects is inspired by concern about impacts on the landscape.

Therefore, we will discuss the aspects that should be elaborated in a CBA of landscape impacts, followed by a description of the inventory of the appreciation of the actual landscape. As a preliminary exercise, we will first summarize the limitations of the existing methods as applied in Zwaneveld et al. (2009). Thereafter, we demonstrate and discuss the potential of the Hotspotmonitor.

Aspects of CBA of landscape impacts

For a CBA of landscape impacts to take place, three basic aspects need to be elaborated before the final monetarization can take place. Firstly, the effects of the construction of new urban developments and railways on landscape perception should be identified; we should also know how the actual landscape is perceived and appreciated; and thirdly, effects on the actual landscape values need to be quantified. These three aspects will be considered separately in the following three sub-sections.

Landscape perception research in the Netherlands shows that most man-made structures have some negative impact on the appreciation of landscapes. Large-scale infrastructure has the biggest impact on the appreciation of landscapes; landscapes with industrial or commercial buildings, or large barns within sight are, on average, appreciated one-third less than the same landscape without these man-made structures (Van der Wulp et al. 2009). The impact of new housing construction ranges from only slightly negative to no effect whatsoever.

Nevertheless, impacts differ according to context: they are more severe the larger they are and when they are situated in highly appreciated landscapes (De Vries et al. 2012). The limitations of the application of these results for this cost–benefit analysis can be illustrated by a survey of yacht owners who sail on the IJmeer and Markermeer. Half of the owners mentioned spontaneously that new house construction along the shores or beside lakes is a serious threat to the landscape values of IJmeer and Markermeer. Building long bridges or dams is also perceived as a threat by 25 per cent of respondents, which is equal to the number of respondents who oppose new wind parks. The results of the Dutch

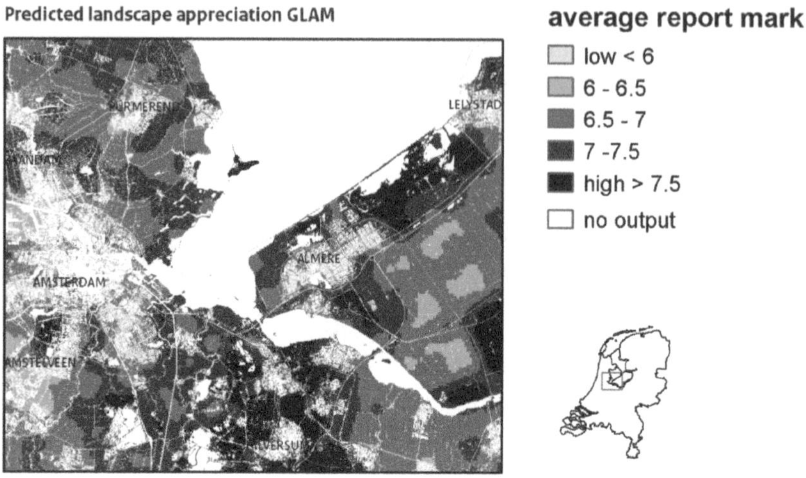

Figure 8.7 Predicted landscape appreciation using GLAM.

landscape perception research therefore do not seem to be entirely valid for the case of Almere.

First estimates of actual landscape appreciation

Actual landscape values were estimated in the first instance using GLAM (see above). The estimations of the actual landscape values of the region are shown in Figure 8.7. Larger nature areas and forests are appreciated the most, whereas large-scale open agricultural land and urban areas are appreciated less. The limitations of this result are that:

- Appreciation differs only slightly between nature areas or between urban areas.
- Geographical resolution is rather low, so it is not possible to estimate the appreciation of urban parks or smaller building sites.
- It was not possible to estimate the appreciation of the large water landscapes that are affected by new railroad tracks and one building site.
- Only the preferences of inhabitants within a 15 km range from home are modeled in GLAM. So there are no indications of appreciation at the national and local scales. The landscape values of the former Southern Sea, used frequently by sailing enthusiasts from throughout the Netherlands, or urban parks visited intensively, are not considered.

We have concluded that the first estimate using GLAM falls short in the valuation of the actual landscape appreciation of the housing project at Pampus Buiten and the new railroad tracks. The three other locations for building houses have nearly the same actual landscape appreciation.

Actual landscape qualities in the Amsterdam–Almere region with Hotspotmonitor

In order to achieve a better estimate of landscape appreciation of areas affected by building new housing and railroad tracks, a Hotspotmonitor enquiry (version 1.9) was performed with 1,286 inhabitants of the region Amsterdam–Almere so as to establish the local, regional and national hotspots. The national survey (version 1.2) of 3,859 respondents was also applied to establish national hotspots in the region.

Respondents' markers were aggregated into regions delineated by both the clustering patterns on the one hand, and their relevance for the impact assessment on the other. Figure 8.8 shows the HotSpotIndex (HSI) for each region; the HSI is normalized according to the number of respondents who were allowed to place a marker in the area of concern. At the national scale, this number is the same for all regions but at regional and local scales this number differs as proximity (within 20 or 2 km) to the preferred area differs. The number of respondents allowed to put a marker in a certain area for the regional scale ranges from 386 to 1,286, which is sufficient for a reliable valuation. On the local scale, however, the inquiry was too small for a reliable estimation for all regions. Regions with fewer than 12 potential respondents were excluded from our study.

We find for the local level that daily to weekly recreational activities, large urban parks and some nature areas within the urban fringe are appreciated the best. The most appreciated area is Twiske, a 25-year-old recreational park with woodlands and swamps. Three areas are appreciated the most for regional, monthly recreational activities: the Amsterdam Forest, which is an 80-year old woodland park; Twiske; and most surprising, a large natural marshland, Oostvaardersplassen, which is closed to the public. Another noteworthy result is the rather high appreciation of waterfronts (Almere-Beach, Port Almere, Muiden, Durgerdam, and Volendam). This is in contrast to the appreciation of landscapes in the Amsterdam–Almere region, which is low compared to the national hotspots, such as the coastal zone and the extensive woodlands, heathlands and sand dunes of the Veluwe. The most appreciated areas on a national scale within the region are Lake Markermeer, a vast section of the former Southern Sea, in addition to Oostvaarderplassen, Amsterdam Forest, Vondelpark, and the inner city of Amsterdam.

The great diversity of highly appreciated areas on all scale levels as revealed by Hotspotmonitor illustrates the limited value of the GLAM results for an impact analysis on a local to regional scale. The substantial differences in appreciation of the same areas at different spatial levels, indicates that a distinction between national, regional and local impact populations is indeed relevant. The rather high appreciation of water landscapes at regional and national scales, moreover, demonstrates the need to estimate the appreciation of water landscapes.

Estimating landscape impacts of Plan Almere 2.0

Our third step in the impact assessment entails the quantification of the landscape impacts. As shown earlier, there is scant knowledge of the magnitude of effects

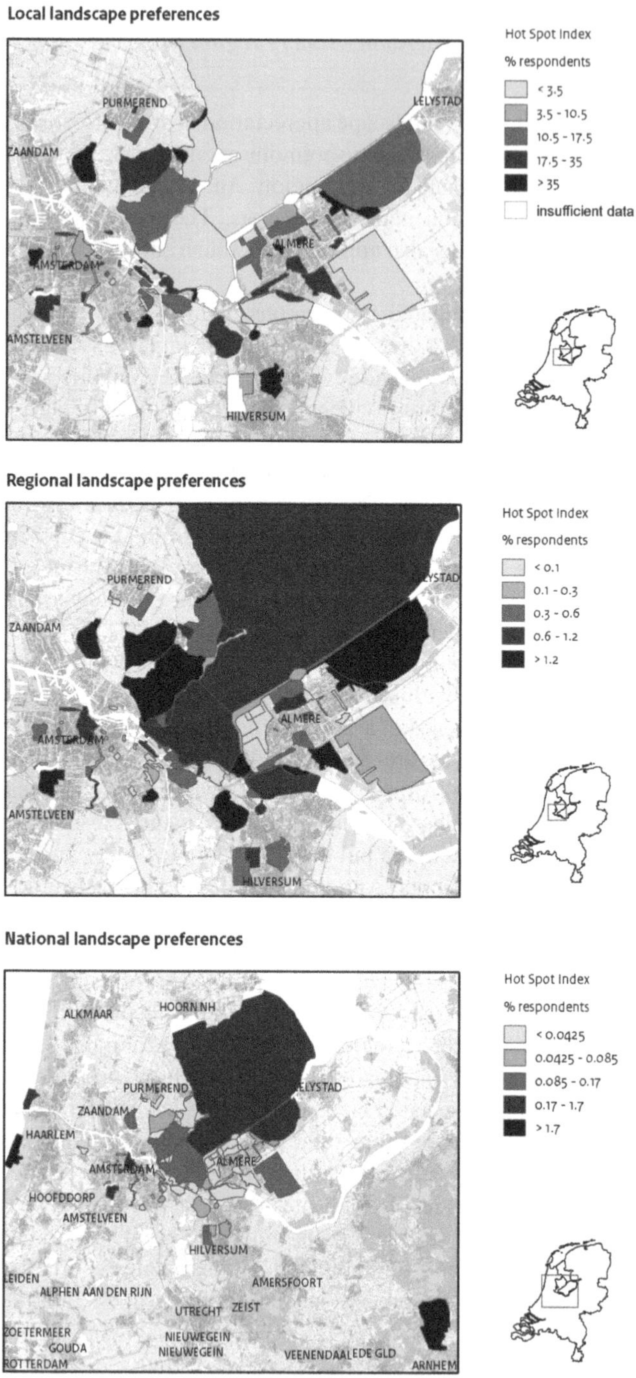

Figure 8.8 HotSpotIndex of selected regions normalized by number of respondents at a local scale (top), a regional scale (middle) and a national scale (bottom).

Table 8.2 Calculated HotSpotIndex (% of respondents who placed a marker) for areas potentially affected by building new houses and railroad tracks

Area	*HSI local*	*HSI regional*	*HIS national*	*Impacts Almere 2.0*
IJmeer	0.00	0.70	0.12	Construction and visual impact dam/bridge
Markermeer	0.00	0.88	0.32	Construction and visual impact new land with buildings
Weerwater	24.79	0.57	0.03	Construction new buildings
Polderstad	2.86	0.26	0.10	Construction new buildings
Shores and waterfronts IJmeer	26.74	0.56	0.01	Visual impact of dam/bridge and building
Almere Pampus	4.58	0.03	0.00	Construction new buildings
Reference maxima:				
National: National Park Hoge Veluwe			3.27	
Regional: Oostvaardersplassen		13.22		
Local: Twiske	112.50			

of building new railroad tracks and housing. So for the moment we can only describe landscape impacts as a range of magnitude with the maximum impact being equal to the actual appreciation. Table 8.2 summarizes the areas that may be affected by Plan Almere 2.0 and the degree to which they are appreciated.

The water landscape of Markermeer, which is affected by the building site Pampus Buiten, is appreciated the most on both national and regional scales. The appreciation of landscape of this lake is about one-tenth of that of the most appreciated landscape in the Netherlands, the national park Hoge Veluwe. The shores, and especially the waterfronts of the IJmeer, which are well appreciated areas at a regional scale, represent five per cent of the most appreciated landscape in the Almere–Amsterdam region (the Oostvaardersplassen). But these areas are visually influenced by the construction of a new bridge. The three alternative locations for building houses on the shores and waterfronts are hardly appreciated, except for Almere Weerwater at a regional scale.

To restate our results, we find that the most explicit impacts of Plan Almere 2.0 are to be expected on the shores and waterfronts of Lake IJmeer (local and regional scale), Weerwater (local scale) and Markerwaard (national scale). The new house building sites have few impacts except for the site Pampus Buiten. Finally, to put these results in another perspective, the actual most appreciated area of the IJmeer shore, a recreational area, is situated just below the (oldest) bridge that connects

Almere with the old land. The implication here is that what is seen as a significant, possibly negative, change in the plans (new infrastructure) may in fact not be a thorny issue for landscape appreciation valuation.

Discussion

We think that we have added something of significance to the *economist's* toolbox, but perhaps we should be more precise and say to the *evaluator's* toolbox. The Hotspotmonitor is designed to bring increased spatial precision, greater spatial consistency, and more standardized quantification of landscape preferences. Our approach may provide comprehensible factual information to analysts from different disciplines. Given the multi-disciplinarity and complexity of the analysis of land-use changes (Brouwer and Van der Heide 2009; Rounsevell et al. 2012), this integrative potential may be seen as a welcome opportunity. We wrote above that, ideally, the HSI and ATWA measurements are used within an MCCBA approach in which monetarization is less essential than standardized measurement on interval or ratio scales, but what have we added to the standard CBA toolbox for capturing scenic beauty?

This CBA toolbox is well-known and contains several revealed preference tools and stated preference tools (Hanley and Barbier 2009; Boardman et al. 2011). With regard to landscape issues, the most used and discussed revealed preference tools are hedonic pricing and travel cost approaches (Palmquist 2005; Herriges and Kling 2008) while for stated preferences, contingent valuation and discrete choice models are most evident in the literature (Kapper 2004; Palmquist 2005; Powe 2007; Hanley and Barbier 2009; Boardman et al. 2011). What then have we added to this literature?

We have neither used any of the aforementioned tools nor have we invented a new monetary valuation tool. Our contribution to the economist's toolbox of stated preference techniques and revealed preference techniques is of a different kind. Boardman et al. (2011) distinguish nine basic steps in a CBA. The fifth step involves monetizing all impacts; while discounting, calculating net present values, carrying out sensitivity analysis, and making recommendations follow thereafter. The HotSpotIndex and ATWA calculations have added nothing to this fifth step and thus logically neither helps the steps thereafter. The merits of our approach in the field of CBA lie primarily in the steps preceding monetarization. Boardman et al. set out four steps: (1) Specify the set of alternative projects; (2) Decide whose benefits and costs count (standing); (3) Catalog the impacts and select measurement indicators (units); and (4) Predict the impacts quantitatively over the life of the project. It is in step (3) that the Hotspotmonitor contributes most to the toolbox of the economist. The Hotspotmonitor provides an improved and systematic understanding of *current* landscape appreciation; in effect, it gives greater insight into the welfare of the landscape at stake. The Hotspotmonitor answers the most basic question in this stage: Do the alternatives affect the welfare of the impact population, in as much as this welfare relates to landscape appreciation? Furthermore, it offers units of measurement for assessing the relative size of the possible impacts, by defining areas which are highly appreciated, as well as by measuring the relative intensity of the attractivity of these areas: the HSI and ATWA measurements. As we observed above, when we discuss the current practice in Dutch CBA case studies, this is an essential aspect of any CBA, but Dutch CBA analysts have thus far been left empty handed to seriously address this issue.

As to step (2), the "standing" issue, the Hotspotmonitor approach is especially designed to satisfy the economist's basic concern in relation to landscape preferences. Obviously the Hotspotmonitor does not help in choosing standing as such, but its spatial layers and its spatial-consistency make it easy for the analyst to assess impacts on the welfare of the people who do have standing (the people whose welfare counts), while the Hotspotmonitor is also able to identify their landscape preferences at local, regional and national levels (Boardman et al. 1993; Bateman 2009).

These measurements can be used for step (4), which is the prediction of the impacts, quantitatively, over the lifetime of the project. However, the Hotspotmonitor will only help in as much as changes in current preferences are concerned. If the project alternatives that are considered in the CBA "only" consider destroying existing landscape qualities, the Hotspotmonitor will be very useful. If for instance we reconsider the two alternatives in the Randstad urbanization CBA discussed above as case (i), then using the Hotspotmonitor measurements will clarify the landscape impact – using current preferences. The limitations may be clear from the CBA of a new business park development at Hoeksche Waard, discussed as case (ii) above. Here, the Hotspotmonitor measurement would have obviously been helpful in answering the basic question: 'Do the alternatives affect the welfare of the impact population in as much as this welfare relates to landscape appreciation?' and if so, carried out the assessment of the relative size of the impacts compared to other areas. However, this CBA also had to assess what effect landscaping efforts might have concerning building new landscape elements to enhance attractivity. Here the Hotspotmonitor would not have been helpful, since it merely evaluates the relative attractivity of existing places.

For the monetarization step – the fifth of nine basic steps in a CBA – the Hotspotmonitor measurements of HSI and ATWA may be indirectly helpful. They may provide the units of measurement for monetary valuations of landscape impacts. Although the Hotspotmonitor measurement has its limitations due to its focus on current preferences for existing places, we still feel that in as much as a specific CBA can be limited to welfare changes relating to those types of preferences, the explicit monetary valuations of changes in HSI and ATWA will help in the search for better results from stated preference techniques (Powe 2007). This is especially so because of its systematic and non-project-specific measurement units and its spatial consistency (compare Table 8.2), the two elements that seem most to hamper many monetary stated preference studies and benefit transfer approaches (Sugden 2005; Bateman et al. 2006; Bateman 2009).

And to conclude, Hotspotmonitor data may also be used to make revealed preference approaches more convincing (Sijtsma and Brouwer 2011). This point is interesting, especially in as much as these data can give a clearer view on the causality between, for instance, recreational expenditures or real estate prices on the one hand, and landscape qualities on the other, which then can be seen as a combination of stated and revealed preferences (Earnhart 2001).

References

Aoki, Y. (1999) 'Review article: Trends in the study of the psychological evaluation of landscape', *Landscape Research*, 24(1): 85–94.

Bateman, I.J., Lovett, A.A. and Brainard, J.S. (2005) *Applied Environmental Economics – A GIS Approach to Cost–benefit Analysis*. Cambridge: Cambridge University Press.

Bateman, I.J. (2009) Bringing the real world into economic analyses of land-use value: Incorporating spatial complexity, *Land Use Policy*, 26: S30–S42.

Belton, V. and Stewart, T.J. (2002) *Multiple Criteria Decision Analysis: An Integrated Approach*. Dordrecht: Kluwer Academic Publishers.

Berg, A.E. van den (1999) *Individual Differences in the Aesthetic Evaluation of Natural Landscapes*. Groningen: Rijksuniverstiteit Groningen, PhD thesis.

Boardman, A.E., Greenberg, D., Vining, A. and Weimer, D. (2011) *Cost–benefit Analysis: Concepts and Practice*. Boston, Mass.: Pearson Education.

Boardman, A.E., Vining, A. and Waters, W.G., II (1993) 'Costs and benefits through bureaucratic lenses: Example of a highway project', *Journal of Policy Analysis and Management*, 12(3): 532–555.

Boer, T.A. de and Groot, M. de (2010) *Belevingswaardenmonitor Nota Ruimte 2009 – Eerste Herhalingsmeting Landschap en Groen in en om de Stad*. Wageningen: Wettelijke Onderzoekstaken Natuur & Milieu, report no. 109 (in Dutch).

Brouwer, F. and Heide, C.M. van der (eds) (2009) *Multifunctional Rural Land Management – Economics and Policies*. London: Earthscan.

Brown, G., and Raymond, C. (2007) 'The relationship between place attachment and landscape values: Toward mapping place attachment', *Applied Geography*, 27(2): 89–111.

Brown, G. and Weber, D. (2011) 'Public participation GIS: A new method for national park planning', *Landscape and Urban Planning*, 102(1): 1–15.

Buijs, A.E., Arts, B.J.M., Elands, B.H.M. and Lengkeek, J. (2011) 'Beyond environmental frames: The social representation and cultural resonance of nature in conflicts over a Dutch woodland', *Geoforum*, 42(3): 329–341.

Buijs, A.E. and Van Kralingen, R. (2003) *Meten van Beleving. Inventarisatie van Bestaande Indicatoren en Meetmethoden*. Wageningen: Alterra, report 782 (in Dutch).

Coeterier, J.F. (2000) *Landschapsbeleving – Toepassing van de Meetmethode Landschapsbeleving in Vier Gebieden in Nederland*. Wageningen: Alterra, report 209 (in Dutch).

Dakin, S. (2003) 'There's more to landscape than meets the eye: Towards inclusive landscape assessment in resource and environmental management', *Canadian Geographer*, 47(2): 185–200.

Daniel, T.C. (2001) 'Whither scenic beauty? visual landscape quality assessment in the 21st century', *Landscape and Urban Planning*, 54(1–4): 267–281.

Earnhart, D. (2001) 'Combining revealed and stated preference methods to value environmental amenities at residential locations', *Land Economics*, 77(1): 12–29.

ECORYS (2005) *Maatschappelijke Kosten en Baten IBO Verstedelijking (Inclusief Bijlage Rapport)*. Rotterdam: ECORYS Nederland BV (in Dutch).

ECORYS (2006) *Optimalisatie KKBA Bereikbaarheid en Landschappelijke Inpassing Bedrijventerrein Hoeksche Waard – Eindrapportage*. Rotterdam: ECORYS Nederland BV (in Dutch).

Glaser, B.G. (2008) *Doing Quantitative Grounded Theory*. Mill Valley, CA: Sociology Press.

Hanley, N. and Barbier, E.B. (2009) *Pricing Nature – Cost–benefit Analysis and Environmental Policy*. Cheltenham, UK: Edward Elgar.

Herriges, J. and Kling, C.L. (eds) (2008) *Revealed Preference Approaches to Environmental Valuation – Volumes I & II*. Aldershot: Ashgate.

Kapper, T. (2004) 'Bringing beauty to account in the environmental impact statement: The contingent valuation of landscape aesthetics', *Environmental Practice*, 6(4): 296–305.

Kyttä, M. (2011) 'SoftGIS methods in planning evaluation'. In: A. Hull, E. Alexander, A. Khakee and J. Woltjer (eds), *Evaluation for Participation and Sustainability in Planning*. pp. 334–354, London: Routledge.

Kyttä, M. and Kahila, M. (2011) 'Softgis methodology', *GIM International*, 25(3). http://www.gim-international.com/issues/articles/id1677-SoftGIS_Methodology.html.

Palmquist, R.B. (2005) 'Property value models'. In: K.-G. Mäler, and J. Vincent (eds), *Handbook of Environmental Economics – Volume 2*. pp. 763–819, Amsterdam: North-Holland.

Pearce, D.W., Atkinson, G. and Mourato, S. (2006) *Cost–benefit Analysis and the Environment: Recent Developments*. Paris: Organisation for Economic Co-operation and Development (OECD).

Pomerol, J.C. and Barba-Romero, S. (2000) *Multicriteria Decision in Management – Principles and Practice*. Boston: Kluwer Academic Publishers.

Powe, N.A. (2007) *Redesigning Environmental Valuation – Mixing Methods within Stated Preference Techniques*. Cheltenham, UK: Edward Elgar.

Roos-Klein Lankhorst, J., Vries, S. de, Buijs, A E., Bloemmen, M.H.I. and Schuiling, C. (2005) *BelevingsGIS Versie 2; Waardering van het Nederlandse Landschap door de Bevolking op Kaart*. Wageningen: Alterra, report 1138 (in Dutch).

Rounsevell, M.D.A., Pedroli, B., Erb, K.-H., Gramberger, M., Busck, A.G., Haberl, H., Kristensen, S., Kuemmerle, T., Lavorel, S., Lindner, M., Lotze-Campen, H., Metzger, M.J., Murray-Rust, D., Popp, A., Pérez-Soba, M., Reenberg, A., Vadineanu, A., Verburg, P.H. and Wolfslehner, B. (2012) 'Challenges for land system science', *Land Use Policy*, 29(4): 899–910.

Sijtsma, F.J. and Brouwer, A.E. (2011) 'Book review: J. Herriges and C. L. Kling, Revealed preference approaches to environmental valuation – Volume I & II', *European Spatial Research and Planning*, 1(1): 105–106.

Sijtsma, F.J., Heide, C.M. van der and Hinsberg, A. van (2011) 'Biodiversity and decision-support: Integrating CBA and MCA'. In: A. Hull, E. Alexander, A. Khakee and J. Woltjer (eds), *Evaluation for Participation and Sustainability in Planning*. pp. 197–218, London: Routledge.

Sijtsma, F.J., Kruitwagen, S. and Farjon, H. (2008) *Landschapseffecten in Maatschappelijke Kosten-baten Analyses van Integrale Gebiedsontwikkeling – Deel 2b van het PBL Onderzoek naar het Gestandaardiseerd, Systematisch en Precies Meten van 'Lastig Inpasbare Effecten' in MKBA's*, Unpublished manuscript (in Dutch).

Sijtsma, F.J. (2006) *Project Evaluation, Sustainability and Accountability – Combining Cost–benefit Analysis and Multi-criteria Analysis*. Groningen: Stichting REG, PhD thesis.

Stobbelaar, D.J. and Pedroli, B. (2011) 'Perspectives on landscape identity: A conceptual challenge', *Landscape Research*, 36(3): 321–339.

Stolp, A. (2006) *Citizen Values Assessment – An Instrument for Integrating Citizens Perspectives into Environmental Impact Assessment*. Leiden: Leiden University, PhD thesis.

Sugden, R. (2005) Anomalies and stated preference techniques: A framework for a discussion of coping strategies. *Environmental and Resource Economics*, 32: 1–12.

Vries, S. de, Groot, M. de and Boers, J. (2012) 'Eyesores in sight: Quantifying the impact of man-made elements on the scenic beauty of Dutch landscapes', *Landscape and Urban Planning*, 105(1–2): 118–127.

Vries, S. de, Lankhorst, J.R. and Buijs, A.E. (2007) 'Mapping the attractiveness of the Dutch countryside: A GIS-based landscape appreciation model', *Forest Snow and Landscape Research*, 81(1–2): 43–58.

Wulp, N.Y. van der (2008) *Belevingswaardemonitor Nota Ruimte 2006 – Nulmeting Landschap naar Gebieden*. Wageningen: Wettelijke Onderzoekstaken Natuur & Milieu, report no. 75 (in Dutch).

Wulp, N.Y. van der, Veeneklaas, F.R. and Farjon, J.M.J. (2009) *Krassen op het Landschap: Over de Beleving van Storende Elementen*. Wageningen: Wettelijke Onderzoekstaken Natuur & Milieu, paper no. 1 (in Dutch).

Zube, E.H., Sell, J.L. and Taylor, J.G. (1982) 'Landscape perception: Research, application and theory', *Landscape Planning*, 9(1): 1–33.

Zwaneveld, P., Romijn, G., Renes, G. and Geurs, K. (2009) *Maatschappelijke Kosten en Baten van Verstedelijkingsvarianten en Openbaarvervoerprojecten voor Almere*. The Hague: Centraal Planbureau (CPB) (in Dutch).

9 Rural landscape and optimal agricultural land-use

Iddo Kan, David Haim, Mickey Rapaport-Rom and Mordechai Shechter

Introduction

As a supplement to the supply of agricultural products, agricultural lands provide a range of amenity services that are valuable for society, including recreation areas, cultural heritage and rural landscapes. These amenities are the driving force behind policies established in some developed countries to reward farmers for the positive external benefits they generate (EC 2003; OECD 2000, 2003; Peterson et al. 2002), and by this means to slow urbanization and other economic processes that shrink agricultural lands. Contrary to agricultural products that can be shipped to a distance, rural landscape is a non-transportable resource, and therefore it is in the interest of local communities to preserve their agricultural open spaces. Various studies have evaluated the willingness to pay for maintaining agricultural lands, including Halstead (1984), Bergstrom et al. (1985), Beasley et al. (1986), Bowker and Didychuk (1994), Hackl and Pruckner (1997), Ready et al. (1997), Ready and Abdalla (2005) and Fleischer and Tsur (2003). In McConnell (1989), Lopez et al. (1994), Brunstad et al. (1999) and Fleischer and Tsur (2009) the allocation of land between urban and rural uses is analyzed in view of the agricultural landscape amenity values.

Drake (1992), Brunstad et al. (1999) and Fleischer and Tsur (2009) evaluate agricultural landscape amenity values while taking into account the variability across agricultural land-uses with respect to their contribution to the rural landscape. Recently, Kan et al. (2009) utilized the landscape amenity-value function estimated by Fleischer and Tsur (2009) in the evaluation of the potential for increasing social welfare in Israel by changing land allocation among crops, where the tradeoff between farming profit and the landscape amenity value associated with each crop is considered.

The objective of this chapter is to explore, both theoretically and empirically, how this welfare increase is affected by two factors. The first factor is the size of the regional area for which a policy to encourage (discourage) the expansion of land allocated to crops with high (low) landscape amenity values is specified. The second factor is related to the question of whether a marginal increase in land devoted to crops, particularly those associated with non-traditional rural landscapes such as greenhouses, can decrease the landscape amenity value of an entire regional agricultural area.

The chapter continues as follows. The next section develops a theoretical model describing the urban-rural and intra-agricultural land allocations under market equilibrium versus the socially optimal allocations; this model is utilized to analyze the impact of the two aforementioned factors on the welfare increase associated with internalization of the landscape amenity value. Then a section follows that presents the empirical model and the policy analysis with application to the case of Israel. The final section concludes.

Theory

The model developed in this section builds on the theoretical framework developed by Fleischer and Tsur (2009) for the analysis of the socially optimal (SO) rural-urban and intra-agricultural land allocations versus allocations under the farming-sector's profit maximization (PM) solution.

The model

Consider a small open economy, such that all agricultural input and output prices are constant. The economy's land that is available for agriculture and housing uses amounts to $\bar{L}$ hectares. The population is composed of N identical households, of which N^f are identical farmers, where both populations are uniformly spatially distributed across area $\bar{L}$. The economy's land is administratively separated into regions, where the community of residents in each region has the authority to set its own land policy, i.e. to create incentives for growing socially preferable crops. It is assumed that, while designing such a policy, each community ignores landscape amenities provided by other regions, as well as amenities to residents of other regions; that is, only the landscape amenities provided by the regions' agricultural lands to the local residents are considered. Moreover, our analysis is limited to the case in which neither internal nor external immigration is allowed; this restriction can be attributed to the attachment of the population to its residential location, which entails high immigration costs. This assumption implies that in each region there is a separate regional land market, in which the regional equilibrium rural-urban land allocation is determined. In each region this land market is "free" such that ownership allocation has no impact on welfare assessments.

Define a sub-region in this economy, with an area of $\alpha\bar{L}$ hectares, available for agricultural and urban uses, αN households and αN^f farmers, where $1 \geq \alpha \geq 0$. Farmers allocate their lands among I crops, where l_i denotes the land devoted to crop i, $i = 1, \ldots, I$, by a typical farmer. The region is characterized by some exogenous factors such as climatic conditions, soil quality, etc. which are represented by a composite parameter, g. For the theoretical analysis, it is assumed that there is no spatial variability in g throughout the economy (whereas our empirical model takes into account the heterogeneity in g). Let $\pi_i(l_i; g)$ be crop i's net profit, excluding agricultural land rental cost, which is denoted r.

The prices of all production factors other than land are constant; these factors are set to their optimum, and expressed as a function of l_i and g. The marginal profit of l_i is therefore $\pi_i'(l_i; g)$. It is assumed that $\pi_i(l_i; g)$ is strictly concave and that $\pi_i'(l_i; g)$ can be inverted such that, given the agricultural rental rate r, the demand for land for crop i is given by $l_i(r; g) = \pi_i'^{-1}(r; g)$, where $l_i(r; g) \geq 0$. The regional land allocated to crop i is therefore $L_i(r; g) = \alpha N_f l_i(r; g)$, and the derived demand for land allocated to crop i is expressed as $\pi_i'(L_i/\alpha N_f; g)$. By summing the demands for I crops we get the regional demand for land for the purpose of agricultural production, $L(r; g) = \alpha N_f l(r; g)$, where $l(r; g) = \sum_{i=1}^{I} l_i(r; g)$ is the per-farmer land-use. The regionally derived demand for agricultural land, denoted $\Pi'(L/\alpha N_f; g)$, is obtained by horizontally summing the aggregate crop-demand functions $\pi_i'(L_i/\alpha N_f; g)$ across crops.

Households residing in the region derive utility from the consumption of a composite private good, z, housing land l_h, and landscape amenities generated by the regional agricultural land allocated among the I crops; this land allocation is represented by $\mathbf{L} = (L_1, L_2, \ldots, L_I)$. The household's utility function is assumed to be of the form

$$u(z, l_h, \mathbf{L}) = u_p(z, l_h) + u_a(\mathbf{L}), \tag{9.1}$$

where $u_p(z, l_h)$ and $u_a(\mathbf{L})$ are the utilities derived from the private goods, z and l_h, and the landscape amenities associated with the agricultural land allocation, $\mathbf{L}$, respectively. Let the household's budget constraint be $e = p_z z + r_h l_h$, where e is the household's exogenous income, r_h is the urban-land rental rate, and p_z is the price of z, which is assumed constant. Maximizing the utility subject to the budget constraint yields the demands for the private goods, $z(r_h; e)$ and $l_h(r_h; e)$, where p_z is suppressed as an argument. By inverting $l_h(r_h; e)$, we get the household's housing-land demand, $D_h(l_h; e)$, such that, given r_h, there is $D_h(l_h; e) = r_h$. It is further assumed that $D_h(l_h; e)$ is down-sloped with respect to l_h, meaning that the marginal willingness to pay for housing lands reduces with the size of the housing areas. Denote by $L_h \equiv \alpha N l_h$ the regional urban land consumption, and by expressing the housing demand in terms of L_h, we get $D_h(L_h/\alpha N; e) = r_h$.

The indirect utility function, $v(\mathbf{L}; e) = u_p(z(r_h; e), l_h(r_h; e)) + u_a(\mathbf{L})$, is obtained by setting r_h as a constant and substituting the demands for the private goods into equation (9.1). Denote by $a(\mathbf{L}; e)$ the landscape amenity value of the agricultural lands, which is the per-household willingness to pay for the preservation of the agricultural lands due to the rural landscape they provide. This value is equivalent to the income increase needed to compensate a household for the loss of landscape amenities associated with the disappearance of the agricultural areas, L, when these areas are allocated among crops according to $\mathbf{L}$, such that $v(e + a(\mathbf{L}; e), 0) = v(\mathbf{L}; e)$. The per-household marginal landscape amenity value (MLAV) of the land devoted to crop i, conditional on the land allocation $\mathbf{L}$, is given by $\partial a_i(\mathbf{L}; e)/\partial L_i$.

Empirical estimations (e.g. Lopez et al. 1994; Fleischer and Tsur 2009) show that the MLAV decreases with L_i, and at a particular crop i's land size, denoted L_i^0, it may become zero: $\partial a_i(\mathbf{L}; e)/\partial L_i|_{L_i=L_i^0} = 0$. This means that for each crop there is a land size beyond which individuals' willingness to pay for an additional hectare assigned to this crop is zeroed. For further analyses, we let $\mathbf{L}^0 = (L_1^0, L_2^0, \ldots, L_I^0)$ be the regional land allocation among crops under which there is $\partial a_i(\mathbf{L}^0; e)/\partial L_i = 0$ for all i, $(i = 1, \ldots, I)$, and $L^0 = \sum_{i=1}^{I} L_i^0$ is the associated regional agricultural area.

One question that comes to mind is whether the MLAV can become negative, i.e. can devoting an additional hectare to a crop reduce the amenity services provided by the entire agricultural area? In other words, would individuals ask for compensation if the land allocated to low-amenity crops became too large? Since the aforementioned amenity evaluation studies do not answer this question, it is left to our interpretation of the estimated amenity-value function. In the following, we distinguish between the case of non-negative MLAV with respect to all crops, and the case in which some crops, specifically those grown in greenhouses, can exhibit negative MLAVs.

A regional rural-urban market-equilibrium land allocation emerges when the housing and agricultural rental rates are equal, such that

$$D_h(L_h/\alpha N; e) = \pi_i'\left(L_i/\alpha N_f; g\right), i = 1, \ldots, I. \tag{9.2}$$

We denote by L_h^m and $\mathbf{L}^m = (L_1^m, L_2^m, \ldots, L_I^m)$ the land allocation which constitutes a solution to the I+1 conditions set by equation (9.2), subject to the regional land constraint, $L_h + L \leq \alpha \overline{L}$, where the corresponding total agricultural area is $L^m = \sum_{i=1}^{I} L_i^m$. Note that the intra-agricultural land allocation among crops under this market equilibrium represents the farming sector's PM solution. Using the aggregate agricultural demand function, $\Pi'(L/\alpha N_f; g)$, and assuming a binding regional land constraint, we get

$$r^m = D_h\left((\alpha\overline{L} - L^m)/\alpha N; e\right) = \Pi'\left(L^m/\alpha N_f; g\right), \tag{9.3}$$

where r^m is the market-equilibrium rental rate. Given a specific land allocation $(L_h, \mathbf{L})$, the representative household's total willingness to pay for housing and agricultural landscape amenities is $\int_0^{L_h/\alpha N} D_h(l_h; e)dl_h + a(\mathbf{L}; e)$, and $\sum_{i=1}^{I} \pi_i(L_i/\alpha N_f; g)$ is the profit of the representative farmer. Summation of these across households and farmers yields the regional social-welfare function

$$W(L_h, \mathbf{L}, \alpha) = \alpha N\left[\int_0^{L_h/\alpha N} D_h(l_h; e)dl_h + a(\mathbf{L}; e)\right] + \alpha N_f \sum_{i=1}^{I} \pi_i\left(L_i/\alpha N_f; g\right). \tag{9.4}$$

The socially optimal (SO) land allocation, denoted $\left(L_h^s, \mathbf{L}^s\right)$, $\mathbf{L}^s = \left(L_1^s, L_2^s, \ldots, L_I^s\right)$, which maximizes equation (9.4) subject to the regional land constraint, $L_h + L \leq \alpha\overline{L}$, satisfies

$$D_h\left(L_h/\alpha N; e\right) = \pi_i'\left(L_i/\alpha N_f; g\right) + \alpha N \partial a\left(\mathbf{L}; e\right)/\partial L_i = \lambda, i = 1, \ldots, I, \quad (9.5)$$

where $\lambda \geq 0$ is the shadow price of the land constraint; i.e. the benefit which would be obtained from a marginal increase in the region's total land. We define $L^s = \sum_{i=1}^{I} L_i^s$ as the SO total agricultural land. The difference between the SO and market-equilibrium land allocations is attributed to the landscape amenity value. The public-good nature of the amenities is reflected by the multiplication of the per-household MLAV, $\partial a(\mathbf{L}; e)/\partial L_i$, by the number of households residing in the region, αN.

Let $D(L; g, e)$ represent the regional social demand for agricultural lands, which incorporates both farmers' agricultural profits and households' landscape amenities. This function can be derived as follows. Let $\hat{L}$ be an auxiliary regional agricultural land constraint, where $\hat{L} \in [0, \alpha\overline{L}]$. Define $L_i(\mu; g, e)$, $(i = 1, \ldots, I)$, as the cropland allocation that maximizes the contribution of the agricultural lands to the social welfare, $\alpha N a(\mathbf{L}; e) + \alpha N_f \sum_{i=1}^{I} \pi_i(L_i/\alpha N_f; g)$, subject to the land constraint $\sum_{i=1}^{I} L_i \leq \alpha\hat{L}$, where μ is the constraint's shadow value. By setting $\mu = r$ and aggregating across crops, we get $L(r; g, e) = \sum_{i=1}^{I} L_i(r; g, e)$. The function $D(L; g, e)$ is defined as the inverse of $L(r; g, e)$, which satisfies $D(L(r; g, e); g, e) = r$. Under the SO allocation, $(L_h^s, \mathbf{L}^s)$, and assuming a binding regional land constraint, $\alpha\overline{L} = L^s + L_h^s$, we get

$$r^s = D_h\left(\left(\alpha\overline{L} - L^s\right)/\alpha N; e\right) = D\left(L^s; g, e\right). \quad (9.6)$$

When the rural landscape amenities are ignored, the market-equilibrium land allocation, $(L_h^m, \mathbf{L}^m)$, appears. If $(L_h^m, \mathbf{L}^m) \neq (L_h^s, \mathbf{L}^s)$, then, the market equilibrium entails a welfare loss. We let $\Gamma(\alpha) = \dfrac{W(L_h^s, \mathbf{L}^s, \alpha) - W(L_h^m, \mathbf{L}^m, \alpha)}{\alpha\overline{L}}$ be the per-hectare welfare loss (increase) associated with the market-equilibrium (SO) land allocation in a region with an area of $\alpha\overline{L}$ hectares.

Effects of regional size and interpretation of the MLAV

As agricultural amenities appear to be a public good exhibiting a decreasing return to scale, a regional policy aimed at increasing the landscape amenities is expected to be sensitive to the region's size. This implies that the level of a nation's land partitioning into regions can affect the regional welfare loss entailed by ignoring the landscape amenities, and therefore impact the contents of the associated local policies. In some cases, it may cause the welfare loss to vanish, thereby eliminating the need for a policy altogether. Our interest is in studying the effect of regional area $\alpha\overline{L}$ on both the SO rural-urban land allocation and the intra-agricultural land allocation among crops. We commence with the first.

Rural-urban land allocation

To facilitate the analysis, assume a single crop, such that the contribution of the agricultural lands to social welfare becomes $\alpha Na(L;e) + \alpha N_f \pi \left(L/\alpha N_f; g\right)$. Let $0 \leq \beta \leq 1$ be an auxiliary parameter representing the weight assigned to the landscape amenity value in the social-welfare function:

$$W(L_h, L, \alpha, \beta) = \alpha N \int_0^{L_h/\alpha N} D_h(l_h; e)\, dl_h + \beta\alpha Na(L; e) + \alpha N_f \pi \left(L/\alpha N_f; g\right), \tag{9.7}$$

such that by maximizing equation (9.7) under $\beta = 0$ and $\beta = 1$, we get the market equilibrium and the socially optimal land allocations, respectively. By maximizing (9.7) subject to the land constraint, we get the SO agricultural land, $L^{s\beta}$, and the SO urban land, $L_h^{s\beta} = \alpha\overline{L} - L^{s\beta}$, which satisfy the condition:

$$D_h\left(\left(\alpha\overline{L} - L^{s\beta}\right)/\alpha N; e\right) - \pi'\left(L^{s\beta}/\alpha N_f; g\right) - \beta\alpha Na'\left(L^{s\beta}; e\right) = 0. \tag{9.8}$$

Given β, the maximum per-hectare regional social welfare is:

$$\frac{W\left(L^{s\beta}, \alpha, \beta\right)}{\alpha\overline{L}} = \left[N \int_0^{\left(\alpha\overline{L} - L^{s\beta}\right)/\alpha N} D_h(l_h; e)\, dl_h + N\beta a\left(L^{s\beta}; e\right) + N_f \pi\left(L^{s\beta}/\alpha N_f; g\right)\right] \Big/ \overline{L}. \tag{9.9}$$

Now suppose that while adjusting $L^{s\beta}$ such that the optimality condition in equation (9.8) is preserved, we increase the regional area by increasing α, and at the same time change the landscape amenity-value weight β, such that the value of the optimal per-hectare social welfare, as in (9.9), remains constant. Note that due to the assumed uniform and constant spatial distribution of population and environmental conditions, these factors are not affected by the increase in α. Forming the complete differential, and employing the envelope theorem – a theorem that is used to solve optimization problems in microeconomics – we get

$$\frac{d\beta}{d\alpha} = \frac{\left[D_h\left(\left(\alpha\overline{L} - L^{s\beta}\right)/\alpha N; e\right) - \pi'\left(L^{s\beta}/\alpha N_f; g\right)\right] L^{s\beta}}{-\alpha^2 Na\left(L^{s\beta}; e\right)}. \tag{9.10}$$

Assume $a(L^{s\beta}; e) > 0$. In view of condition (9.8), for the case of $a'(L^{s\beta}; e) > 0$ the numerator is negative, so that $\frac{d\beta}{d\alpha} < 0$; i.e. the larger the region (α), the smaller the weight that should be put on the landscape amenities (β) in order to keep the regional optimal per-hectare welfare unchanged. This indicates that, in larger regions, the share of the landscape amenity value in the per-hectare welfare is larger, and hence, the per-hectare welfare loss associated with ignoring the landscape amenity value is higher.

Note that for large enough α, the market-equilibrium land allocation, L^m, may become larger than L^0. In the case of non-negative MLAVs, this implies $a'(L^m; e)|_{L^m \geq L^0}. = 0$, and utilizing equation (9.8) we get $L^m = L^{s\beta}$ for every $L^m \geq L^0$. It then follows from equations (9.3) and (9.10) that $\frac{d\beta}{d\alpha} = 0$ for every $L^m \geq L^0$. On the other hand, if the single crop under consideration is allowed to exhibit a negative MLAV, $a'(L^{s\beta}; e) < 0$, then, given $a(L^{s\beta}; e) > 0$, conditions (9.8) and (9.10) imply $\frac{d\beta}{d\alpha} > 0$ for the case of $L^m \geq L^0$; i.e. in this case the welfare loss associated with ignoring the landscape amenity value is lower as regions become larger.

Intra-agricultural land allocation

We now hold the rural-urban land allocation constant, and focus on the impact of regional size on the intra-agricultural allocation of land among crops. Assume that the land devoted to agriculture is set exogenously as equal to $L = \gamma\alpha\bar{L}$, where $0 \leq \gamma \leq 1$. Thus, changes in land allocation among crops affect only the social welfare derived from the agricultural lands. For simplicity, we limit the number of crops to two ($I = 2$). Following the same analytical approach used above, we utilize the auxiliary parameter β in formulating a social-welfare function of the form:

$$W(L_1, L_2, \alpha, \beta) = \beta\alpha N a(L_1, L_2; e) + \alpha N_f \left[\pi_1\left(\frac{L_1}{\alpha N_f}; g\right) + \pi_2\left(\frac{L_2}{\alpha N_f}; g\right)\right]. \tag{9.11}$$

By maximizing (9.11) subject to the land constraint, we get the set $\mathbf{L}^{s\beta} = (L_1^{s\beta}, L_2^{s\beta})$ which satisfies the SO condition:

$$\pi_1'\left(L_1^{s\beta}; g\right) - \pi_2'\left(L_2^{s\beta}; g\right) = \beta\alpha N\left[\frac{\partial a\left(L_1^{s\beta}, L_2^{s\beta}; e\right)}{\partial L_2} - \frac{\partial a\left(L_1^{s\beta}, L_2^{s\beta}; e\right)}{\partial L_1}\right]. \tag{9.12}$$

Expressing the per-hectare welfare function, $\frac{W}{\alpha\gamma\bar{L}}$, and calculating the $\frac{d\beta}{d\alpha}$ relation that keeps the per-hectare optimal social welfare unchanged, we get

$$\frac{d\beta}{d\alpha} = \frac{\frac{L_1}{\alpha^2}\left[\pi_1{}' - \pi_2{}'\right] - \gamma\bar{L}\beta N \frac{\partial a(\cdot)}{\partial L_2}}{Na(\cdot)}. \tag{9.13}$$

In view of (9.12), it is concluded that the sign of $\frac{d\beta}{d\alpha}$ is indeterminate; hence, the impact of a region's size on the welfare loss is vague. The empirical analysis presented in the next section illustrates the sensitivity of the regional-size effect to the assumption regarding the limitations put on the sign of the MLAV. Nevertheless, in the following theoretical analysis we show that, under the assumption of non-negative MLAV with respect to all crops, for regions with agricultural land large enough such that $L > L^0$, the likelihood of the landscape amenities having no impact on the SO agricultural land allocation increases with regional size.

Consider again the agricultural land being determined by $L = \gamma\alpha\bar{L}$. Given γ, let $L_i^m(\alpha, g)$ be the regional land allocated to crop i under the market equilibrium. Define $\mathbf{L}^a(\alpha) = \left(L_1^a(\alpha), \ldots, L_I^a(\alpha)\right)$ as the land allocation that maximizes the rural landscape amenity value for the specific regional size $\alpha\bar{L}$. Let $g^{am}(\alpha)$ be the exogenous environmental conditions under which there is $L_i^m(\alpha, g^{am}(\alpha)) = L_i^a(\alpha)$ for all i. For simplicity, assume $g^{am}(\alpha) \equiv g^{am}$; i.e. the set of conditions g^{am} imposing the equality $L_i^m(\alpha, g^{am}) = L_i^a(\alpha)$ does not vary with α. Recall our assumption that g is homogeneous throughout the economy, which implies that altering a region's area does not change the environmental conditions there. Now, if the economy under consideration happens to be characterized by g^{am}, then the regional intra-agricultural land allocation under the market equilibrium is the same one that maximizes the landscape amenity value, and hence this land allocation also constitutes the SO one, $\mathbf{L}^s$. Note that in this case, both sides of equation (9.12) are zeroed.

Let us again restrict the number of crops to two, such that $\mathbf{L}^0 = \left(L_1^0, L_2^0\right)$ and $L^0 = L_1^0 + L_2^0$. Suppose first that the regional size $\alpha\bar{L}$ is small enough such that $\gamma\alpha\bar{L} < L^0$. In this situation, there is $L_i^a(\alpha) < L_i^0$, $i = (1, 2)$. Assume $g \neq g^{am}$, and consider the case in which $L_1^m(\alpha, g) < L_1^a(\alpha)$. This implies that increasing the land devoted to Crop 1 would increase the landscape amenity value, albeit at the expense of some reduction in the profitability associated with this crop. Given that $L_1^a(\alpha) = \gamma\alpha\bar{L} - L_2^a(\alpha)$ and $L_1^m(\alpha, g) = \gamma\alpha\bar{L} - L_2^m(\alpha, g)$, there must be $L_2^m(\alpha, g) > L_2^a(\alpha)$; i.e. decreasing the land devoted to Crop 2 would increase the landscape amenity value, while entailing some loss of profit on this crop.

Suppose now that the regional size is large enough such that $\gamma\alpha\bar{L} > L^0$. In this case there is $L_i^a(\alpha) \geq L_i^0$ for all $i = (1, 2)$. This implies that, given g, for each crop i there is a range of land sizes, $\left[L_i^0, \gamma\alpha\bar{L} - L_j^0\right]$, $i = (1, 2)$, $i \neq j$, such that if for crop i there is $L_i^0 \leq L_i^m(\alpha, g) \leq \gamma\alpha\bar{L} - L_j^0$, then, for the other crop, j, there is $L_j^0 \leq L_j^m(\alpha, g)$. As long as $L_i^0 \leq L_i^m(\alpha, g) \leq \gamma\alpha\bar{L} - L_j^0$, the market-equilibrium land allocation, $\mathbf{L}^m$, is the same one that maximizes the landscape amenity value, $\mathbf{L}^a$, and therefore $\mathbf{L}^m = \mathbf{L}^s$. In this case, there is a support of exogenous conditions, $G^{am} = \left[\underline{g}^{am}, \bar{g}^{am}\right]$, under which there is $\mathbf{L}^m = \mathbf{L}^s$. Note, now, that the larger the regional size, α, the larger the support G^{am}, and hence, the larger the probability that the economy is characterized by a set of conditions, g, with $g \in G^{am}$.

Empirical analysis

The empirical analysis is applied to the state of Israel, based on regional-scale aggregated data on land allocation among 45 crops (i.e. $I = 45$) as obtained from the Israeli Central Bureau of Statistics (ICBS 2002). A stern farmland-protection policy in Israel prevents the emergence of rural-urban land market equilibrium (Alterman 1997; Feitelson 1999), and developments of rural areas are subject to authorization by official institutes. Thus, intra-agricultural policies are not expected to significantly affect the allocation of land between urban and rural designations. Therefore our analysis focuses on the intra-agricultural land allocation – and thus not on the rural-urban land allocation – while the total regional agricultural area is assumed constant.

Recalling the sensitivity of the SO solution to the regional size, the decision about the resolution according to which the country is separated into regions is expected to affect the results. We therefore compare two types of partitions, known as Natural Zones and Ecological Regions; both are frequently used by governmental institutes for data analyses and design. Figure 9.1 presents these two geographic allocations. Our analysis focuses on the heavily populated northern part of the country, which lies above the bold lines in Figure 9.1. Overall, we analyze a partition of this area into 43 natural zones in comparison to 16 ecological regions. The average total area of a natural zone and an ecological region is 20,270 and 54,300 hectares, respectively; the agricultural and open-space areas, which are relevant for the landscape amenity values, amount to 24,200 and 9,030 hectares, respectively. Under both allocations we study the implications of the assumption regarding the negativity/non-negativity of the MLAV in terms of welfare and agricultural land policy.

The programming model

The analysis is based on the positive-mathematical-programming (PMP) approach (Howitt 1995). First, the model is calibrated such that it reproduces the land allocation observed in each region – this allocation is considered a result of PM farming behavior. Then, in a second stage, the objective function is reformulated such that it incorporates both the farmers' profits and the landscape amenity value for the local residents; i.e. it represents the regional social welfare. Then, the model computes the SO agricultural land allocation and calculates the welfare increase relative to the PM solution.

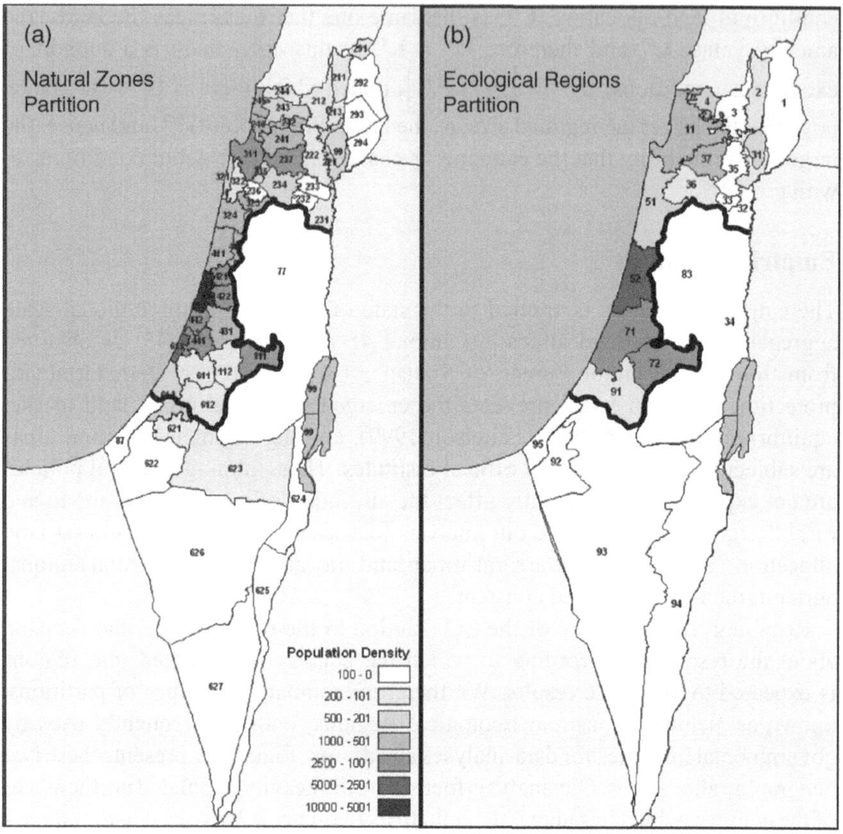

Figure 9.1 Division of Israel according to (a) natural zones and (b) ecological regions.

Currently, farmers are not rewarded for the landscape amenities they create. Hence, in the calibration stage it is assumed that in each region k ($k = 1, \ldots, 16$ and $k = 1, \ldots, 43$ under the ecological-region and natural-zone partitions, respectively), the observed land allocation is an outcome of the PM behavior, denoted $\mathbf{L}_k^m = (L_{1k}^m, \ldots, L_{Ik}^m)$, which solves the problem:

$$\max_{L_{1k},\ldots,L_{Ik}} \Pi_k = \sum_{i=1}^{I} L_{ik}\left[p_i y_i - \left(\gamma_{ik} + \frac{1}{2}\delta_{ik} L_{ik}\right)\right] \; s.t. \; \sum_{i=1}^{I} L_{ik} \le L_k, \qquad (9.14)$$

where L_{ik} (ha) is the land allocated to crop i in region k, Π_k (\$ yr^{-1}) is the annual profit of the vegetative agricultural sector in that region, p_i (\$ ton^{-1}) is crop i's statewide output price, y_i (ton ha^{-1} yr^{-1}) denotes the (statewide average) per-hectare annual yield, and L_k (ha) is the total agricultural land constraint of the region. The element $\gamma_{ik} + \frac{1}{2}\delta_{ik}L_{ik}$ (\$ ha^{-1} yr^{-1}) represents the per-hectare cost associated with the production of crop i, which is expressed as a linear function of land allotted to the crop, L_{ik}. This dependency is used to indirectly reflect the impact of a range of unobserved factors considered by farmers while contemplating their land allocation among crops, including spatial variability of soil quality, agronomic and marketing risks, input constraints, know-how limitations, traditional habits, et cetera. As a result, the total cost is a quadratic function of L_{ik}, which allows the optimal land allocation calculated by this PMP model to be smoothly altered in response to exogenous shocks.

In the first PMP stage, the parameters γ_{ik} and δ_{ik} are calibrated by the procedure developed by Howitt (1995), as is briefly described: Let crop 1 ($i = 1$) be the one exhibiting the lowest observed average per-hectare profit. Thus, the dual value of the regional total land constraint, λ_k^1, is given by $\lambda_k^1 = p_1(y_1 - \Delta y_1) - c_{1k}$, where Δy_1 is crop 1's observed maximum yield reduction below the average, y_1, and c_{1k} (\$ ha^{-1} yr^{-1}) is the observed average per-hectare production cost. Then λ_{ik}^2, the crop i's dual value of a land-calibration constraint $L_{ik} \leq L_{ik}^m$, is given by $\lambda_{ik}^2 = p_i y_i - c_{ik} - \lambda_k^1$. Using the observed ($\mathbf{L}_k^m$assumed) land allocation, and using the equality $c_{ik} = \gamma_{ik} + \frac{1}{2}\delta_{ik}L_{ik}^m$, we get $\delta_{ik} = 2\lambda_{ik}^2 / L_{ik}^m$ and $\gamma_{ik} = c_{ik} - \lambda_{ik}^2$ for every crop i, $i = 1, \ldots, I$.

Information on prices, yields and production costs was obtained from various reports published by the Israeli Ministry of Agriculture and Rural Development (2002) for the 45 crops under consideration. Production costs were calculated by Rapaport-Rom (2006), reflecting variation among regions due to differences in precipitation and surface-water constraints. All monetary terms are in 2002 dollars.

Having the calibrated model based on the observed PM situation, the regional amenity value, $N_k a(\mathbf{L}_k; e_k)$, is introduced into the objective function to form the social-welfare function. The problem expressed in equation (9.14) becomes:

$$\max_{L_{1k},\ldots,L_{Ik}} W_k = \sum_{i=1}^{I} L_{ik}\left[p_i y_{ik} - \left(\gamma_{ik} + \frac{1}{2}\delta_{ik}L_{ik}\right)\right] + N_k a(\mathbf{L}_k; e_k) \ s.t. \sum_{i=1}^{I} L_{ik} \leq L_k, \tag{9.15}$$

where W_k denotes the regional social-welfare value.

Following Kan et al. (2009), we make use of the per-household landscape amenity-value function estimated by Fleischer and Tsur (2009) for the northern part of Israel – our relevant studied area. Fleischer and Tsur (2009) found that individuals can distinguish among the landscape amenities generated by three groups of land-uses: (1) orchards and citrus, (2) vegetables, field crops and preserved open spaces, and (3) greenhouses. We index each group, respectively, by n, $n = 1, 2, 3$. The land devoted to a crop assigned to group n is denoted L_{ik}^n.

The total regional land devoted to group n's crops is given by $L_k^n = \sum_{i=1}^{I^n} L_{ik}^n$, where I^n denotes the number of crops in that group. Note that, while the model allocates land among 45 crops, the landscape amenity value is calculated based on the total land devoted to the three amenity-related groups of crops.

Using the double-bounded dichotomous-choice elicitation method (for an explanation of this model, see Mitchell and Carson 1989), Fleischer and Tsur (2009) estimated the quadratic function:

$$a(\mathbf{L}_k; e_k, A_k) = \sum_{n=1}^{3} \left(\phi_k^n L_k^n + {}^1\!/\!_2 \varphi^n \left(L_k^n\right)^2 \right) + \eta^{12} L_k^1 L_k^2 + \eta^{13} L_k^1 L_k^3 + \eta^{23} L_k^2 L_k^3, \tag{9.16}$$

where ϕ_k^n incorporates the impacts of the income e_k, and age A_k, of the region's representative household, calculated by the relation $\phi_k^n = \phi^n + \phi_e^n e_k + \phi_A^n A_k$. In addition to the crops' land allocations, the data required for calculating the landscape amenity value for each region include the preserved open-space areas, which are assumed constant in the analysis and were obtained from Frenkel (2001), and the average per-head-of-household age and income, as taken from the ICBS (2002).

Table 9.1 presents the crops included in each group and the estimated parameters of the landscape amenity-value function.

Table 9.1 Landscape-amenity-influential groups of crops and the estimated parameters of the landscape-amenity-value function

Crop group	*Crops*	*Parameters*
Orchards and citrus	Orange, grapefruit, lemon, apple, pear, peach, plum, table grape, wine grape, banana, olive non-irrigated, olive irrigated, almond, avocado, palm	$\phi^1 = 2.2 \times 10^{-2}$· 1 $\phi_e^1 = 7.5 \times 10^{-7}$ $\phi_A^1 = -2.1 \times 10^{-4}$·1 $\varphi^1 = -9.5 \times 10^{-6}$ $\eta^{12} = -3.9 \times 10^{-7}$ $\eta^{13} = 1.3 \times 10^{-5}$
Vegetables, field crops and open spaces	Potato, tomato open-field, eggplant, vegetable marrow, onion, carrot, lettuce, cabbage, cauliflower, celery, radish, artichoke, garlic, bean, wheat, barley, cotton, chickpea, corn, pea, groundnut, sunflower, winter forage, summer forage	$\phi^2 = 4.9 \times 10^{-3}$ $\phi_e^2 = 1.2 \times 10^{-7}$ $\phi_A^2 = -8.0 \times 10^{-5}$. $\varphi^2 = -1.5 \times 10^{-7}$ $\eta^{23} = 2.5 \times 10^{-6}$
Greenhouses	Watermelon, sugar-melon, tomato greenhouse, cucumber, pepper, strawberry	$\phi^3 = 7.9 \times 10^{-2}$·r $\phi_e^3 = -1.7 \times 10^{-6}$ $\phi_A^3 = -1.0 \times 10^{-5}$ $\varphi^3 = -2.2 \times 10^{-4}$

The parameter φ^n, $n = (1, 2, 3)$, represents the own-quadratic effect, and is found to be negative for all n. The cross-effect parameters, η^{12}, η^{13} and η^{23}, can be of either sign, where a negative (positive) parameter represents substitution (complementary) relationships between the corresponding crop groups. Due to the negativity of some of these own- and cross-effect parameters, the MLAV with respect to L_k^n, $n = (1, 2, 3)$, may diminish with L_k^n, and even become negative when L_k^n exceeds L_k^{n0}, where L_k^{n0} represents the area of group n under which the MLAV is zeroed. We let L_k^{nf} be the land area of group n which is considered effective with respect to the landscape amenity value. In the case of non-negative MLAV for all crops, L_k^{nf} is calculated such that

$$L_k^{nf} = \begin{cases} L_k^n \text{ } if \text{ } L_k^n < L_k^{n0} \\ L_k^{n0} \text{ } else \end{cases}, n = 1, 2, 3. \tag{9.17}$$

In the case in which some groups of crops are allowed to exhibit negative MLAV, we let $L_k^{nf} = L_k^n$ for all $L_k^n \geq 0$ only for those groups.

The landscape amenity value expressed in equation (9.16) is calculated based on the vector of effective lands $\mathbf{L}_k^f = (L_k^{1f}, L_k^{2f}, L_k^{3f})$. Note that the computation of $\mathbf{L}_k^f$ is non-trivial, since the cross-effects in the landscape amenity-value function make the levels of L_k^{n0}, for each group of crops n, $n = 1, 2, 3$, dependent on $\mathbf{L}_k^f$; an iterative calculation procedure is used to overcome this complexity.

For later discussions, consider the MLAV curves presented in Figure 9.2. These curves were generated as follows: first, the land allocation among the three groups of crops that maximizes the landscape amenity-value function in equation (9.16) was calculated; we denote this land allocation $\mathbf{L}^a = \left(L^{1a}, L^{2a}, L^{3a}\right)$ (which is equivalent to $\mathbf{L}^a(\alpha)$ in the sub-section on "Effects of regional size and interpretation of the MLAV"). The $\mathbf{L}^a$ allocation was computed subject to increasing levels of regional land constraints. Then, the associated MLAV for each group of crops was calculated, and plotted against the group's corresponding land. The resultant curves can be viewed as the crop-groups' landscape amenity demand curves under the maximal amenity levels. Note the relatively low elasticity of the landscape amenity demands for the groups "orchards and citrus" and "greenhouses", in comparison to the elastic demand for the group "vegetables, field crops and open spaces". Horizontal summation of these maximum-amenity demand curves yields the demand curve of the entire region. The landscape amenity-value function exhibits a decreasing return to scale, such that at a regional (agricultural and preserved open space) land size of 22,500 ha, the MLAVs of all groups of crops are zeroed; this is the regional size denoted L^0, $L^0 = \sum_{n=1}^{3} L^{n0}$, where the corresponding land-allocation vector is $\mathbf{L}^0 = \left(L^{1,0}, L^{2,0}, L^{3,0}\right) =$ (1,754, 20,070, 676). That is, under this land allocation, the region's residents would not be willing to pay for the landscape amenities derived by the expansion of the land allocated to any of these groups of crops.

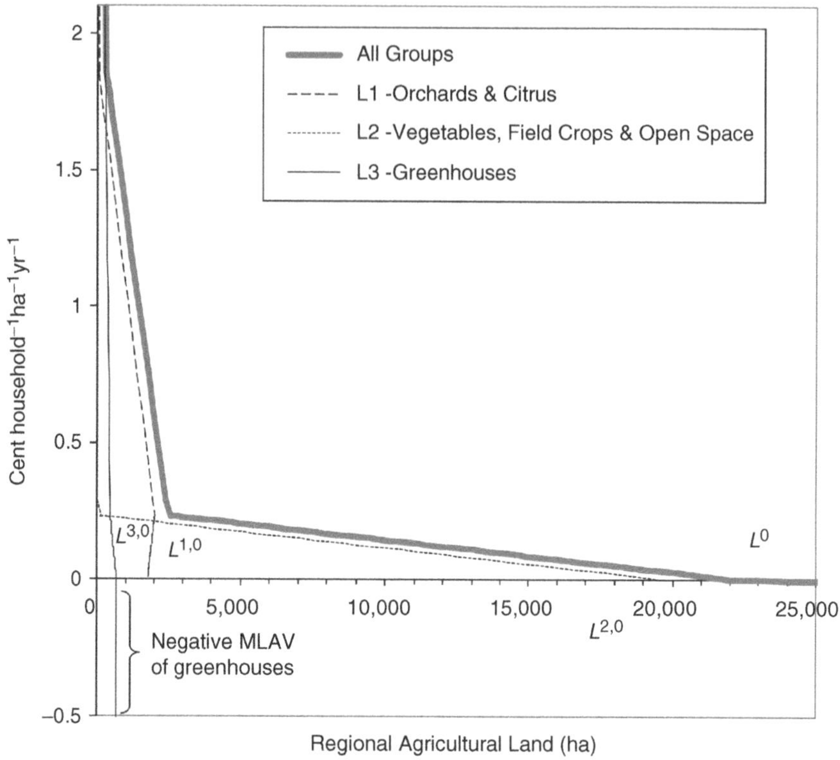

Figure 9.2 Marginal-landscape-amenity-value (MLAV) curves under the maximum-landscape-amenity land allocation.

Similar to the amenity-effective land allocation, $\mathbf{L}_k^f$, let us denote by $\mathbf{L}^{af}$, $\mathbf{L}^{af} = \left(L^{1af}, L^{2af}, L^{3af}\right)$, the maximum-landscape-amenity-effective land allocation. In regions with agricultural areas smaller than 22,500 ha $\mathbf{L}^{af} = \mathbf{L}^a$, whereas in regions with an agricultural area larger than 22,500 ha, if the MLAV is assumed to be non-negative with respect to all crops, $\mathbf{L}^{af} = \mathbf{L}^0$. In the case in which the MLAV of greenhouses is allowed to be negative, as depicted in Figure 9.2 for this group, there is $\mathbf{L}^{af} = \mathbf{L}^a \neq \mathbf{L}^0$.

The programming model was built on an Excel worksheet and run by a Premium Solver Platform V6.5 instrument. To overcome the non-convexity that results from the structure of the amenity-value function, the program seeks the global optimum (the model and the entire dataset are available from the authors upon request).

Analysis

Consider a policy according to which the farming sector in each region is compensated for the entire profit loss associated with switching from the PM land allocation to the SO one. Table 9.2 presents the nationwide values of the welfare elements and land allocations of the PM and SO solutions under the two regionalization partitions, and under the non-negative MLAV and negative-greenhouse MLAV assumptions. For each of the four combinations, the PM solution represents the current situation, in which no policy is implemented. Farmers allocate their land among crops in order to maximize their profits, which amount to \$456.5 million per year. The associated landscape amenity value under the non-negative MLAV is \$212.4 and \$96.1 million per year for the 43 natural zones and 16 ecological regions, respectively. The lower landscape amenity value under the lower resolution ecological-region partition is attributed to the decreasing-return-to-scale nature of the landscape amenity-value function. Recalling the assumption that communities consider the landscape amenity provided only by the local lands and only to the local residents, then, for instance, if a large region is divided into a few identical sub-regions, the MLAV of each sub-region is larger than that of the original full-size region. Therefore, the sum of the landscape amenity values of the sub-regions must exceed the amenity value when they are merged.

Consider now the landscape amenity value under the PM solution, where a negative-greenhouse MLAV is assumed. Under both partitions, the landscape amenity value is lower relative to the non-negative MLAV case, and in the ecological-region partition it even becomes negative, implying that agricultural landscapes actually reduce social welfare. This outcome is a result of the inelastic landscape amenity demand for greenhouse crops (Figure 9.2); i.e. if the area devoted to greenhouses exceeds the $L^{3,0}$ area (676 ha), then each hectare above the $L^{3,0}$ amount significantly reduces the amenity value. Obviously, the larger the region, the larger the probability that under the PM land allocation there is $L^{3m} > L^{3,0}$, and therefore the national landscape amenity value is lower under the ecological-region partition.

Moving from the PM to the SO solution, Table 9.2 indicates that the resultant profit reductions under the non-negative MLAV assumption amount to \$2.5 and \$0.5 million per year under the natural-zone and ecological-region partitions, respectively. The associated benefits in terms of landscape amenity-value increases are \$18.6 and \$0.8 million per year, respectively, where the changes in land allocation under the ecological-region partition are also smaller. This implies that the larger the regions, the lower the efficiency of a policy to internalize the agricultural landscape benefits. However, this conclusion is reversed in the case of a MLAV that is allowed to become negative with respect to greenhouse crops, where the welfare increase under the ecological-region partition is much larger than under the natural-zone partition: \$215.9 versus \$50.1 million per year. Also here, the landscape amenity demand for greenhouse crops plays a crucial role, as evidenced by the direction of the change in the total land allocated to greenhouse crops. In the case of non-negative MLAV, the land devoted to greenhouse crops is increased, whereas in the case of negative greenhouse MLAV, it is reduced. This effect is further elaborated in Figure 9.3.

Graphs (a)–(f) in Figure 9.3 present the PM ($\mathbf{L}^{nm}$) and SO ($\mathbf{L}^{ns}$) land allocations in the 43 natural zones and 16 ecological regions, plotted against their

Table 9.2 Nationwide values of the welfare elements and land allocations of the PM and SO solutions under the natural-zones and ecological-regions partitions and under the non-negative MLAV and negative-greenhouse MLAV assumptions

	Natural zones			*Ecological regions*		
	Profit maximization (PM)	*Socially optimum (SO)*	*Difference (SO) – (PM)*	*Profit maximization (PM)*	*Socially optimum (SO)*	*Difference (SO) – (PM)*
Welfare elements (10^6 $/yr)						
Non-negative MLAV						
• Farming profits	456.5	454.0	−2.5	456.5	456.0	−0.5
• Amenity value	212.4	231.0	18.6	96.1	96.9	0.8
• Social welfare	668.9	685.1	16.2	552.6	552.8	0.2
Negative greenhouse MLAV						
• Farming profits	456.5	450.9	−5.6	456.5	447.7	−8.8
• Amenity value	174.1	229.8	55.7	−128.4	96.4	224.7
• Social welfare	630.6	680.7	50.1	328.1	544.0	215.9
Land allocation (ha)						
Non-negative MLAV						
• Orchards and citrus	61,179	61,659	480	60,991	60,446	−545
• Vegetables, field crops and open spaces	310,794	309,002	−1,791	310,122	310,320	198
• Greenhouses	16,189	17,579	1,390	16,096	16,443	347
Negative greenhouse MLAV	61,179	62,641	1,462	60,991	62,218	1,227
• Orchards and citrus						
• Vegetables, field crops and open spaces	310,794	312,157	1,363	310,122	316,854	6,732
• Greenhouses	16,189	13,443	−2,747	16,096	8,137	−7,959

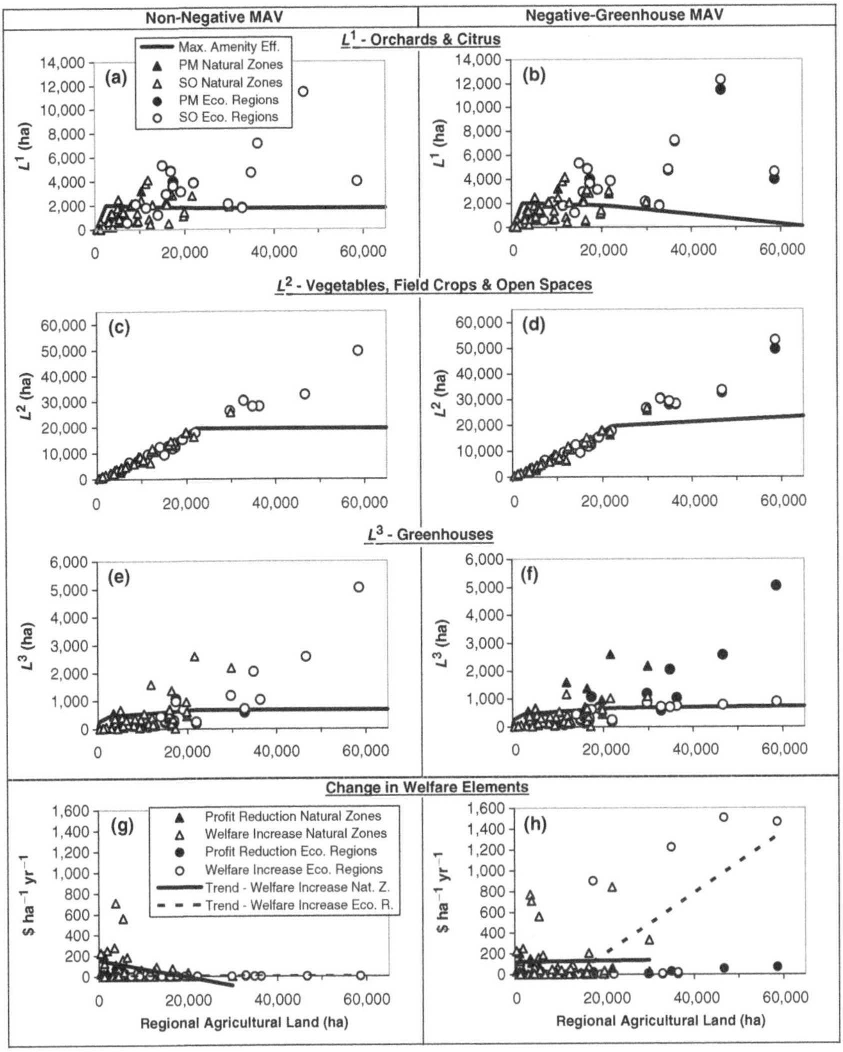

Figure 9.3 (a)–(f) Variation of the maximum-landscape-amenity-effective (L^{naf}), PM (L^{nm}) and SO (L^{ns}) land allocations of the three groups of crops ($n = 1,2,3$) in the natural zones and ecological regions, plotted against the regional total agricultural land, and (g) and (h), the associated profit reduction and welfare increase under non-negative MLAV and negative-greenhouse MLAV.

total regional agricultural areas, under the non-negative MLAV and negative-greenhouse MLAV assumptions. The switch from PM to SO land allocation can be motivated by the aforementioned compensation policy. The solid curves are the maximum-landscape-amenity-effective land allocations, $\mathbf{L}^{naf}$, corresponding to the MLAV curves in Figure 9.2. Note that in regions with total agricultural land of up to 2,400 ha, $L^{2af} = 0$ (Figure 9.3c, d); i.e. the maximum landscape amenity is achieved when no land is allotted to Group 2 (vegetables, field crops and preserved open space). Figure 9.2 indicates that within that range of regional agricultural sizes, Group 2's associated MLAV is lower than those of the other two groups. Yet, when the total regional agricultural land grows beyond 2,400 ha, the maximum-landscape-amenity allocation proceeds such that every additional hectare is devoted to Group 2.

Consider the case of non-negative MLAV. As expected, when the land allocation is changed from PM to SO, the directions of the changes are toward the maximum-landscape-amenity-effective ($\mathbf{L}^{naf}$) curves. Note, however, that the changes are minor. Focusing on graph (e), we note that most of the greenhouse areas in both the natural zones and ecological regions fall below the maximum-landscape-amenity-effective curve. Therefore, switching from the PM to SO allocation involves an increase in the lands allocated to greenhouse crops in these regions. On the other hand, in regions where the greenhouse-crops' PM land allocation is above the maximum-landscape-amenity-effective curve, the MLAVs with respect to greenhouse crops are zeroed; therefore, changes in the land allocated to Group 3 (greenhouses) are driven by the situation of the other two groups of crops with respect to the PM allocation versus the maximum-landscape-amenity-effective allocations. Also note that, in line with the trend predicted by the theoretical analysis in the sub-section on "Effects of regional size and interpretation of the MLAV", in almost every region with a total regional agricultural size larger than the L^0 size (22,500 ha), the land allocated to each of the three crop groups under the PM solution is above the corresponding group's maximum-landscape-amenity-effective curve, and therefore the MLAVs with respect to all groups are zeroed. Hence, in these regions the PM solution also constitutes the SO one. This explains why, at the national level, there is an increase in Group 3's land, as indicated by Table 9.2.

The negative-greenhouse MLAV scenario presents a different picture: whenever the area devoted to the group of greenhouse crops under the PM allocation exceeds the maximum-landscape-amenity-effective area (Figure 9.3f), moving to the SO allocation entails a cut in Group 3's land; in some regions the reduction in Group3's area is so sharp that the lands allocated to the other crop groups are increased above and beyond the maximum-landscape-amenity-effective curve. This explains the nationwide reduction in Group 3's land in the case of negative-greenhouse MLAV (Table 9.2). Moreover, since Group 3's maximum-landscape-amenity-effective curve is almost unchanged with regional size, the larger the regions, the larger the probability for such an occurrence. Indeed, 37 per cent of the ecological regions exhibit a greenhouse-crop-PM land which exceeds the maximum-landscape-amenity-effective area, whereas only 14 per cent of the natural regions show this incident. Therefore, under the ecological-region partition the nationwide reduction in Group 3's area amounts to 50 per cent of the PM allocation, versus only 14 per cent under the natural-zone partition.

The welfare implications of the two geographical partitions can be further studied by using Graphs (g) and (h) in Figure 9.3, where the per-hectare annual profit reduction and the increase in social welfare, caused by the shift from PM to SO land allocation, are plotted against the total regional areas. Linear trend lines were fitted by OLS regressions (Ordinary Least Squares) to the welfare-increase outputs associated with the natural-zone and ecological-region partitions. The natural-zone trend lines indicate that under the non-negative MLAV assumption, the welfare-increase declines with regional size, whereas it is not influenced by regional size in the negative-greenhouse MLAV case. On the other hand, the ecological-region partition trend lines are horizontal and increase under the non-negative and negative-greenhouse MLAV assumptions, respectively. Thus, the regional size's impact depends on both the resolution of the geographic partition and our interpretation of the landscape amenity value function with respect to greenhouse crops. Policies aiming at internalizing the external benefits ensued by rural landscapes can therefore yield welfare changes that are sensitive to scale effects and to the methodology used for evaluating the economic value of landscape.

Concluding remarks

In this chapter we investigated factors affecting a regional scale agricultural land-use crop-discriminating policy, which is aimed at increasing social welfare by accounting for rural landscape benefits. Our empirical analysis for the specific case of Israel indicates considerable sensitivity to the regional-partitioning resolution and to the features of the landscape-amenity-value function: based on Table 9.2, the increase in social welfare varies between 0.05 and 68 per cent.

Solving the geographical-resolution dilemma may be associated with a wide range of considerations, such as the role played by existing regional authorities, policy-implementation costs, the landscape contribution to tourism and to the welfare of residents of neighboring regions, and the extent to which local residents are exposed to agricultural landscape. The latter factor depends on the joint spatial distribution of a region's population and agricultural areas, which may vary in time due to population growth and development. A possible indication of the landscape exposure may be provided by the commuting-distance measurement. Crane (2007) estimated, for the entire United States in 2005, an average one-way commuting distance of 18.8 and 22.6 km for women and men, respectively. Compared to the year 1985, these distances have increased by 30 and 22 per cent, respectively. For the case of Israel, Koslowsky et al. (1996) estimated an average commuting distance of 15.0 km in the Tel-Aviv area, implying that the commuters are spread over an area of about 35,000 hectares, i.e. a region whose size lies between the average sizes of the natural zones and ecological regions considered in our analysis. Yet, given the steady increase in commuting distances in Israel (Blumen and Kellerman 1990), in the long run, the ecological-region partition may be a more representative resolution for the case of agricultural landscape policies.

If so, then our results indicate the importance of the question of whether non-traditional agricultural-landscape crops can generate negative values of landscape amenities. On the other hand, the ongoing extensive development of rural areas in Israel (Fleischer and Tsur 2003) shrinks the size of the rural areas viewed by commuters along the way, and therefore the natural-zones resolution may be favored as a basis for policy design. Hence, in order to be useful for landscape-preservation policies, the methodologies applied for evaluating landscape values should accurately represent the exposure of the population to the landscape under consideration.

References

Alterman, R. (1997) 'The challenge of farmland preservation: lessons from a six-nation comparison', *Journal of the American Planning Association*, 63(2): 220–243.

Beasley, S., Workman, W. and Williams, N. (1986) 'Amenity values of urban fringe farmland: a contingent valuation approach', *Growth and Change*, 17(4): 70–78.

Bergstrom, J., Dillman, B. and Stoll, J. (1985) 'Public environmental amenity benefits of private land: the case of prime agricultural land', *Southern Journal of Agricultural Economics*, 17(1): 139–149.

Blumen, O. and Kellerman, A. (1990) 'Gender difference in commuting distance, residence and employment: Metropolitan Haifa 1972 and 1983', *Professional Geographer*, 42(1): 54–71.

Bowker, J. and Didychuk, D. (1994) 'Estimation of nonmarket benefits of agricultural land retention in eastern Canada', *Agricultural Resource Economics Review*, 23(2): 218–225.

Brunstad, R.J., Gaasland, I. and Vårdal, E. (1999) 'Agricultural production and the optimal level of landscape preservation', *Land Economics*, 75(4): 538–546.

Crane, R. (2007) 'Is there a quiet revolution in women's travel? Revisiting the gender gap in commuting', *Journal of the American Planning Association*, 73(3): 298–316.

Drake, L. (1992) 'The non-market value of the Swedish agricultural landscape', *European Review of Agricultural Economics*, 19(3): 351–364.

EC (2003) *Council Regulation No. 1782/2003*. http://eur-lex.europa.eu/LexUriServ/site/en/consleg/2003/R/02003R1782-20060101-en.pdf (accessed 1 May 2012).

Feitelson, E. (1999) 'Social norms, rationales and policies: reframing farmland protection in Israel', *Journal of Rural Studies*, 15(4): 431–446.

Fleischer, A. and Tsur, Y. (2003) 'Measuring the recreational value of open space', *Journal of Agricultural Economics*, 54(2): 269–283.

Fleischer, A. and Tsur, Y. (2009) 'The amenity value of agricultural landscape and rural-urban land allocation', *Journal of Agricultural Economics*, 60(1): 132–153.

Frenkel, A. (2001) *Open versus Development: The Loss of Agricultural Land in Israel in the 2000's*. Haifa: The Technion.

Hackl, F. and Pruckner, G.J. (1997) 'Towards more efficient compensation programs for tourists' benefits from agriculture in Europe', *Environmental and Resource Economics*, 10(2): 189–205.

Halstead, J. (1984) 'Measuring the nonmarket value of Massachusetts farmland', *Journal of Northeastern Agricultural Economics*, 13(Apr): 12–19.

Howitt, R.E. (1995) 'Positive mathematical programming', *American Journal of Agricultural Economics*, 77(2): 329–342.

ICBS (2002) *Statistical Abstract of Israel No. 53*. Jerusalem: The State of Israel.

Kan, I., Haim, D., Rapaport-Rom, M. and Shechter, M. (2009) 'Environmental amenities and optimal agricultural land-use: The case of Israel', *Ecological Economics*, 68(6): 1893–1898.

Koslowsky, M., Aizer, A. and Krausz, M. (1996) 'Stressor and personal variables in the commuting experience', *International Journal of Manpower*, 17(3): 4–14.

Lopez, R.A., Shah, F.A. and Altobello, M.A. (1994) 'Amenity benefits and the optimal allocation of land', *Land Economics*, 70(1): 53–62.

McConnell, K.E. (1989) 'Optimal quantity of land in agriculture', *Northeastern Journal of Agricultural and Resource Economics*, 18(2): 63–72.

Ministry of Agriculture and Rural Development (2002) *Instructions for Growers. The Agricultural Extension Service, Israel*. http://www.shaham.moag.gov.il/ (accessed 1 May 2012).

Mitchell, R.C. and Carson, T.R. (1989) *Using Surveys to Value Public Goods – The Contingent Valuation Method*. Washington, D.C.: Resources for the Future.

OECD (2000) *Multifunctionality: Towards an Analytical Framework*. Paris: OECD.

OECD (2003) *The Future of Rural Policy – From Sectoral to Place-Based Policies in Rural Areas*. Paris: OECD.

Peterson, J.M., Boisvert, R.N. and Gorter, H.de (2002) 'Environmental policies for a multifunctional agricultural sector in open economies', *European Review of Agricultural Economics*, 29(4): 423–443.

Rapaport-Rom, M. (2006) *Economic Assessment of Climate-change Impacts on the Vegetative Agriculture in Israel*. Haifa University, Israel: MSc thesis.

Ready, R. and Abdalla, C. (2005) 'The amenity and disamenity impacts of agriculture: Estimates from a hedonic pricing model', *American Journal of Agricultural Economics*, 87(2): 314–326.

Ready, R., Berger, M. and Blomquist, G. (1997) 'Measuring amenity benefits from farmland: Hedonic pricing vs. contingent valuation', *Growth and Change*, 28(1): 43–58.

Part III

Worldwide applications and detailed case studies – integration of practices

10 An economic evaluation of the grassland landscape in Aso Kuju National Park, Japan

Kenji Okubo

Introduction

Recently, Japanese people have begun to display an increasing interest in the natural environment and ecosystem services related to the multifunctionality of agriculture and forestry. This growing interest is directed toward issues such as biological diversity preservation, the nurturing of water resources, landscape formation, and recreational experiences. As a result, in Japan, tourism demand for agricultural and rural regions with abundant natural resources and rural landscapes has increased among people inhabiting urban areas.

Aso-Kuju National Park is exemplary of the above characteristics. This National Park is situated in the Aso area, which is located in the northeast of Kumamoto prefecture in the center of Kyushu Island – the southernmost of the four major Japanese islands (Figure 10.1). Aso-Kuju is a traditional national park characterized by a unique terrain, geology, biodiversity, and a beautiful landscape. In particular, the park is famous for having the largest caldera – a volcanic feature usually formed by the collapse of land following a volcanic eruption – in the world formed by a central cone and somma; the caldera is covered with grassland. Furthermore, the ecosystem of the park is the result of a close relationship between wildlife and grasslands. Apart from the National Park, the Aso area is also known for its traditional cuisines, hot spring towns, and riverheads with clear water. It is a popular tourist destination visited by over 18 million tourists annually, and is also one of the largest livestock farming areas in Japan.

Historically, the grassland landscape of the area has been shaped not only by nature (i.e. wildlife) but also by the daily activities of farmers or dwellers who have used and maintained the land in commonage for their livelihood over the past 1,000 years. Livestock farmers, for example, utilize the grassland for grazing cattle and gathering roughage. In the past, dwellers also used the grasses as fuel, compost, and thatched roof material. Therefore, grasslands in the Aso area can be characterized as semi-natural grasslands.

After the liberalization of beef imports in 1991, the fall in the price of veal calves led to the worsening of the conditions for livestock farm management, and particularly led to the cessation of livestock farming by the small-scale farmers, who experienced a drastic decrease in their profits. As a consequence,

Figure 10.1 Location of Aso-Kuju National Park.

the maintenance of the grasslands became difficult due to the ageing of the rural population and a decrease in the number of livestock farmers who were dependent mainly on the grassland. This crisis of grassland preservation in the Aso area has been popularized through the mass media, and support for this preservation has been garnered not only from the public at large, but also from fundraising campaigns by local newspapers and systematic volunteer activities for grassland preservation by non-profit organizations.

Efforts at grassland preservation have been continued based on a firm belief that the grasslands must be preserved. These protection activities, however, have been conducted without explicitly considering the economic aspects of the use of the grasslands in the Aso area. If the protection activities are continued, it is necessary to evaluate the economic functions and value of the grasslands in the Aso area, so that effective decisions concerning grassland preservation can be made. Therefore, in this chapter, we focus on the grassland landscape and characteristics of the Aso area as a famous tourist destination and employ correspondence analysis and the travel cost method to estimate the economic value of the Aso area for tourists.

This remainder of this chapter is as follows. In the following section, we elaborate on the relationship between the use and preservation of the grasslands in the Aso area. Subsequently, the on-site survey that has been conducted in the area is summarised. The section thereafter presents the survey responses and the response patterns on various questions. This is followed by an estimation of the economic value of the grassland landscape by using the travel cost method. First, we introduce the method, then we spend a few words on the travel cost demand model and its associated data, and finally we discuss and analyse the valuation results and test some hypotheses. The chapter finishes with some conclusions.

Utilization and preservation of grasslands

In the Aso area, grasslands have been mainly used as pastureland and meadows in commonage. Traditionally, a number of Brown Cattle, which are breeding cattle, were put to pasture. However, as mentioned above, after the liberalization of beef imports in 1991, a fall in veal calf prices and the subsequent decrease in breeders and the number of cattle for breeding have had a significant effect on grassland preservation in the Aso area. If there is a decrease in the number of breeders who have the right to use the common grounds, the group exercising this right may lose its incentive to make positive efforts toward preserving the grassland (Otaki 1999).

Figure 10.2 illustrates the usage and maintenance schedule of the grassland in the Aso area over a year. One of the characteristics of the grassland in this area is that it is semi-natural and gradually changes into bush if maintenance operations are not conducted.

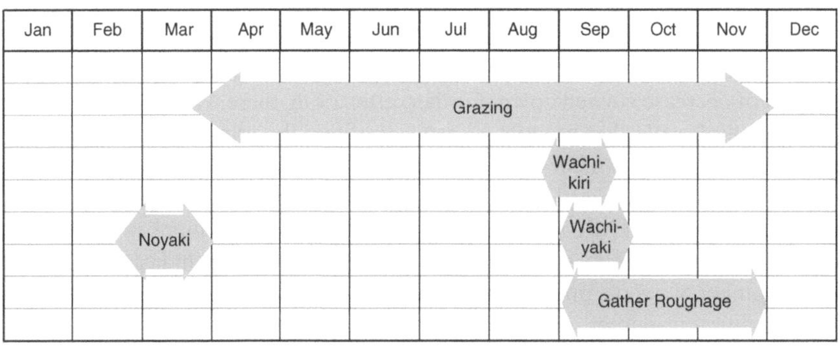

Figure 10.2 Utilization and management of grassland in the Aso area.

Source: Reported by the Ministry of Environment of Japan (Aso Design Center 1996).

Some of the main operations necessary for maintaining grasslands are the creation of a firebreak belt and field burning. A firebreak belt is necessary to prevent the spread of fire to any adjacent forest when grassland is burnt. The creation of a firebreak belt can be divided into two processes known as "Wachikiri" and "Wachiyaki" in Japanese. "Wachikiri" is the process of cutting grasses in an area approximately 8 m wide at the border of the forest. This operation is conducted from the end of August to mid-September. After several days of "Wachikiri", "Wachiyaki" is conducted in the same area. In this process, dry grasses that were cut in the process of "Wachikiri" are burned; this completes the creation of a firebreak belt. As the period between the end of August and mid-September is the hottest season in the Aso area and both the abovementioned operations are conducted on grassland with a steep slope, conducting these operations is hard work, particularly for old people. The process of burning the grassland is termed "Noyaki" in Japanese. "Noyaki" is conducted from the end of February to the end of March in order to burn the dry grasses of the preceding year, inhibit the growth of bushes, and hasten the growth of new grasses.

In the case that the number of non-livestock farmers – for example, employees from the non-agricultural sector exercising the right of common – exceeds the number of livestock farmers, these operations are conducted only during holidays. Whether or not these operations are conducted is dependent on the weather. If these operations cannot be conducted according to a regular schedule because of bad weather, it is likely that grassland utilization and preservation will encounter serious problems. Furthermore, a decrease in the number of cattle that are put to graze on the grassland will cause an increase in overgrown grasses, which may hinder the operations that are conducted for maintaining the grassland. Moreover, if overgrown grasses that are not eaten by cattle are burned in the "Noyaki" operation, a huge fire may occur, thereby making the process of "Noyaki" more dangerous.

Figure 10.3 illustrates the change in the number of breeders and cows for breeding in Aso area from 1986 to 2005. It is evident that until 1991, the number of cows for breeding was over 25,000; however, during the period 1992–1996, this number fell rapidly to 17,000. This rapid decrease was caused by the cessation of livestock farming along with poor or inappropriate farm management. Thereafter, the rate of decrease slowed down. Further, after 1996, there was a marked change in the type of cattle that brought in better profits – the number of Brown Cattle decreased while the number of Black Cattle, which fetches a higher price relative to Brown Cattle, increased. In 2005, the ratio of Black Cattle was one-third of the total number of cattle in the Aso area. Furthermore, the number of breeders decreased steadily during the period 1986–2005. These trends indicate a decrease in the number of people undertaking the utilization and preservation of grasslands as agricultural resources.

Figure 10.4 presents the change in the total grassland area in the Aso area during the period 1986–2005. The grasslands in the area can be classified into artificial grasslands (farmlands) and semi-natural grasslands. If we look

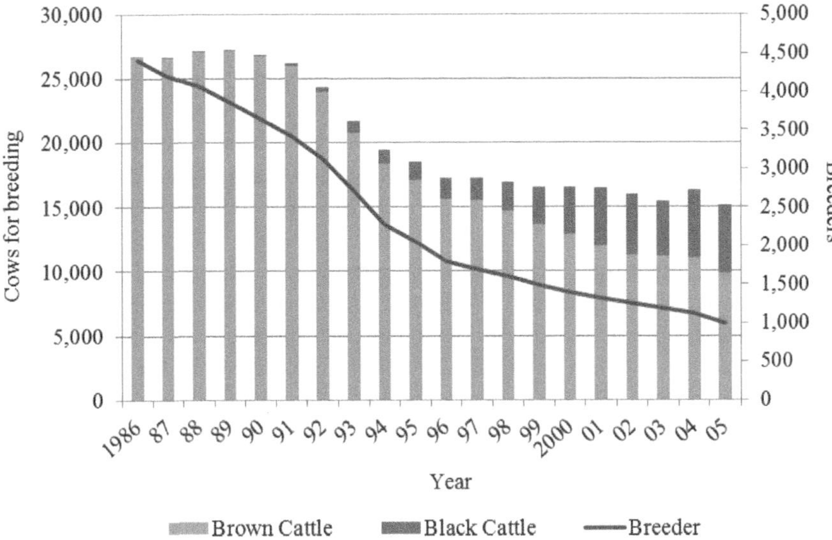

Figure 10.3 Change in the number of cows for breeding and breeders in the Aso area.

Source: Livestock Statistics of Kumamoto Prefecture, Japan, 1986–2005.

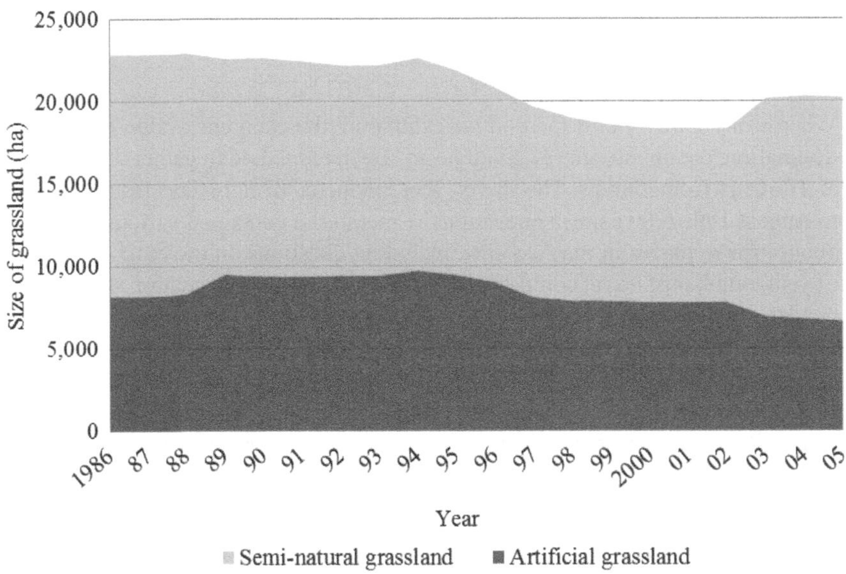

Figure 10.4 Change in the size of grassland (in ha) in the Aso area.

Source: Livestock Statistics of Kumamoto Prefecture, Japan, 1986–2005.

at the change in gross grassland area, then we see that until 1994 the gross grassland area was maintained at over 22,000 hectares. However, during the period 1995–2000, there was a rapid decrease in this area to approximately 18,000 hectares. Thereafter, there was a recovery to approximately 20,000 hectares. Further, the artificial grassland area was maintained at over 9,000 hectares during the period 1989–1995, followed by a steady decline after 1996. The semi-natural grassland area, on the other hand, declined steadily by 4,000 hectares during the period 1986–2002 and recovered to over 13,000 hectares after 2003.

Taking these facts into consideration, it is imperative for rational preservation of the grassland landscape that a sufficient number of cattle are put to graze on the grassland. Indeed, one of the major factors responsible for the decrease in grassland area is a fall in grassland utilization. In other words, the preservation of the grasslands solely by livestock farmers who have right of common has become increasingly difficult during the last decade. However, there has been a recent popularization of grassland preservation among Japanese people, and support activities have been firmly established for grassland preservation. For example, a volunteer organization has been formed, comprising members who recognize the supreme value of grasslands and are willing to participate in this support activity. They have been trained and organized for grassland preservation by Aso Green Stock, which is a foundation for natural resource preservation in the Aso area. This support activity has made progress towards the recovery of lost grasslands since 2003.

Summary of the on-site survey

When using a travel cost method for estimating the economic value of a tourist destination, the on-site survey could be an effective method to gather data quickly and directly from tourists. The survey was conducted in eight days from July 24th to August 10th 2000 using a questionnaire method. It contained various questions about trips to places in the Aso area, including questions on travel time, purpose of visit, number of travel companions, etc. Demographic information such as age, gender, income, and family was also collected through the questionnaire. The form was completed by 214 tourists.

The sample was drawn from three representative locations in the Aso area: Kusasenri (central Aso), Daikanbo (north Aso), and Tawarayama (south Aso). These sites are famous for their grassland landscapes. Kusasenri is the most famous site and is known for its beautiful grassland dotted with cows put to pasture. Moreover, this site is also frequently used for poster or TV advertisements. Further, Daikanbo is located in the somma of north. The place offers tourists a magnificent view over the entire rural landscape of Aso City and the north side of the central cone. Tawarayama is located in the somma of south Aso and tourists have a view from this site overlooking the entire south Aso area and the south side of the central cone.

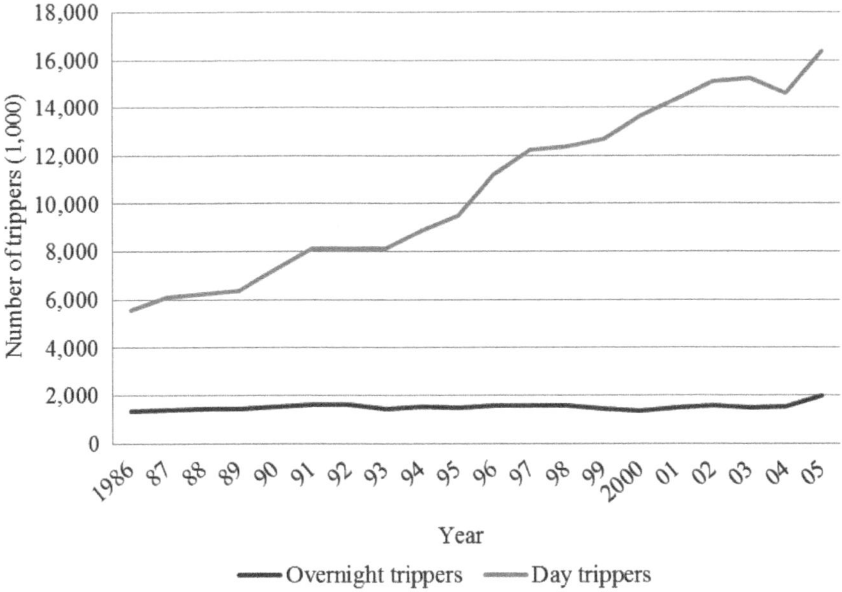

Figure 10.5 Change in the number of overnight trippers and day trippers to the Aso area.
Source: Tourism Statistics of Kumamoto Prefecture, Japan, 1986–2005.

The characteristics of overnight and day trippers to Aso area

Change in the number of overnight and day trippers

Figure 10.5 presents the change in the number of overnight and day trippers to the Aso area from 1986 to 2005. According to this figure, approximately 1.3 million overnight trippers visited the area in 1986, and this number increased to approximately 1.6 million in 1991. From 1992 to 2004, the number of overnight trippers was approximately 1.4–1.6 million per year. In 2005, the number of overnight trippers increased to 1,995,000. Further, the number of day trippers was approximately 5.5 million in 1986. Thereafter, there was an increase in the number of day trippers. In 2005, the number of day trippers to the Aso area was 16.4 million. So, in conclusion, the number of both overnight and day trippers to the Aso area in 2005 was approximately 18.4 million.

Next, we examine the attributes of travel to the Aso area, such as travel mode, travel distance and travel time, based on the results of the on-site survey. The average travel frequency to Aso area in the last 1 year – i.e. from August 1999 to July 2000 – was 2.4 times, and that in the last 5 years – from August 1995 to July 2000 – was 6.0 times. The average travel frequency for overnight trippers was 66.7 per cent and that for day trippers was 34.3 per cent. Moreover, regarding traveling companions: 57.1 per cent was "member of the family", 34.9 per cent was "friend", and 7.5 per cent was categorized as "other", including alone. The average number of people with traveling companions was 3.2.

Trippers traveled to the Aso area for different purposes. To be more concrete, 59.3 per cent of the interviewees traveled to the area for a panoramic view from a high point, 72.9 per cent for landscape, 52.8 per cent for open spaces, 24.3 per cent for wildlife, 25.2 per cent for visiting the crater, 26.6 per cent for spas, 16.4 per cent for local foods, and 15.4 per cent for hot springs. Further, the proportion of trippers who responded that the grassland landscape is an essential element of the Aso area was 87.4 per cent.

Difference in travel purposes

After the above descriptive overview, in this sub-section we apply a correspondence analysis to clarify the differences in tourist destinations between different types of trippers and travel purposes. Correspondence analysis is an exploratory data analysis technique used for the graphical display of contingency tables; it displays various frequency distributions of variables that are assessed on a nominal scale. The scales are shown in rows and columns of a rectangular data matrix in corresponding units so that each row and column can be displayed in the same low-dimensional space (Hoffman and Franke 1986). We use correspondence analysis to indicate the relationship between types of trippers and tourist destinations and obtain a graphical display that can help us detect structural relationships among the various categories.

Trippers to the Aso area can be classified into the following four groups.

DK: Day trippers from Kumamoto prefecture.
DO: Day trippers from outside Kumamoto prefecture.
TK: Overnight trippers from Kumamoto prefecture.
TO: Overnight trippers from outside Kumamoto prefecture.

Further, the variables for tourist destinations in the Aso area are as follows: grassland landscape, panoramic view from a high point, open space for recreational use, wildlife, visiting the crater, traditional cuisines, hot spring town, and riverheads with clear water.

Figure 10.6 presents the result of the correspondence analysis. The four abovementioned groups are displayed in different quadrants and are situated relatively far from each other; this indicates that visitors have different purposes for overnight or day trips to the Aso area.

DK and open space are correlated; this result implies that day trippers from Kumamoto prefecture displayed a strong tendency to relax in the open space of the grassland and enjoy the recreational use of it. Further, DO, grassland landscape, panoramic view from a high point, and wildlife were displayed in proximity. This result implies that day trippers from outside Kumamoto prefecture mainly enjoy the grassland landscape, commune with the nature and wildlife of a national park, and enjoy the panoramic view of the entire Aso area. Moreover, TK is displayed in close proximity to visiting the crater and traditional cuisines,

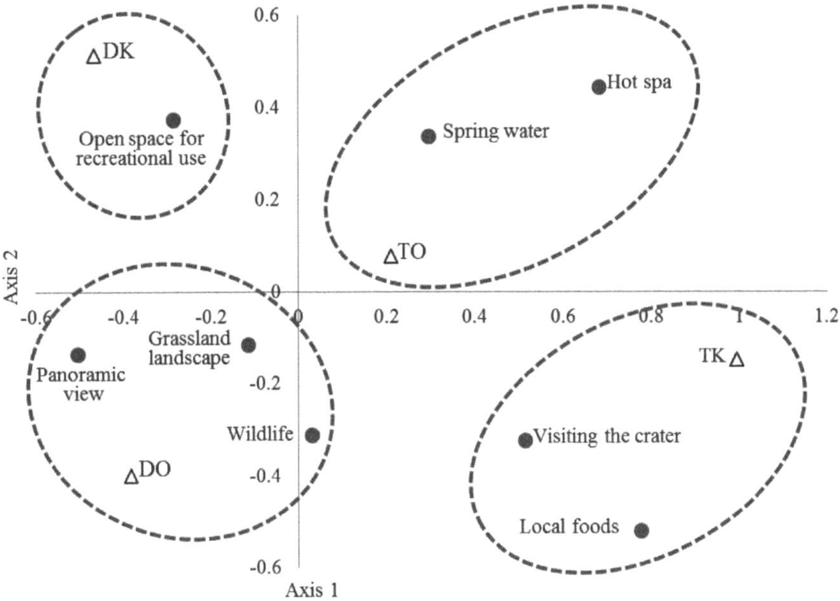

DK: Day trippers from Kumamoto prefecture.
DO: Day trippers from outside of Kumamoto prefecture.
TK: Overnight trippers from Kumamoto prefecture.
TO: Overnight trippers from outside of Kumamoto prefecture.

Figure 10.6 Result of correspondence analysis.

implying that trippers from Kumamoto prefecture showed strong preferences to visit the volcanic crater in the central cone and eat local foods in the Aso area. And finally, TO, hot spring town, and riverheads with clear water were displayed in immediate proximity, indicating that trippers from outside Kumamoto prefecture were very much inclined to visit the hot spring town to bathe, or the riverheads to experience clear water in the Aso area.

These results suggest that day trippers to the Aso area have a tendency to visit visually appealing areas such as open spaces, areas with panoramic views, and grassland landscapes. Overnight trippers to Aso area, on the other hand, have a tendency to value regional attraction resources such as spas, hot springs, the volcanic crater, and traditional foods.

Estimation of the economic value of the grassland landscape

Travel cost method

The travel cost method (TCM) is used to evaluate the benefits of recreational sites such as the Aso area with an estimation of the travel cost demand model. The travel cost demand model is estimated using the following variables: frequency of visits, travel expenditure, socio-economic attributes, and other variables that influence their decision-making regarding the travel destination. Thus, TCM is a revealed preference method, which is a method that identifies the best possible option on the basis of consumer behavior. TCM has an advantage over the contingent valuation method (CVM) in that the result of estimation obtained by employing TCM is not affected by the results of the questionnaire because the data that is used for TCM is based on actual human behavior. The estimated value obtained through TCM is not a compensating or equivalent surplus, but a consumer surplus, which is an approximate value of the compensating and equivalent surplus.

The travel cost demand model can be divided into two types: the zone travel cost method (ZTCM) and the individual travel cost method (ITCM). ZTCM is useful for estimating the travel cost demand function based on aggregated tourist data for every equal travel cost zone. ITCM is particularly appropriate for estimating the travel cost demand model by using individual data as a socio-economic attribute.

When we conducted the on-site survey in 2000, the number of overnight and day trippers was approximately 1.34 million and 13.63 million, respectively. Thus, it is evident that over 90 per cent of the tourists to the Aso area are day trippers. In other words, overnight trippers from distant regions are less-frequent visitors. Therefore, ITCM is a more appropriate method than ZTCM as the latter method causes a loss of information regarding each respondent.

Furthermore, the methods used for the evaluation of the benefits of recreational areas must be used separately due to the differences in quality of pleasure, decision-making processes, expenditures for overnight trips and day trips, and so on. Therefore, in this study, three types of travel cost demand models are estimated: full model (amount of trippers), overnight trippers' model, and day trippers' model. We also apply the likelihood ratio (LR) test in order to statistically examine the difference between an overnight trip and day trip. If the null hypothesis that an overnight trip and day trip are equal goods is rejected by the likelihood ratio test, there is a necessity to classify trips into overnight trip and day trip for the estimation of the travel cost demand model.

Travel cost demand model and data

In this study, the frequency of a trip is a dependent variable that is a non-negative integer and non-zero because the data has been collected from trippers in the on-site survey. Thus, we apply a zero-truncated Poisson regression model for estimating the travel cost demand model. Greene (2000) presented the probability function of the zero-truncated Poisson regression value, which is given by equation (10.1):

$$Prob(Y_i = y_i | Y_i > 0) = \frac{e^{-\lambda}\lambda_i^{y_i}}{y_i!\left(1 - e^{-\lambda_i}\right)}. \tag{10.1}$$

$$y_i = 1, 2, 3, \ldots \quad i = 1, 2, 3, \ldots, n,$$

where y_i *is* frequency of trips; and λ_i is variance and average of distribution.

In equation (10.1), λ_i is the dependent variable, y_i is the independent variable, which is the socio-economic attribute, and β is the coefficient vector of y_i. Then, equation (10.2) is expressed in the following manner:

$$\ln \lambda_i = \beta x_i. \tag{10.2}$$

Further, β – a coefficient vector – can be estimated by maximum likelihood. Thus, equation (10.3), which is a likelihood function, is expressed in the following manner:

$$\ln \mathrm{L} = \sum_{i=1}^{n} \left[-\lambda + y_i \beta x_i - \ln y_i! - \ln \left(1 - e^{\lambda^i}\right) \right]. \tag{10.3}$$

In the sample, the number of day trippers was 64 and that of overnight trippers was 71; the destination of both types of trippers was the Aso area. The variables used to estimate the travel cost demand model were the frequency of trips to Aso area, travel cost, age, coming with friends (dummy variable), and coming with family (dummy variable). The travel cost is given by equation (10.4):

$$\mathrm{TC}_i = \frac{d_i \times {}^{g_i}/_{f} + I_i}{P}, \tag{10.4}$$

where TC_i is travel cost; d_i is distance covered; g_i is cost of fuel (100 yen/1 liter); d_i is fuel consumption (10 km/1 liter); I is other costs; and P is number of fellow trippers.

The total travel cost, which is one of the dependent variables, is expressed as the sum of TC and cost of transit time. The cost of transit time is calculated in the following manner. According to a report by the Ministry of Health, Labor and Welfare of Japan (2000), the average wage was 315,901 yen and average work time was 149.8 hours in August 2000. The average wage divided by the average work time yields the hourly wage (2,109 yen). The cost of transit time is calculated as one-third of the abovementioned hourly wage (Cesario 1976).

Estimation of the travel cost model and consumer surplus

Table 10.1 presents the estimation of the parameters of a zero-truncated Poisson regression model. First, the likelihood ratio test is applied in order to examine the statistical difference between overnight and day trips. The likelihood ratio is given by equation (10.5):

$$\mathrm{LR} = -2\left[\mathrm{L}_1 - (\mathrm{L}_2 + \mathrm{L}_3)\right], \tag{10.5}$$

where LR is likelihood ratio; L_1 is the logarithmic likelihood ration of Model 1, which is estimated by all samples; L_2 is the logarithmic likelihood ration of Model 2, which is estimated by overnight trippers' samples; and L_3 is the

Table 10.1 Results of the travel cost demand model

	Model 1 (All trippers)		*Model 2 (Overnight trippers)*		*Model 3 (Day trippers)*	
	Coefficient	*t-value*	*Coefficient*	*t-value*	*Coefficient*	*t-value*
Constant	−0.874	−3.228***	3.319	3.765***	−0.286	−0.854
Travel cost	-4.015×10^{-5}	−6.952***	-7.737×10^{-5}	−3.687***	-2.819×10^{-4}	−5.976***
Income	7.535×10^{-8}	2.729***	-1.482×10^{-7}	−2.167**	1.823×10^{-6}	5.190***
Age	0.022	4.419***	0.0260	2.818**	0.154	2.910**
Coming with family	1.048	8.306***	−1.522	−5.318**	0.823	5.637***
Coming with friends	1.206	7.421***	−0.959	−2.473**	0.881	4.807***
Number of sample	135		64		71	
Log L	−313.466		−68.684		−174.033	
Likelihood Ratio test			141.498***			

*** = significant at 0.01; ** = significant at 0.05.

logarithmic likelihood ration of Model 3, which is estimated by day trippers' samples.

The distribution of the likelihood test statistic, which is given by (10.5) is chi-squared. The result of the likelihood ratio test was 141.498, strongly rejecting the null hypothesis that overnight trippers are equal to day trippers. This result indicated that, based on empirical observations, the estimation of the travel cost demand model should be done separately for overnight and day trippers.

Next, we examine the estimation result of the zero-truncated Poisson regression of Models 2 (overnight trippers) and 3 (day trippers). In the estimation result of all models, the total travel cost coefficients were statistically significant at the 1 per cent level, and had a negative effect on overnight and day trips. Further, the income coefficients were statistically significant at the 5 per cent and 1 per cent levels in Models 2 and 3, respectively. In Model 3, the income coefficient had a positive effect on day trips. However, in Model 2, the income coefficient had a negative effect on overnight trips. These results indicated that tourists with higher incomes prefer day trips to overnight trips. In the estimation result of all models, the age coefficients were statistically significant at the 1 per cent level, and had a positive effect on both overnight and day trips. In the estimation result of all models, the effect of "coming with family" was statistically significant at the 1 per cent level. This effect was positive for day trips and negative for overnight trips. Further, the coefficient related to "coming with friends" was statistically significant at the 5 per cent and 1 per cent levels in Models 2 and 3, respectively. But also here is a difference between Model 2 and Model 3: "coming with friends" had a positive effect on the number of day trips, but a negative effect on overnight trips.

Furthermore, Table 10.2 presents the consumer surplus that was calculated from the reciprocal of the absolute value of total travel cost-related coefficients in Models 2 and 3 for overnight and day trippers. The result of the calculation of consumer surplus for overnight trippers was 12,925 yen per trip; therefore, the consumer surplus for trippers in a year (13.4 million people in 2000) was 17.3 billion yen. The consumer surplus for day trippers was 3,547 yen per day trip and thus the consumer surplus for day trippers in a year (in 2000: 13.6 million people) was 48.3 billion yen.

Conclusion

The main aim of this study was to evaluate the economic value of the grassland landscape that is used and maintained by livestock farmers or dwellers in commonage. In particular, we focused on the grassland landscape and

Table 10.2 Estimated consumer surplus (CS)

	Overnight trippers	*Day trippers*
CS/person (in yen)	12,925	3,547
Total CS/year (in billion yen)	1.73	48.34

characteristics of the Aso area as a famous tourist destination. The following are the main conclusions drawn from this study. First, the result from the correspondence analysis indicates that the main purpose of visiting the area varies according to the tourists' itinerary and place of residence. In particular, overnight trippers indulge in experiences such as hot spa, local foods, and visiting the crater. On the other hand, day trippers especially enjoy viewing the grassland landscape and using the open spaces for recreational activities.

Second, the result from the estimated travel cost demand model indicated the behavioral characteristics of overnight and day trippers to the Aso area. It appeared, for example, that tourists with higher income prefer day trips rather than overnight trips and that age has a major positive impact on both day and overnight trips. Further, it turned out that according to the travel cost model estimates, the consumer surplus for overnight and day trippers – after aggregating individual estimates – was calculated to be 17.3 billion yen and 48.3 billion yen annually, respectively. However, it is must be noted that these estimated values include not only revenue from the grassland area but also that from other tourist destinations in the area. Thus, it was necessary to focus on calculating the consumer surplus for those day trippers whose major purpose in visiting the Aso area was enjoying the landscape and open spaces that are formed by grasslands. It was found that the estimated consumer surplus for day trippers is much more than the budget for national park management in Japan (approximately 140 billion yen in 2000), according to the Ministry of Environment of Japan (2008). From the perspective of total costs and benefits, a simple comparison of these two major determinants of economic performance suggests that the grasslands of the Aso area have an enormous positive external effect on day trippers.

It is evident from the results of this study that grassland landscapes are of great value in attracting tourists, and the value of the grassland landscape that was estimated by TCM may be considered as an economic basis for the preservation of grasslands. Because in our analysis, we focused only on the recreational use value of the Aso landscape, and did not include any estimate for non-use values (which can be derived for example by the contingent valuation method (CVM)), we feel it is safe to conclude that the value of the grassland landscape will be even higher than the abovementioned estimates. In fact, our TCM outcome provides a lower bound for the value of the Aso landscape, rather than an exact answer to the question: What is the economic value of landscape?

All in all, in view of the fact that the current environment of grasslands is fraught with difficulties and that the value of grassland landscapes in terms of attracting tourists is huge, it is imperative that a grassland preservation system is created with the cooperation of people who enjoy such landscapes.

References

Aso Design Center (1996) *Report of Participatory Management Project of National Park*. Tokyo: The Commission of the Ministry of Environment of Japan (in Japanese).

Cesario, F.J. (1976) 'Value of time in recreation benefit studies', *Land Economics*, 52: 32–41.

Greene, W.H. (2000) *Econometric Analysis*, 4th edition. Upper Saddle River, New Jersey: Prentice-Hall.

Hoffman, D.L. and Franke, G.R. (1986) 'Correspondence analysis: Graphical representation of categorical data in marketing research', *Journal of Marketing Research*, 23: 213–227.

Ministry of Environment of Japan (2008) *National Park System*. Tokyo (in Japanese).

Ministry of Health, Labor and Welfare of Japan (2000) *Explanation of Monthly Labor Survey of August, 2000*. Tokyo (in Japanese).

Otaki, N. (1999) 'Aso grassland for a thousand years. Burning stopped, grassland endangered', *Environmental Research Quarterly*, 114: 31–36 (in Japanese).

11 From an "integrated" to a "dismantled" landscape

María D. Domínguez García and David Soto Fernández

Introduction

Landscapes are the result of the interactions between humans and natural factors and, as a result, are subject to continuous transformation: visually, socio-economically, ecologically and culturally. Landscape is a complex concept that has been studied by many disciplines and perspectives. From a subjective (or perceptive) perspective, a landscape can be analysed as a socio-cultural representation created by a particular society. From an objective (or "neutral") perspective, a landscape can be considered as the visible or perceptible part of an ecosystem, the approach often adopted by landscape ecologists (Forman and Gordon 1986; Burel and Baudry 2001). In this chapter, we approach changing landscape values by focusing on changes in social metabolism, which we define as 'the particular form in which societies establish and maintain their material inputs from, and outputs to, nature; the modes in which they organise the exchange of matter and energy with their natural environment' (Fisher-Kowalsky and Haberl 1994: 3). To describe and illustrate this approach, we focus on the dismantling of the organic agro-ecosystem in the region of Galicia (Spain). Our study looks at the concept of an organic agro-ecosystem, its dismantling, the interrelations between the socio-economic and environmental aspects of agro-ecosystems and the analysis of energy and material flows (González de Molina and Guzmán Casado 2008; Tello et al. 2008).

Galicia in the northwest of Spain has been traditionally a rural region with a strong reliance on the primary sector. Its rural landscape has been profoundly shaped by farming activities. For centuries, the organic agro-ecosystem performed well and, although it was not static, the main features of this agro-ecosystem remained in place until well into the second half of the 20th century. The organic agro-ecosystem describes the Galician peasant-like agrarian system; its management is characterised by a combination of different land-use systems, such as crops, cattle and *monte* (areas of forest, scrub and bushes). This land-use combination was harmonious and well balanced because the healthy functioning of one system determined the healthy functioning of the others. In this agro-ecosystem, material and energy flows were highly efficient and the nutrient cycles were closed. This is because traditional agriculture is dependent on solar energy and therefore subject to strong limitations, forcing it to be highly efficient in the use of locally available resources (Wrigley 1993).

This agro-ecosystem was progressively dismantled due to different institutional, economic and social factors, and began to develop into a more industrial system of farming. One important result of this has been a change in land-use patterns, and particularly the abandonment of productive lands, which has entailed a significant change in the landscape. The process of dismantling the organic agro-ecosystem, and the transition to a new agro-industrial model, led to the functions of the *monte* becoming redundant. The transformation entailed a change from a landscape in which the socio-economic, ecological, cultural and symbolic realms were strongly related and interconnected, into one where parts of these interconnections have been lost. Furthermore, it also implies a transformation of the interactions between the constituent elements of the landscape and society. We focus on this aspect, and we compare and try to explain the differences between the integrated landscape, linked to an organic management system and the current landscape, linked to an industrial production system – which is more intensive, specialised and homogeneous.

This chapter is structured as follows. After this introduction, the next section gives some theoretical approaches to understanding how to approach the process of the transformation of the organic agro-ecosystem and the subsequent changes in landscape that this gives rise to. The concept of social metabolism is introduced, along with other key concepts: sustainability, co-evolution and common goods. Then, the following section presents the main outcomes of our research, based on the case study area, providing a brief description of the Galician organic agro-ecosystem and its structure, highlighting its transformation and the changes in land-use patterns that have had an impact on the landscape. In the last section, we make some concluding remarks.

Social metabolism, sustainability and common goods

The concept of social metabolism is a way of explaining the process through which each society reproduces its material conditions of existence (Fischer-Kowalsky 1998; Fischer-Kowalsky and Hüttler 1999; Toledo and González de Molina 2006). The analysis of social metabolism focuses on energy and material flows and on the relation between nature and society which can be understood as a process of co-evolution, driven by mutual determination (Toledo and González de Molina 2006) between humans and nature. Over recent years several methodologies have emerged for developing different indicators that reveal the features of this process, such as material and energy flows, the Human Appropriation of Net Primary Product (HANPP, see Schandl et al. 2002) and the balance of nutrients (Garrabou and González de Molina 2008).

Social metabolism views the landscape from an integrative perspective which considers the interactions between cultural, social, economic and environmental elements. The landscape is a socially and culturally constructed space, but it is important to understand the relation between its different bio-physical components

(Guzmán Casado and González de Molina 2009). A historical or dynamic perspective helps us to understand the interactions between society and nature, and the resultant landscape. To valorise the current functions of the landscape and to identify sustainable ways of management, it is also essential to understand the functionality of the landscape at different times, its historical transformations and how these transformations have influenced its current spatial configuration.

Sustainability is a central concept here. This is generally understood as the capacity of a social ecosystem to meet the basic needs of the current and future generations without degrading the natural resource basis. We use sustainability as a criterion for evaluating and comparing the landscapes that result from different types of farming. A number of different parameters can be used to measure sustainability in general, or in terms of a specific social metabolism. These include productivity, stability, resilience, autonomy, and equity in the distribution of goods and services produced (Conway 1985; Conway and Barbier 1990; Lopez Ridaura et al. 2000). Sustainability is not just a measure but also an objective and it is important to be able trace progress and changes over different temporal and spatial scales.

From a historical perspective it is obvious that there is not just one single equilibrium in nature, but a number of equilibria. Ecosystems move from one equilibrium to another (González de Molina and Guzmán Casado 2006). The dynamics associated with such change can bring further unexpected changes, as a result of the interactions between many factors. The focus on sustainability has increased awareness of the importance of resilience and stability (as opposed to a sole focus on productivity). When assessing sustainability, it is useful to analyse the use and the way an agro-ecosystem is organised in comparison with other agro-ecosystems over a long period of time.[1]

The tools of social metabolism (energy flows, Human Appropriation of Net Primary Production and nutrient balances) provide a useful way of analysing environmental history and clearly show that organically managed pre-industrial farming systems were more sustainable than the current day industrial model (González de Molina and Guzmán Casado 2006; Garrabou et al. 2006; Tello et al. 2009; González de Molina et al. 2009; Krausmann and Cunfer 2009). This is largely because of the strong interrelations between the different components of the landscape, the volume and type of energy and material flows[2] and method of fertilisation. These arguments have also been made in a more qualitative way for the case study region (Bouhier 1979; Balboa 1990; Simón 1995; Soto 2006; Dominguez 2007). Galician peasant organic farming was highly sustainable, its energy and matter flows were within the natural carrying capacity and it allowed for a high level of biodiversity. However, there was also a "land cost", meaning that land had to be used for different and complementary purposes, which left an imprint on the landscape (Guzmán Casado and González de Molina 2009).[3]

For peasant organic farmers this land cost meant that they had to keep aside a large part of their land for use as pasture and forest. The extension of cropping lands was only possible to the extent that it did not affect the supply of materials needed to sustain the needs of livestock, energy needs and other requirements (timber for building etc.). This argument highlights another interesting aspect of landscape management, since much land given over to forests and pastures in peasant organic societies was under common ownership, with access to the resources guaranteed by common rights.[4]

Elinor Ostrom's seminal work (1990) on the commons reveals a number of factors that can contribute to the sustainable management of common property resources. These include: well defined limits of common property, the adaptation of rules to local conditions, channels for user participation, monitoring of the systems and scaled sanctions, mechanisms for conflict resolution, a certain degree of autonomy from external powers, and an ordered structure of relationships with wider systems. Her analysis has been applied by several social historians. Van Zanden (1999) and de Moor (2009) have applied this scheme and shown its value in analysing the commons between the 16th and 18th centuries in the Netherlands. Lana Berasain (2008) used this approach to explain how commons institutions survived several liberal revolutions in Spain and contine to this day. These works also show that an exclusively institutional analysis is not sufficient to understand the role of the commons and attention also needs to be paid to how environmental and institutional aspects interrelate. This idea coincides with recent developments in the literature about the commons, which stresses the importance of considering complexity, uncertainty and institutions (Berkes 2002; Laerhoven and Ostrom 2007). In other words, it is not enough to just analyse common rights of use and the rules that regulate them. Consideration also needs to be given to the ways in which resources are managed. Ecosystems need to be managed in an iterative way (Holling 1993), considering the dynamics of natural and anthropic changes. This approach highlights the adaptive character of traditional management, in which cultural and biological diversity interrelate (Berkes and Folke 1998; Toledo and Barrera-Bassols 2008).

Adapting this approach to the historical analysis, it is clear that a particular institutional arrangement regarding the commons might have a different effect upon sustainability, depending on the specific historical context. In the case of Galicia, we hypothesise that the existing forms of property and common rights were well adapted to supporting the functioning of the organic agro-ecosystem. In our case study the maintenance of common property allowed the survival of small sized farms, at least as long as the system maintained its organic features (see the following section) and was sustainable in environmental and social terms. This balance was derived from the interrelations between the different animal, crop and land systems, which provided a central element in the landscape. Furthermore, the maintenance of common property resources also supported social equity and democratic participation.

The following section examines how the integrated management of the Galician landscape, which remained in force throughout the 19th century and the first half of the 20th century, has been dismantled and replaced (for the location of Galicia in Spain, see Figure 11.1). This has been largely due to the progressive dismantling of the organic agro-ecosystem that existed during this period. The dismantling progressively disconnected the different components

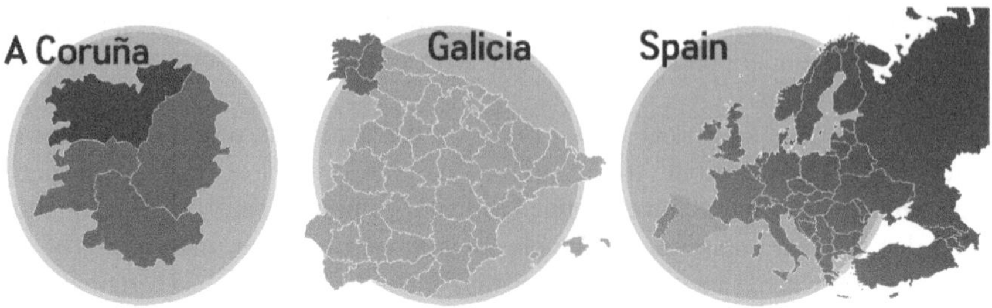

Figure 11.1 Galicia in Spain and the province of A Coruña.

of the agro-ecosystem from each other and disarticulated them.[5] This is particularly evident when looking at the *monte*, described in more detail below, which was to a large extent managed under a common property regime.

Case study: landscape changes in Galicia caused by the dismantling of the organic agrosystem

Galicia has a richly diverse landscape, with high mountains in the eastern part, thousands of rivers, old granitic rocks and highly diverse weather conditions, due to its proximity to the Atlantic (Loureiro and Barrio 2009). Throughout the second half of the 20th century the diversity of this landscape has been diminished by a process of homogenisation which has simplified and degraded the landscape (Lois 2005). A large part of this transformation is related to changes in land-use that have taken place in agriculture. This case study explores the transformation of Galician farming: the progressive dismantling of the organic agro-ecosystem and the subsequent increase, since the 1960s, of agro-industrial farming.

To better understand the transition from an organic to an industrial agro-ecosystem and the impact of this on the landscape, we first analyse the main features of the organic agro-ecosystem and how it operated. Firstly we focus on the social metabolism of this agro-ecosystem in one of the four provinces of Galicia (A Coruña), the only such analysis that has thus far been elaborated. We then briefly summarise the institutional and socio-economic factors that contributed to the progressive dismantling of the organic agro-ecosystem, before going on to analyse the subsequent changes in land-use patterns. This analysis demonstrates the significance of the dismantling of the organic agro-ecosystem in recent times. Finally we discuss two other, more recent, factors related to the abandonment of productive lands that have an impact on landscape; forest fires and property arrangements (the commons).

The functioning of the organic agro-ecosystem

Until well into the 20th century, most of the Galician population lived in rural areas and worked in farming and related activities. The organisation of the

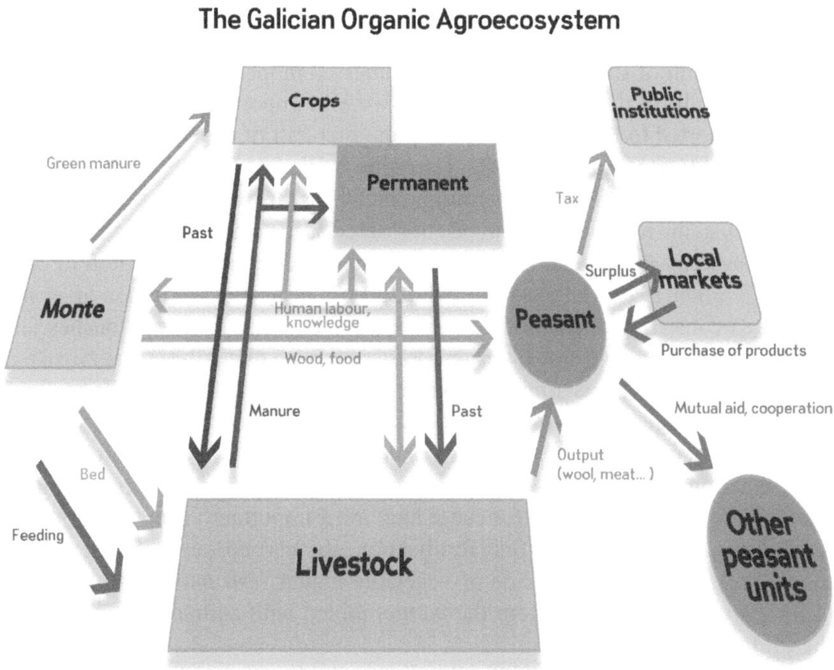

Figure 11.2 The Galician organic agro-ecosystem.

Source: Simón (1995: 217).

farming space, which optimised a combination of different ecosystems (crops, permanent meadows, cattle and *monte*) played a central role in maintaining small peasant units of production. Community and local relations were very important in maintaining this organisation and in keeping farm units working.

Figure 11.2 summarises the functioning of the organic agro-ecosystem in Galicia. It shows the flows of materials, the socio-economic relations between the different parts of the system, and how this (sub) system engaged with the broader economic system. The arrows indicate the directions of these flows and the relations between (the different parts of) the systems. Ovals indicate components of the system that are start or end points, rectangles show activities and circles represent units of connection and linkages. The boxes labelled "public institutions" and "local markets" indicate external, but integral activities.

From a management perspective, the organic agro-ecosystem achieved a high level of integration and mutual interaction between animals, crops and meadows and *monte*. There was a delicate and fine tuned balance between these sub systems in terms of the exchange of energy (labour and animal traction, fertilisation), matter (cattle fodder, wood, etc.) and information flows (applied technology, "learning by doing" processes, institutional arrangements, etc.). The agro-ecosystem was characterised by spatial heterogeneity, genetic and biological biodiversity, but also by poor economic conditions. The peasant farm unit was not isolated but constantly interacted with the transformed and natural and social environments. It was not a unit of subsistence, but was immersed in local and national markets: since the middle of the 19th century, cattle were an important commercial output, initially exported to England and later, in the first half of the 20th century, to the rest of Spain (Carmona Badía 1982; Carmona Badía and Puente 1988).

The health of the soil was a determining factor for the functioning of the organic agro-ecosystem and the *monte* can be considered as the cornerstone of the organic agro-ecosystem. In Galicia, the term *monte* has a different connotation to other places (Bouhier 1979; Balboa 1990; Soto 2006). The general definition of *monte*, as an area consisting of non-arable land covered by trees, scrub and bushes falls half way between two English terms: forest, and bush or scrub (*monte baixo*). In Galicia the *monte* consists more of bushes and scrub than of forest. It has been traditionally subject to an intense process of domestication, or, in other words, has been intensively farmed. As such it is not strictly non-arable land. It had three main functions: supplying cereals (by means of the *rozas* or *estivadas* culture),[6] supplying pasture and forage for cattle and, most importantly, supplying natural fertilisation (Bohuier 1979). Soil fertilisation was carried out by using organic manure (*esterco* or dung), made of vegetable matter (*estrume*), especially *toxo* (*Ulex Europeus*), obtained from the *monte*, mixed with animal excrement from the cattle stalls (Soto 2006).

The *monte* played a crucial role in supplying inputs that sustained the reproduction of family farm units, especially those with insufficient land to support themselves through cropping and pasture alone. The domestication of the *monte* compensated for the small size of the farms[7] – as we will show later, when discussing communal and/or individual access to the *monte*. It also provided wood and other products such as fruits, mushrooms and medicinal herbs for domestic consumption (Marey Pérez et al. 2003).

The organic agro-ecosystem continued to function until the middle of the 20th century (Bouhier 1979), although it was subject to some changes; an intensification of land-use by changing the rotation system, an "improvement" of the autochthonous cattle breeds, a selection of crops that were better adapted to the soil features and the introduction of chemical fertilisers in the first third of the 20th century, etc. There was also a parallel process of intensification in the uses of the *monte*, with increased collection of bushes, the transformation of more areas into pasture and an extension of the forest area. Despite these changes, the organic agro-ecosystem continued to operate for about 200 years without substantively changing: to use the words of Soto (2006: 81) 'without any structural fracture that altered substantially the interrelations among the three main elements of the agro-ecosystem.'

Table 11.1 provides some relevant key data of flows of materials that show the agrarian metabolism in the Galician province of A Coruña in 1922. Around this time important transformations were taking place in agriculture: the privatisation of part of the commons, the introduction of chemical fertilisers, the consolidation

Table 11.1 Biomass flows in the province of A Coruña, 1922

	Unit	*Quantity*
Domestic extraction (DE) cropland	000 tons dry matter	1,233
Domestic extraction (DE) *montes*	000 tons dry matter	1,113
Domestic extraction (DE) total	000 tons dry matter	2,345
Cropland	ha	242,803
Montes	ha	513,197
Other land	ha	34,279
Total	ha	790,279
Land productivity, cropland	kg dry matter/ha	5,077
Land productivity, *montes*	kg dry matter/ha	2,168
DE used for seed	per cent	1.0
DE used for food	per cent	6.1
DE used for feed	per cent	84.9
DE used for fuel and raw materials	per cent	7.9
Livestock units 500 kg	n°	262,176
Livestock units/km^2	n°	33.2
Manure production	000 tons dry matter	789
Meat, milk and wool	000 tons dry matter	32
Consumption of chemical fertilisers	000 tons dry matter	6.5

Sources: Ministerio de Fomento (1921, 1922, 1923). Methodological guidelines in Schandl et al. (2002) and González de Molina and Guzmán Casado (2006). It has not been possible to assess the flows of imports and exports of biomass.

of small peasant properties, etc. (Soto 2006). The data show a very intensive farming system, producing up to five tons of biomass per cultivated hectare. This puts Galician farming well above the Spanish average which was around 1.6 tons per hectare at the time (Soto et al. 2010) and that of Mediterranean area, which was between 1 and 1.5 tons per hectare (González de Molina et al. 2009).

These data imply that a significant share of domestic extraction (see the biomass flows in Table 11.1) was destined for cattle fodder, since Galicia was highly specialised in cattle production. The cattle density (at 33 livestock units per km^2) was much higher than in most other places in Spain in that time (eight livestock units per km^2 in some Andalucian and Catalonian villages – González de Molina et al. 2009; Tello et al. 2009) at the end of the 19th century. This high productivity, combined with cattle specalisation, in Galicia was only possible because of an organic farming system that maintained a well-balanced management of the land, especially the *monte*, of which a significant part was dedicated to sustaining cattle and producing organic fertilisation. In A Coruña almost 65 per cent of the area was used for pasture, bushes or forest (*monte*). While there was significant, although limited, use of chemical fertilisers (6,500 tons, all phosphate) cattle played an important role not only as a source of income but also in maintaining and enhancing soil fertility.

Table 11.2 illustrates the land-use structure at that time. The cropping area occupied only a small part of the territory, with very intensive rotations of mixed farming. Likewise the area dedicated to pasture was small. Most of the area was

Table 11.2 Land-use patterns in Galicia and its four provinces, 1922 (hectares)

	A Coruña	*Lugo*	*Ourense*	*Pontevedra*	*Galicia*
Horticulture	3,052	0	3,392	1,200	7,644
Orchard		1,000	6,014		7,014
Olive grove		110	104		214
Vineyard	864	4,800	17,100	7,610	30,374
Industrial crops	172	400	755	200	1,527
Roots, tubers and bulbs	30,559	91,000	39,757	6,500	167,816
Cereals and pulses	164,575	90,200	80,332	92,300	427,407
Permanent artificial meadows	43,181	12,500	0	14,000	69,681
Useful Agricultural Area (UAA)	242,403	200,010	147,454	121,810	711,677
Natural meadows	37,157	52,000	11,200	0	100,357
Scrubland	281,100	72,000	305,080	181,912	840,092
Forest and pasture	155,300	40,000	19,500	89,900	304,700
Forest without pasture	39,590	25,000	203,388	20,348	288,326
Monte unspecified (probably scrubland)		410,000			410,000
Monte	513,147	599,000	539,168	292,160	1,943,475
Unproductive	10,572	72,000	1,862	21,162	105,596
Total (UUA+*monte*+ Unproductive)	766,122	871,010	688,484	435,132	2,760,748

Source: Ministerio de Fomento (1923).

scrubland (part of the *monte*) which played a decisive role in fertility replacement and livestock feeding. Thus, most of the forest area was under pluri-functional management with a low tree density, that made it compatible with pasturing. More importantly, and a typical feature of the organic agro-ecosystem, no productive land was left unexploited. This is in stark contrast with the situation today where much land has fallen out of productive use and lies abandoned (López Iglesias 1996) – as will be shown in the next section. All the landscape elements were interconnected by having different simultaneous functions: economic, ecological and social.

Factors that triggered the progressive dismantling of the organic agro-ecosystem

A range of different factors – political, demographic and socio-economic – contributed to the progressive dismantling of the Galician organic agro-ecosystem in the second half of the 20th century (Dominguez 2007). Two of them were of vital importance: afforestation and the modernisation of farming that took place after 1960.

In the period 1936–1970 the Franco regime encouraged afforestation in Galicia. This particularly affected *monte* areas and entailed significant changes in land-use, which led to progressive changes in landscape. The afforestation of the *monte* areas enlarged the forest areas and diminished the *monte baixo* (scrub and bushes) which played a central role in providing different inputs and functions to the Galician organic agro-ecosystem. Moreover, forest enlargement was largely achieved by planting non-native varieties which significantly contributed to spreading forest fires.

The process of afforestation carried out under the National Forest Plan (NFP) of 1941, was different from the afforestation that took place at the end of the 19th century. These earlier changes mostly affected *monte*s under private ownership and did not entail any real change to the forestry use but more a diversification of uses (Soto 2006). By contrast, the government expropriation under Franco's dictatorship had more impact on *communal monte*s and dramatically altered the way in which the Galician organic agro-ecosystem operated: the *monte* area was taken out of a pattern of integrated farm uses and replaced with an intensive monoculture of fast growing forest species.

The aim of the government was to develop self-sufficiency in wood supplies and the government's programme was subsequently reinforced by private speculative afforestation (Rico Boquete 1995). In the end, both had the result of taking the *monte* out the agricultural system by shifting the emphasis towards wood production (Bouhier 1979). When the *communal monte* were returned to the original owners by the legal reforms that occurred between 1950 and 1970 (Grupo dos Comúns 2006), the farmers did not have the same use for this resource. Even in those areas where the *monte* had not been expropriated, it was no longer integrated within the farming system because the socio-demographic and economic conditions had drastically changed as agriculture followed a more agro-industrial trajectory.

The modernisation of farming was the second important factor that triggered the dismantling of the organic agro-ecosystem. Despite a late start, the Galician farming sector was transformed dramatically from 1960 onwards via a process of homogenisation, specialisation and intensification in which production became strongly oriented to dairy and poultry farming, which was increasingly disconnected from the land, locally available resources and highly dependent on a range of external inputs. This disconnection and the associated process of land demobilisation is described in the next section

Dismantling the organic landscape: land demobilisation

Changes in land-use patterns can be used as a way of analysing social metabolism. Here we will show the main changes in the land structure and how these have been linked to changes in agrarian activities. This offers a good insight into the reasons underlying the landscape changes. An analysis of land-use pattern dynamics, changes in use and particularly land demobilisation highlights the dismantling of the organic agro-ecosystem.

Over the last 50 years or so, the structure of Galician farming has changed significantly. Probably the most important change has been the disappearance of

Table 11.3 Dynamics of the structural basis of the Galician agrarian sector, 1962–1999

	1962	*1972*	*1982*	*1989*	*1999*	*Net change 1962–1999*
Number of farms	421,294	385,832	360,436	358,886	268,995	−152, 259
Total Area (TA)	2,424,352	2,456,592	2,241,968	2,217,145	2,041,799	−382, 554
UAA	629,718	687,815	655,043	675,039	696,690	66,972
TA/number of farms	5.75	6.37	6.22	6.18	7.56	1.8
UAA/number of farms	1.49	1.78	1.82	1.88	2.58	1.1
Number of plots	9,544,981	7,289,176	5,454,555	5,445,081	4,717,093	−4, 827, 888
Plots per farm	22.66	18.89	15.13	15.17	17.5	−5.2
TA/number of plots	0.25	0.34	0.41	0.41	0.43	0.2
UAA/number of plots	0.07	0.09	0.12	0.12	0.15	0.1
UAA/TA (percentage)	26	28	29.2	30.4	34.1	8.1

Source: INEa (various years) UAA: Useful Agricultural Area (including land for crops and pasture); TA: Total Area (including UAA, bush and scrub, forest area and other areas).

Table 11.4 Dynamics of the structural basis of the Galician agrarian sector, 1987–2005

	1987	*1997*	*2003*	*2005*	*Net variation 1987–2005*
Number of farms*	227,359	114,781	101,537	91,452	−135, 907
TA	1,329,935	1,070,598	1,130,386	1,112,643	−217, 292
UAA	710,514	621,552	724,624	732,759	22,245
UAA/number of farms	3.1	5.4	7.1	8.0	4.9
TA/number of farms	5.8	9.3	11.1	12.2	6.3
UAA/TA	53.4	58.1	64.1	65.9	12.4

Source: INEb (various years).

*Farms with UAA.

farms: estimated at 35–55 per cent – depending on the source and time period (see Tables 11.3 and 11.4).[8] Yet at the same time two classical structural features of the Galician agrarian sector have remained relatively unchanged: the small size of farms and the high number of small plots per farm. The reduction in the number of farms and the persistence of small sized farms suggests a high level of land abandonment. In fact, the Useful Agricultural Area (UAA) has slightly increased while the Total Area (TA) has declined significantly. This trend continues – even in the most recent period (between 1987 and 2005), Table 11.4 shows that there has been a strong reduction in the TA (−16 per cent) and a small increase in the UAA (3.1 per cent). These dynamics explain the change in the ratio of UAA/TA from 53.4 per cent in 1985 to 65.9 per cent in 2005. Historically in Galicia this ratio has been much lower and has been often used to highlight the constraints caused by a shortage of productive land (López Iglesias 2000).

However, some aspects of the UAA/TA ratio need further explanation. Firstly, this ratio has been traditionally low in Galicia because the *monte* areas, which helped articulate the organic agro-ecosystem, were not been statistically considered as agricultural land (and thus not included within the UAA). Secondly, the

Table 11.5 Evolution of the number of farms, the UAA and the TA between 1987 and 2005. UAA groups by size (> and < 10 ha)

	Net variation 1987–2005						
	Number of farms	*% farms**	*UAA*	*% UAA***	*TA*	*% TA****	*UAA/TA*****
< 10 ha	−144,568	−18	−285,614	−41.1	−557,393	−35.6	−0.1
> 10 ha	8661	18	307,859	41.2	340,101	35.6	15.3
Total	−135,907	63.3	22,245	3.1	−217,292	−16.3	

Source: INE (1987 and 2005).

*Percentage of farms that have disappeared.
**Change in UAA (in percentage).
***Change in TA (in percentage).
****UAA/TA (in percentage).

increase in this ratio in recent decades corresponds with a slight increase in the UAA, and with a strong decline in the TA. This is because those lands that were previously used by the farming system for agricultural purposes, without being part of the UAA, no longer appear in agrarian statistics; i.e., land that had been productive has now been abandoned or demobilised. Thus the higher ratio in recent years is not due to any increase in average farm size but because of the abandonment of *monte* areas, and the loss of their productive functions.

The sharp decline in the number of farms might have been expected to generate some transfer of farming land, with land that has been abandoned or neglected being transferred from one farm to another through renting, selling, inheritance, etc. However in Galicia these mechanisms are rarely used, although the use of land has been changing, for instance from cropping or forest to pasture. These changes are shown in Table 11.5, which gives the net changes in UAA and TA for the period 1987–2005 for small (less than 10 ha) and large farms (more than 10 ha). The increase in the UAA on the bigger farms outweighs the loss on the smaller farms (by 22,245 hectares). This shows that overall there has been an increase in land that is categorised as UAA in this period, highlighting another form of mobilisation.

Land dynamics can also be analysed by considering the different land-uses. The Agrarian Census (several years) shows a number of different processes that contributed to the evolution of land-use during the industrialisation of Galician farming (see Table 11.6). Firstly, there was a change in the structure of the UAA (crops, and meadows and pastures), with a reduction of the area used for crops and an increase in the area used for pasture. This change is consistent with the specialisation in livestock that occurred as Galician farming industrialised. This change also led to spaces that formerly had a productive function within the organic agro-ecosystem (e.g. the scrublands – *monte baixo*) to become abandoned or degraded. Furthermore, despite the policy of afforestation since the mid-1930s onwards, the forest areas have not increased significantly. On the contrary, they

Table 11.6 Change in land-use 1962–1999 (in hectares)

Year	*Crops*	*Meadows and pasture*	*Forest*	*Scrubland*	*Unproductive*	*Total*
1962	476,499	152,938	616,742	1,089,399	88,429	2,424,007
1972	444,121	243,694	743,127	935,437	90,217	2,456,596
1982	330,798	324,245	714,296	843,754	28,875	2,241,968
1989	245,977	429,063	nd**	nd	nd	nd
1999*	258,878	437,812	605,161	648,892	91,056	2,041,799

Source: INEa (various years).

*For the year 1999, the "unproductive" category includes land with eiral and esparto and "other areas".
**nd – no data.

have regularly been affected by forest fires, an emerging problem in Galicia since the 1960s. The increase in forest fires is related to the withdrawal of the *monte* areas from the agrarian system (and the organic agro-ecosystem) and the loss of its functionality for the local communities (see next section).

The data from Tables 11.5 and 11.7 show that there was an overall increase in UAA (22,245 ha) between 1987 and 2005. This did not come about because of transfers of land from small farms that disappeared, but rather from a change in the use of some lands within the WEB category (Wasteland, Esparto grass, and Bushes/or scrub) or forest, which declined by 230,301 ha and 23,874 ha respectively (Table 11.7). On the basis of historical patterns we might conclude that most land-use changes would come from the forest category, as this land usually has more fertile soils than the WEB category and is closer to, or more accessible from, farms (López Iglesias 1996).

However, it is evident that, between 1987 and 2005, the category of WEB lands declined the most (both in small and big farms). These changes were the most significant contributor to the net decline in Total Area (217,292 hectares between 1987 and 2005). Some allowance should be made here for methodological changes in statistical collection/definitions and the possibility that a small part of those lands could be given over to forest use (therefore, falling outside the agricultural survey).Even making such allowances we can state that around 200,000 hectares of land disappeared from the survey in the 18 years to 2005 as they no longer fulfilled any farming function. This land was demobilised as the traditional uses were abandoned.

In summary, the recent dynamics of the territorial structure of Galician farms highlights the dismantling of the organic agro-ecosystem as a result of the industrialisation of farming. There have been changes in the structure of the UAA, a reduction of the area dedicated to crops, an increase in pasture areas (because of specialisation in livestock), while spaces (i.e. scrublands and, to a lesser extent, forests) that previously had a productive function in the organic agro-ecosystem become abandoned or degraded. This has had a detrimental effect on the landscape, leading to an increase in forest and *monte* fires (especially since the 1960s) and has further reduced the value of these areas for local communities.

Table 11.7 Land-use dynamics in the Galicia agrarian sector, 1987–2005 (in hectares)

	1987	*2005*	*Net variation 1987–2005*
Crop lands	334,621	166,268	–168,353
Permanent pasture	375,893	566,491	190,600
Meadows	324,067	445,021	120,954
Others	51,834	121,469	69,635
UAA	710,514	732,759	22,245
Other lands*	619,421	379,884	–239,537
WEB**	355,136	124,835	–230,301
Forest area	253,360	229,486	–23,874
• Non-commercial	103,235	91,181	–12,054
• Commercial	150,125	138,305	–11,820
Other areas***	10,925	25,563	14,638
Total area****	1,329,935	1,112,643	–217,292

Source: Own calculations based on EEEA (1987) and (2005) (only farms with UAA, and based on data from 1,130,922 farms).

*The category "Other lands" includes: WEB, forest area (commercial and non-commercial), and "other areas" (including those lands that, being part of the TA of the farm, are neither UAA, nor WEB, nor forest land).

**Wastelands, Esparto grass land and Bushes.

***Other areas: those for non-farming or productive purposes.

****Total area = the sum of crop lands, permanent pasture (UAA) and Other lands.

Forest fires, abandonment and the commons

The Galician organic agro-ecosystem used fire (*rozas* or *estivadas*) as an agricultural method of shifting cultivation. Knowledge about how to do this was perfected over the centuries, limiting the possibilities of starting catastrophic fires. The dismantling of the agro-ecosystem and the subsequent abandonment of the areas, especially those covered by scrub, altered the culture of fire. As Lage (2003) highlights, forest fires have become the symptom of a traumatic and conflict-ridden change in the way land-use is organised. The data speak for themselves: between 1961 and 1997, 132,233 fires burned 1,411,384 hectares, almost half (47.9 per cent) of the total area of Galician area. Between 1970 and 2006 around 1,700,000 hectares have burned: 60 per cent of these being "non-tree areas", mostly brush and bushes – vegetation forms that dominate the WEB category (Conselleria do medio rural 2006).

The problem is far more than localised, its effects spread well beyond the areas where the fires take place, damaging not only the soil and vegetation, but also the water, atmosphere and animal habitats. Fire is one of the main causes of soil erosion in Galicia (as it denudes soil). Fires destroy vegetation – when they are frequent, the only vegetation that remains is brush. They also affect fauna – the most affected species are insects and invertebrates, small mammals, reptiles, etc., but large vertebrates also often cannot escape them due to fences to control livestock and hunting. They also destroy habitats and feeding resources, forcing many species to migrate to other areas. Aquatic species are also affected by the sediment caused by subsequent soil erosion and the destruction of river vegetation. Moreover, fires are also an important source of atmospheric pollution, emitting oxides (nitrate and sulphur) as well as contributing to climate change because of CO_2 emissions.

There are two main reasons why the *monte* has been so badly affected by fires in recent decades, both directly related to changing agricultural activities. The first reason is that public and private afforestation has mostly planted tree varieties, which are highly flammable,[9] such as eucalyptus and pines. Second, and more importantly, the *monte* burns because it is abandoned. As a consequence of the dismantling of the organic agro-ecosystem nobody "cleans out" the area any more; the brush and scrub accumulate there, contributing to extending the fires. The *monte* is also kept out of use because of structural problems: too many widely scattered plots, and the lack of good long term planning and adequate institutional control. This is still an important issue for agriculture, since as we have shown, farms are still in general rather small and the use of suitable *monte* lands for pasture could enlarge the territorial basis of those farms and close the gap between the UAA and the Total Area. This would help both reduce the fire risk and improve the economic and productive performance of the farms by better managing and increasing the use of locally available natural resources.

One also needs to ask about the impact of fires on the sustainable management of the *monte* and on the relation to the type of land property. As said above, common ownership played a crucial function within the organic agro-ecosystem – ensuring the collective management of part of the territory that was essential for peasant production. The return of the property rights to the local communities at the end of the Franco period and during the transition to democracy did not lead to local communities becoming (re)engaged in the management of this resource, nor did it lead to more sustainable use of the resources. On the contrary, the separation between agriculture and *monte* meant that most communities with *monte* resources had little interest in managing and protecting them. Only a small number of communities sought to build a relationship with the *monte* that combined conservation with income.

Today, the management of private *monte* is more sustainable than that of public *monte* (Grupo dos Comúns, 2006). This does not imply a defence of the line that private property is more sustainable than the commons. On the contrary, the Galician commons followed a pattern of sustainable management during the 19th and first half of the 20th centuries. However, the form of property rights is not enough to guarantee good management of the agro-ecosystem, especially when that management does not seek to integrate the interrelations between the different sub-components of a system, the specific environment where they are located, and the general pattern of resource use.

Concluding remarks

In this chapter, we have sought to extend the more traditional approach of analysing landscape through subjective (perceptual) means and to include the material elements that largely condition the landscape and its performance. We have also extended our analysis beyond the economic, to look at ecological functioning and integrity. To do this we have used social metabolism as a tool to describe the articulation among different land-uses. We started by examining the Galician organic agro-ecosystem that was in force until the middle of the 20th century. In terms of sustainability this can be considered to be a well functioning system. The system, which integrated the *monte* with other ecosystems, helped to create a rural landscape characterised by a high level of biodiversity, heterogeneity and a contextually specific and sustainable pattern of land and soil use.

We provided a first estimation of the agrarian metabolism in the province of A Coruña in the year 1922, when this agro-ecosystem was still operative. There, we see how the use of the *monte* is not due to the underuse of the soil but, on the contrary, due to the need to maximise the efficiency of soil use by using a highly productive area suitable for cattle specialisation. The high production of biomass at this time provides proof of the efficiency and sustainability of this management.

The analysis of the subsequent changes in land-use patterns showed how the process of agricultural industrialisation drastically altered the interrelations among the different land-uses. This process caused a loss of the functionality of many spaces, especially the *monte*, provoking abandonment or an intensive forestry use, far removed from farming.

The loss of the *monte's* function has had several negative impacts on the landscape. Afforestation, aimed at increasing wood production has led to plantations of fast growing species replacing more diverse and more fire-resistant autochthonous vegetation. The incidence of forest fires, practically non-existent five decades ago, has increased as a result of afforestation and the abandonment of the *monte's* functions as a resource base. The result is the increasing disarticulation of the Galician landscape.

This work shows the importance of studying the landscape from a dynamic and historical perspective. Long run studies are useful in highlighting changes that have taken place in territorial organisation and the origin of (and possible solutions for) contemporary problems.

Last but not least, the paper addresses the question of common goods since the *montes* were largely under this sort of property arrangement. No property arrangements can necessarily guarantee that an agro-ecosystem is well managed. Good management also needs to take into account the imperatives of sustainability. Only when (common or private) property is combined with a management that aims at fostering the interrelations among different components of a landscape or different (agro) ecosystems and the environment where these agro-ecosystems are located, can sustainability be increased.

Notes

1 This is not a full discussion of the implications of sustainability. González de Molina and Guzmán Casado (2006) provide a more detailed discussion of this point in relation to traditional and current agrarian systems.

2 This means taking into account the quantity of energy and material flows, the quality of those flows, whether or not they are renewable and whether they are reabsorbed by nature or they are polluting, etc.

3 Crops for human nourishment, pastures for cattle feeding (cattle played a very important role in maintaining soil fertility and for traction) and forests.

4 Lana Berasain (2008) discusses the difference between common property and common rights.

5 Martínez Alier (1992, 1995) and Ortega Santos (2002) discuss the concept of disarticulation, applying it to historical studies of the commons. Disarticulation refers to both the change in the property rights (e.g. from commons to private) as well as to changes in land-use which can produce a rupture between the different parts of an (agro) ecosystem.

6 The *rozas* culture or *estivadas* involved the summer burning of brush and scrub (*matogueira*) that was uprooted in the spring. The ashes were spread over the area to be seeded in the autumn. Usually, only one (or two depending on the quality of the soil) harvests of cereal or gorse (*Ulex Europeus*) was obtained in this manner. After harvesting, the area was again left fallow for several years (from 8 to 30 years), depending on the soil (Bouhier 1979; Balboa 1990; Soto 2006).

7 Access to common land has historically determined the configuration and low ratio of Useful Agricultural Area (UAA)/Total Area (where the UAA includes crops and pasture and the TA includes scrub, forest and other non-farming areas). More information about this is provided in the main text.

8 To cover a longer period, Tables 11.3–11.7 have been constructed using data from two different statistical sources. The Agrarian Census INEa (various years) covers the period 1962–1999, and the *Encuesta sobre la Estructura de las Explotaciones Agrarias* – EEEA (Survey of the Structure of Agrarian Farms) INEb (various years) – the period 1987–2005. The two sources use different samples and methodologies and the absolute figures are not comparable, although they show the same dynamics (Dominguez 2007).

9 The third National Forest Inventory (INF 1997–2006) describes the degree of combustion of the Galician forest as "extreme or high" in 85 per cent of cases. The 2006 classification of the forest systems shows that around 70 per cent of the forest area is occupied by highly flammable species (*Pinus pinaster*, *P. radiata* and *P. silvestris*, and *Eucalyptus globulus*), 24 per cent is occupied by lower and autochthonous lush-trees (*Castanea sativa*, *Quercus robur* and *Q. pyrenaica*) and the rest is occupied by different combinations of these tree-varieties (*Consellereia do medio rural* 2006).

References

Balboa, X. (1990) *O monte en Galicia*. Vigo: Edicións Xerais de Galicia (in Spanish).

Berkes, F. (2002) 'Cross scale institutional linkages: Perspectives from the bottom up'. In: E. Ostrom, T. Dietz, N. Dolsak, P.C. Stern, S. Stonich and E.U. Weber (eds), Committee on the Human Dimensions of Global Change, *The Drama of the Commons*. pp. 293–321, Washington DC: National Academic Press.

Berkes, F and Folke, C. (1998) 'Linking social and ecological systems for resilience and sustainability'. In: F. Berkes (ed.), *Linking Social and Ecological Systems*. pp. 1–25, Cambridge: University Press.

Bouhier, A. (1979) *La Galice. Essay geographique d'annalyse et d'interpretation d'un vieux complex agraire*. La Roche-sur-Yon: Imprimerie Yonnaise (in French).

Burel, F. and Baudry, J. (2001) *Ecología del paisaje, conceptos, métodos y aplicaciones*. Madrid: Mundiprensa (in Spanish).

Carmona Badía, X. and Puente, L. de la (1988) 'Crisis agraria y vías de evolución ganadera en Galicia y Cantabria'. In: R. Garrabou (ed.), *La Crisis Agraria de Fines del Siglo XIX*. pp. 191–211, Barcelona: Crítica (in Spanish).

Carmona Badía, X. (1982) 'Sobre as orixes da orientación exportadora na producción bovina galega. As exportacións a Inglaterra na segunda metade do século XIX', *Grial*, Anexo I, Historia (in Spanish).

Consellería do medio rural (2006) *O monte en cifras*, Xunta de Galicia. http://mediorural.xunta.es/institucional/publicacions/forestal/o_monte_en_cifras/ (accessed 8 September 2011) (in Spanish).

Conway, G.R. (1985) 'Agro-ecosystem analysis', *Serie Agricultural Administration (RU)*, 20: 31–55.

Conway, G.R. and Barbier, E.B. (1990) *After the Green Revolution*. London: Earthscan Publications Ltd.

De Moor, T. (2009) 'Avoiding tragedies: A Flemish common and its commoners under the pressure of social and economic change during the eighteenth century', *Economic History Review*, 61(1): 1–22.

Domínguez, M.D. (2007) *The Way You Do, It Matters*. Wageningen: Wageningen University (PhD Thesis).

Fischer-Kowalski, M. (1998) 'Society's metabolism. The intellectual history of materials flow analysis. Part I, 1860–1970', *Journal of Industrial Ecology*, 2(1): 61–78.

Fischer-Kowalsky, M. and Haberl, H. (1994) 'On the cultural evolution of social metabolism with nature', *Schriftenreihe Soziale Ökologie*, Vienna: Iff, Band 40.

Fischer-Kowalsky, M. and Hüttler, W. (1999) 'Society's metabolism: The intellectual history of materials flow analysis, part II, 1970–1998', *Journal of Industrial Ecology*, 2(4): 107–129.

Forman, R.T.T. and Gordon, M. (1986) *Landscape Ecology*. New York: John Wiley and Sons.

Garrabou, R. and González de Molina, M. (eds) (2008) *La Reposición de la fertilidad en los Sistemas agrarios tradicionales*. Barcelona: Icaria (in Spanish).

Garrabou, R., Cussó, X., Olarrieta, J.R. and Tello, E. (2006) 'Balances energéticos y usos del suelo en la agricultura catalana: una comparación entre mediados del siglo XIX y finales del siglo XX', *Historia Agraria*, 40: 471–500.

González de Molina, M., R. García Ruiz, D. Soto, G. Guzmán Casado, A. Herrera and J. Infante Amate (2009) 'Nutrient Balances and Management of Soil Fertility in Andalusia, Spain (18th and 19th centurias)', Paper presented at 1st World Congress of Enviromental History, Copenhagen (Denmark), August 2009.

González de Molina, M. and Guzmán Casado, G. (2006) *Tras los pasos de la insustentabilidad. Agricultura y Medio ambiente en perspectiva histórica (siglos XVIII-XX)*. Barcelona: Icaria.

González de Molina, M. and Guzmán Casado, G. (2008) 'Transformaciones agrarias y cambios en el paisaje. Un estudio de caso en el sur peninsular (1752–1997)'. In: R. Garrabou, R and J.M. Naredo (eds) *El paisaje en perspectiva histórica. Formación y transformación del paisaje en el mundo mediterráneo*. pp. 199–231, Zaragoza: Sociedad española de historia agraria.

Grupo dos Comúns (2006) *Os montes veciñais en man común: o patrimonio silente. Naturaleza, economía, identidade e democracia na Galicia rural*. Vigo: Xerais.

Guzmán Casado, G. and González de Molina, M. (2009) 'Preindustrial agriculture versus organic agriculture. The land cost of sustainability', *Land Use Policy*, 26(2): 502–510.

Holling, C.S. (1993) 'Investing in research for sustainability', *Ecological Application*, 3(4), Ecological Society of America: 552–553.

INE (1987) *Encuesta sobre la Estructura de las Explotaciones Agrarias (EEEA) 1987*. Instituto Nacional de Estadística, Madrid.

INE (2005) *Encuesta sobre la Estructura de las Explotaciones Agrarias (EEEA) 2005*. Instituto Nacional de Estadística, Madrid.

INEa (various years) *Censos Agrarios 1962, 1972, 1982, 1989 and 1999*. Instituto Nacional de Estadística, Madrid.

INEb (various years) *Encuesta sobre la Estructura de las Explotaciones Agrarias (EEEA), 1987, 1997, 2003, and 2005*. Instituto Nacional de Estadística, Madrid.

Krausmann, F. and Cunfer, G. (2009) 'Sustaining soil fertility: Agricultural practice in the old and new worlds', *Global Environment*, 4: 48–77.

Laerhoven, F. and Ostrom, E. (2007) 'Traditions and trends in the study of the commons', *International Journal of the Commons*, 1(1): 3–28.

Lage, X.A. (2003) 'El monte, el cambio social y la cultura forestal en Galicia', *Revista de Investigaciones Políticas y Sociologicas* 2(1–2): 109–123, Santiago de Compostela: Universidade de Santiago de compostela.

Lana Berasain, J.M. (2008) 'From equilibrium to equity. The survival of the commons in the Ebro Basin: Navarra from the 15th to the 20th centuries', *International Journal of the Commons*, 2(2): 162–191.

Lois, R. (2005) 'Paisaxe and despoboamento rural'. In: V. Paül (coord.) *Paisaxes galegas: unha escolma plural de olladas ás paisaxes de Galiza*. Santiago de Compostela: Xunta de Galicia, Dirección Xeral de Turismo.

López Iglesias, E. (1996) *Movilidad de la tierra y dinámica de las estructuras agrarias*. Madrid: Serie Estudios, MAPA.

López Iglesias, E. (2000) 'El sector agrario gallego a las puertas del siglo XXI: Balance de sus transformaciones recientes', *Revista Galega de Economía* 9 (001): 1–30. Santiago de Compostela: Universidade de Santiago de Compostela.

López-Ridaura, S., Masera, O. and Astier, M. (2002) 'Evaluating the Sustainability of complex socio-environmental systems, the MESMIS framework', *Ecological Indicators*, 2(1): 135–148.

Loureiro, M.L. and Barrio, M. (2009) *Valoración medioambiental, cultural y paisajística de los espacios rurales gallegos: una perspectiva económica*. Santiago de Compostela: CIEF-Fundación Caixa Galicia.

Marey Pérez, M.F., Crecente Maseda, R. and Rodríguez Vicente, V. (2003) *Claves para comprender los usos del monte en Galicia (España) en el siglo XX*. Departamento de Ingeniería Agroforestal, Escuela Politécnica Superior de Lugo. Universidade de Santiago de Compostela, España. http://laborate.usc.es/files/Monte.pdf (accessed 9 July 2010).

Martínez Alier, J. (1992) 'Pobreza y Medio Ambiente. A propósito del Informe Brundtland'. In M. González de Molina and J.A. González Alcantud (eds), *La Tierra: Mitos, Ritos y Realidades*. pp. 295–332, Granada: Anthropos/Diputación Provincial de Granada.

Martínez Alier, J. (1995) 'Political ecology, distributional conflicts and economic incomensurability', *New Left Review*, 211: 70–88.

Ministerio de Fomento (1921) *Materiales fertilizantes empleados en la agricultura*. Madrid: Imprenta de M.G. Hernández.

Ministerio de Fomento (1922) *Censo de la riqueza pecuaria*. Madrid: Servicio de publicaciones agrícolas.

Ministerio de Fomento (1923) 'Avance estadístico de la producción agrícola en españa', Resumen hecho por la Junta Consultiva Agronómica de las memorias de 1922 remitidas por los ingenieros del servicio agronómico provincial, Madrid: Imprenta de M.G. Hernández.

Ortega Santos, A. (2002) *La tragedia de los cerramientos. La desarticulación de la comunalidad en la provincia de Granada*. Valencia: Fundación Instituto de Historia Social.

Ostrom, E. (1990) *Governing the Commons. The Evolution of Institutions for Collective Action*. Cambridge: Cambridge University Press.

Rico Boquete, E. (1995) *Politica Forestal en Repoboacions en Galicia (1941–1971)*. Monografías de la Universidad de Santiago de Compostela, no. 187, Santiago de Compostela.

Schandl, H., Grünbühel, C., Haberl, H. and Weisz, H. (2002) *Handbook of Physical Accounting. Measuring bio-physical dimensions of socio-economic activities MFA – EFA – HANPP*, Viena: Social Ecology Working Paper 73.

Simón, X. (1995) *A Sustentabilidade nos Modelos de Desenvolvemento Rural. Unha análise aplicada de agro-ecosystemas*. University of Vigo: PhD Thesis (not published).

Soto, D. (2006) *Historia dunha Agricultura Sustentábel. Transformacións productivas na agricultura galega contemporánea. Xunta de Galicia, Consellería de Medio Rural*. Colección EIDOS, n. 4.

Soto, D., Infante, J., González de Molina, M. (2010) 'The second wave of socio ecological transition in agriculture. The social metabolism of Spanish agriculture (1900–1936)', Ester Boserup Conference 2010. A Centennial tribute. IFF, Viena, 15–17 de Noviembre de 2010.

Tello, E., Garrabou, R., Cussó, X. and Olarieta, J.R. (2009) 'On the Sustainability of Mediterranean Agricultural Systems: fertilising methods and nutrient balance in Catalonia (Spain), 1850–1936'. Paper presented at *1st World Congress of Environmental History*. Copenhagen Denmark, August 2009.

Tello, E., Garrabou, R., Cussó, X. and Olarieta, J.R. (2008) 'Una interpretación de los cambios de uso del suelo desde el punto de vista del metabolismo social agrario. La comarca catalana del Vallès, 1853–2004', *Revista Iberoamericana de Economía Ecológica*, 7: 97–115.

Toledo, V. and Barrera Bassols, N. (2008) *La memoria biocultural. La importancia ecológica de los saberes tradicionales*. Barcelona: Icaria/Consejería de Agricultura y Pesca de la Junta de Andalucía.

Toledo, V. and González de Molina, M. (2006) 'El metabolismo social: las relaciones entre la sociedad y la naturaleza'. In: F. Garrido, M. González, J.L. Serrano, and J.L. Solana (eds), *El Paradigma ecológico en las Ciencias sociales*. pp. 85–112, Barcelona: Icaria.

Van Zanden, J.L. (1999) 'The paradox of the Marks. The exploitation of commons in the East Netherlands, 1250–1850', *Agricultural History Review*, 47: 125–144.

Wrigley, E. (1993) *Cambio, continuidad y azar. Carácter de la revolución industrial inglesa*. Barcelona: Crítica (in Spanish).

12 What can hedonic analysis tell us about the value of landscapes?

Jan Rouwendal and J. Willemijn Weijschede-van der Straaten

Introduction

One of the main tasks of Dutch spatial planning is the preservation of valuable landscapes. Landscapes, in our definition, are considered to be valuable in the sense that significant changes in these landscapes are experienced by many as a loss. For instance the typically Dutch polder landscape has, according to many, a high aesthetic value that justifies costly measures to protect it like the construction of a tunnel for the high-speed railway that crosses the so-called Green Heart of the Randstad which is dominated by this landscape type. There are other important examples of landscapes that are regarded as valuable in other parts of the country, for instance as the typical "bocage" landscape around Ootmarsum in the Twente region (eastern part of the Netherlands), and the heath field in parts of Drenthe (northern part of the Netherlands). These landscapes are closely connected to particular types of land-use (such as cattle farming in the case of the polders). As long as this specific land-use type is the most profitable one, in the conventional economic sense that no other use of the land gives the owner a higher profit, there is no need for special protection measures. However, if other land-use types become more profitable the landscape would disappear or its quality would decline. And hence land-use planning might become necessary.

Land-use policy is clearly not costless. Prohibiting the switch to more profitable types of land-use imposes an opportunity cost on the owner of the land, while the benefits of preserving the landscape can partly be experienced by others, for instance non-farmers who reside in the area or those who travel through it. Land-use planning that intends to protect landscapes therefore has to deal with the diverting interests of different stakeholders.

From the point of view of economic analysis, the need for and the use of planning arises from an external effect. The way land is used has consequences for the owner of the land, but also for those who travel through, live in or close by an attractive landscape. Since these people do not pay the farmers for this pleasant experience, the farmer cannot be expected to take their loss of benefit into account when he considers a switch to a different type of land-use that would result in a less attractive landscape. It is the task of spatial planning to take into account all effects of changes in land-use and to promote the general interest.

A main argument in the literature concerned with optimal land-use planning is that the proximity of a scenic landscape improves residential living conditions (see, for example, Luttik 2000; Geoghegan 2002; Ward Thompson 2002; Chiesura 2004; Mansfield et al. 2005). The benefits generated by the landscape therefore depend on the number of people who live in that landscape or close to it. This provides an argument for allowing people to live in such landscapes, since this would increase the total amount of benefits. However, if many people are allowed to do so, it is highly likely that this has a negative impact on the quality of the landscape. In the absence of protective measures, valuable landscapes can deteriorate too much: meaning that more benefits could be generated had the landscape been protected.

This "tragedy of the commons" highlights the trade-off that is central to the analysis that follows. In many cases, but especially in the vicinity of urban areas, the most profitable alternative to the agricultural land, that is associated with an attractive landscape, is residential land. And while the benefits of the marginal inhabitant of the attractive area may be larger than the price of agricultural land, the effect of more residential land implies a reduction of the benefits of all the existing residents. Again it is the task of land-use planning to take this effect into account.

In this chapter we analyze this issue by developing a formal model for protecting a valuable landscape. This results in a cost–benefit rule that can be made operational through hedonic price analysis. This will be illustrated by an application of the model to the optimal provision of open space in cities. We will argue that there are important similarities between this issue and the protection of valuable landscapes. We continue with a discussion of the limitations of hedonic analysis for the protection of valuable landscapes.

Model for optimal land-use

In this section, we develop a model of a city in which a benevolent planner provides open space by means of spatial planning policy. The model can be applied to all kinds of open space, for example dunes, forest or, as will be done in the next section, parks and public gardens in cities. The model is outlined in the next sub-section and the policy evaluation question is considered in the sub-section thereafter. The third sub-section discusses the suggested cost–benefit rule and its applicability in real world circumstances.

The model

Our model has four elements: a utility function and a budget constraint for households, a space constraint and a welfare function. We start with the first two, and then determine the outcome of the model in the absence of spatial planning. Then, we take the space constraint into consideration. In the next section we introduce a spatial planning policy that is based on a social welfare function.

Households derive utility from consumption goods, which we treat as a single aggregate commodity, from residential land and from the landscape in which they live. The analysis refers to a single area and assumes that the quality of the landscape depends on the amount of land that is not in residential use.[1] If the total amount of land reduces, the quality of the landscape deteriorates. The model focuses on the amount of land that has the original scenic quality which is identified with the amount of land in agricultural use.[2] We will refer to this variable from now on as open space. Utility is denoted as u and we write:

$$u = u(c, h, S), \tag{12.1}$$

where c denotes consumption of the aggregate commodity, h that of land and S is the amount of open space. For an individual household, the amount of open space is given, but it has to determine the quantities of the other two goods itself. It does so by maximizing utility subjected to the budget constraint:

$$c + ph = y - tx, \tag{12.2}$$

where p denotes the price of residential land, y is income, t transportation cost and x the commuting distance. The price of the aggregate consumption good has been normalized to 1.

In the absence of spatial planning, the price of residential land is equal to its value in agricultural use. The commuting distance has been introduced in this equation because the interaction between labor market accessibility and land-use planning is an important one. This relationship is studied extensively in the monocentric model (see, for instance, Fujita (1989) for an exposition of that model), but we do not adopt that setting here. We interpret x as a general indicator of the nearness of employment, such as the average distance to the nearest 100,000 jobs, which is known to be highly correlated with the value of residential land in the Netherlands. We assume that all households have the same preferences and incomes.

Households attach value for the land in the area we study. This is reflected in the amount of money they are willing to bid for that land. To see how that works, we consider the situation in which the household can reach utility u^* by living in the urban area which is closest to their job. It wants to reach at least the same level of utility if it locates at another place in the area we study. We can therefore rewrite (12.1) as $u^* = u(c, h, S)$, and invert this equation in such a way that consumption c appears on the left-hand side. The interpretation is that c gives the amount that has to be consumed of the aggregate good if the consumer has to reach utility u^* and uses h units of residential land, while the amount of open space is equal to S. We write this relationship as $c = C(u^*, h, S)$.

The household always has to satisfy its budget restriction, and to see what this means for the price of land a consumer is able to pay, we can substitute the function C into (12.2) and rearrange terms to find:

$$p = \frac{y - tx - C(u^*, h, S)}{h}. \tag{12.3}$$

The final step is to realize that the consumer can choose the amount of residential land himself. The maximum amount he is willing to bid for a unit of land, which will be denoted as ψ, can therefore be found by maximizing the right-hand side of (12.3) over h. Put differently:

$$\psi = \max_h \left(\frac{y - tx - C(u^*, h, S)}{h} \right). \tag{12.4}$$

The value of ψ gives the marginal willingness to pay for residential land. As long as the price of residential land in the area is lower than this amount, households will move in. We have noted already that the price of residential land will be equal to that of agricultural land in the absence of spatial planning. Hence households will move in as long as non-farmer households are willing to pay at least as much for the land as the farmers do. If the price of agricultural land is constant, the process will only stop when there are no more potential residents from elsewhere who are willing to pay the price of agricultural land.

To see how the latter mechanism works, return to the utility function (12.1). The consumers attach value to open space, and utility will therefore be higher when S is high. This also implies that the consumer needs to consume a smaller amount of the aggregate consumption good (all else equal) when S is high. And this implies that the maximum bid for land in the area is higher when the amount of open space is higher. When non-farmer households move into the area, open space is gradually converted into residential land, which results in a lower amount of open space S and therefore in a lower marginal willingness to pay for residential land. In other words, the deterioration of the landscape that is caused by the inflow of the non-farmer households makes the area less attractive to households. This may stop the inflow of households before all agricultural land is converted into residential land. The final situation is an area which has a substantially less attractive landscape than it originally had. This is experienced as a loss in the quality of the landscape (and therefore in welfare) by all inhabitants of the area, including the non-farmer households that moved in early.

If the area is close to an important employment center (and especially when employment is growing over time) it may happen that all open space is ultimately converted into residential land. In that situation the proximity to jobs makes the area so attractive as a residential location that even in the absence of open space households are willing to bid more for the land than farmers can do.

By noting that more residential land implies less open space, we have in fact already seen the third element of our model in operation. The total amount of land in the area is fixed, and residential land can only be made available by the conversion of agricultural land. To state this condition formally, denote the total amount of land in the area as L and the amount of residential land as R. The land constraint says that open space and residential space have to add up to the total amount of land:

$$R + S = L. \tag{12.5}$$

The story we have just told on the basis of our simple model has some important relationships with actual developments in land-use. Especially in the proximity of large urban areas, there is continuous pressure on agricultural land from urban workers who attach value to a rural environment, for example for its quietness and scenic quality. The result is that some of that land is converted into residential use, which often results in a gradual deterioration of the original quality. The process stops only when the deterioration reaches the level at which the area is no longer attractive to newcomers. The general impression is that this process is regrettable and that land-use planning should play a role in protecting the original quality of the landscape.

It is also interesting to note the implications of this analysis for the spread of residential land-use outside an urban area. The most attractive rural areas will be chosen first as urban workers have a preference for attractive landscapes. If the landscape in these first settlement areas deteriorates, interest shifts to other rural sites that were originally less attractive. This means that the areas with the most valuable landscapes are those that are most vulnerable to the process just outlined, and that this process tends to result in equalizing the remaining quality of partially residential landscapes in the proximity of the urban areas. This analysis suggests that a better situation (for at least some households involved, while no one would be harmed) could be reached if more people would choose to locate outside the areas with high-quality open space. This would improve the utility of those living there, while those living elsewhere would not be worse of, since nothing is lost by using more low-quality open space for residential purposes.

Optimal provision of open space

Now that we have shown that the unconstrained functioning of market forces may result in a substantial decrease in the quality of valuable landscapes we have to consider how restrictive land-use planning could potentially have prevented this outcome. To do so, we have to specify the relationship between land-use and welfare.

In the urban economic literature, it is conventional to use social surplus as an indicator for social welfare. Social surplus is simply the difference between the value of the land in actual use and that in agricultural use. The value of the land in actual, non-agricultural use, is the maximum bid for its residential use. Social surplus SS is therefore defined as:

$$SS = R(p - p^{agr}). \tag{12.6}$$

An immediate implication of this definition is that the social surplus is equal to zero when there is no residential land-use at all. Our analysis is therefore unlikely to result in the prescription that open space should be maximized by prohibiting all residential land-use. The reason is that in our model only inhabitants of the area attach value to its landscape. In a later section we come back to this assumption.

Equation (12.6) also makes clear that a social surplus can never be realized when the price of residential land equals the price of agricultural land. The optimum should therefore be expected to differ from the market solution sketched in the previous section. Since a formal analysis of the social planner's problem is outside the scope of this chapter, we refer to Rouwendal and Van der Straaten (2008) for the derivations and only list the main result: social surplus is maximized when the marginal willingness to pay (MWTP) for open space of all households is equal to the private marginal willingness to pay for residential land of a single household:

$$N\frac{\partial C}{\partial S} = \frac{\partial C}{\partial h}. \tag{12.7}$$

The left-hand side of this equation gives the product of the number of households in the area (N) and the marginal willingness to pay for open space ($\partial C/\partial S$) which is the total marginal willingness to pay of all households that experience utility from the landscape in which they live. The right-hand side gives the marginal willingness to pay for residential land ($\partial C/\partial h$) which equals the social marginal willingness to pay for such land since only the household that uses the land experiences utility from it. The equation thus states that the marginal social benefits of open space should be equal to the marginal social benefits from residential space.

The expression on the left-hand side of equation (12.7) is the product of the total number of households, N, in the area, and their marginal willingness to pay for open space. The latter is the first derivative of the function C we discussed in the previous sub-section with respect to open space. It is the expenditure on consumption goods a household is willing to give up in exchange for an additional unit of open space. This willingness to pay has to be multiplied by the number of households residing in the landscape, since open space is consumed by all these households. Open space has characteristics of a public good, and the rule (12.7) is therefore sometimes referred to as a Samuelsonian rule, after Paul A. Samuelson who provided a path breaking analysis of such goods. The right-hand side of (12.7) is the partial derivative of C with respect to residential land. It has a similar interpretation to the other partial derivative: it is the consumption expenditure a household is willing to give up in exchange for one additional unit of residential land. Since residential land is a private good, there is no multiplication by the total number of households.

It is clear from (12.7) that, in order to maximize social surplus, we should not protect open space to such an extent that N equals zero, as this would imply that the left-hand side of this equation becomes zero. We have also seen that the price of residential land must be larger than that of agricultural land for the social surplus to be positive. The implication is therefore that there is some protection of open space, but not an absolute one.

To make the implications of the Samuelsonian rule more concrete, we now discuss two special situations. First, suppose that we set the price of residential land equal to the maximum bid and that we start from a position in which all space in the area is open. In this situation the marginal willingness to pay for open space is probably small, as there is plenty of it available. The left-hand side of (12.7) is therefore small if we allow the first household to enter. The right-hand side is probably large, as there is plenty of open space available in the area, which makes it attractive to live in. The left-hand side of (12.7) is therefore smaller than the right-hand side if $N = 1$. If N increases, less open space will remain, and the willingness to pay for an additional unit therefore probably goes up, while the willingness to pay for residential land in the area goes down. Put differently, if N increases, the left-hand side of (12.7) goes up, while the right-hand side goes down. When more and more households are allowed to reside in the area, both sides of (12.7) approach each other. Further moving in of households should be stopped when the two sides of the equation become equal to each other. When this happens, the value of social surplus is maximized.

An alternative way to use the rule is to start from the other extreme in which the price of residential land is initially set equal to the price of agricultural land. In this situation the social surplus is equal to zero. If, at this price the left-hand side of (12.7) exceeds the right-hand side, social surplus can be increased by setting a higher price. Again, the optimum can – under some regularity conditions – be approached by a gradual adjustment of the instrument until condition (12.7) is satisfied.

Condition (12.7) does not imply that in every area some open space should be protected. If the landscape is of a low quality, the marginal willingness for open space is low, even if almost none is left. On the other hand, it may be the case that a landscape has a very high scenic value when it is left almost untouched by residential use, whereas its value deteriorates rapidly even if a small amount of such land-use is allowed. In such a case, condition (12.7) may hold while most of the area is still open space. The exact implications of the condition for the amount of open space to be protected should therefore be expected to differ significantly between areas. This emphasizes the need for making the condition operational. That is the subject of the next section.

Hedonic analysis and the cost–benefit rule

The rule for optimal preservation of open space that we derived in the previous section is simple, but not easy to implement because the marginal willingness to pay for open space and for residential land are not directly observable. Residential land is usually bought jointly with the house on it, and it is not immediately obvious which amount of the total transaction refers to land and which to the house that has been constructed on it. The difficulty is therefore caused by the fact that there is hardly a specific market for residential land. For open space we do not even have to think of a market: it simply does not exist. We must therefore look for other methods to find the willingness to pay for residential land and open space.

There is a rich economic literature on evaluation methods of goods that are not traded in a market. Two major branches of this literature deal with stated and revealed preference techniques, respectively. The former uses statements of survey respondents as the basic information for estimating willingness to pay. Economists have traditionally been reluctant to use such information, although they have become gradually more accepted. The market based information used in the second branch of the literature is often seen as more reliable, mainly because it refers to actual transactions instead of hypothetical ones. Hedonic pricing is probably the most popular of the revealed preference techniques for evaluating non-market goods. Much experience has been gained with this method and it is generally regarded as most useful. This technique can be used to investigate the willingness to pay for residential land as well as for open space.

To explain how this works, consider a household looking for a house. If the number of houses on the market is large, it will be able to compare many houses that differ in all kinds of relevant characteristics. The household may, for instance, compare houses in the area with open space that we want to study, with similar houses elsewhere. If the houses are indeed similar in all characteristics, except the availability of open space in the proximity, then a difference in the willingness to pay for such houses can be attributed to the difference in the availability of open space. In other words: the housing price difference reveals the willingness to pay for living in the proximity of the scenic landscape in our study area instead of the much less attractive landscape elsewhere in the region. Similarly, a comparison of the prices for houses that are similar except for a difference in lot sizes informs the researcher about the household's willingness to pay for residential land. Hedonic analysis is a systematic attempt to attribute price differences to differences in characteristics of housing and its environment.

The estimated hedonic price function gives the housing price P as a function of lot size h, the amount of open space in the proximity S, and other variables, x: $P = P(h, S, x)$. It can now be shown (and we refer again to Rouwendal and Van der Straaten (2008) for elaboration) that the partial derivatives of the hedonic price function are equal to the willingness to pay. That is: $\partial P / \partial h = \partial C / \partial h$ and $\partial P / \partial S = \partial C / \partial S$. The hedonic price method therefore allows us to make the theory discussed above operational. The hedonic method has been used extensively to evaluate all kinds of local external effects such as noise, air pollution, crime, a view of the ocean etc. Since applications to landscape evaluation seem to be absent at this moment, we illustrate this on the basis of a related example: open space in cities.

Application: the benefits and costs of open space in cities

The model, described in the previous section, gives us the cost–benefit rule for the appropriate provision of open space. We have seen that estimation of the partial derivatives of the hedonic price function informs us about the willingness to pay for residential land and open space. With this information in hand, we can investigate the optimality of the actual provision of open space. This is the essence of the empirical analysis which will be done in this section.

The starting point of the discussion here is three hedonic price functions estimated for the large Dutch cities, Amsterdam, The Hague and Rotterdam.[3] Details are provided in Rouwendal and Van der Straaten (2008). The estimated functions include a large number of housing characteristics as well as detailed information about the neighborhood in which the house is located. Because the willingness to pay (WTP) for floor area can be different between households living in apartments and other housing types (as shown for instance by Visser and Van Dam (2006)), we distinguish between households living in single-family houses and households living in multi-family houses. We include therefore the cross-effect of floor area and apartment.

Open space was included in the equation as the percentage of land coverage of parks and public gardens[4] within a distance of 500 m radius circle around the house. Later in this section we report the results of a sensitivity analysis of this specification. In all specifications we also included the cross effect between open space and apartments because households living in this dwelling type may differ in their appreciation of open space from households living in single-family dwellings.

The variable lot size is only applicable for single-family houses. However, there is also a trade-off between residential land and open space in areas with apartment buildings, and we have developed a procedure – to be discussed below – to take this into account. It uses the relationship between the floor area of apartments and the amount of residential land needed per apartment. For this reason, we are also interested in the willingness to pay for floor space in apartments, and will discuss these results below.

The explanatory variables explain approximately 88 per cent of the total variance and almost all coefficients had the expected signs and were statistically significant. The study found, for instance, a significant negative coefficient for the share of land in industrial use, which confirms the presence of a negative external effect of industrial land-use on residential areas in their proximity, which is often mentioned as a justification for the spatial separation of different types of land-use enforced by zoning.

The coefficients for open space and lot size are of special interest in this analysis, and the results are shown in Table 12.1.

The first column of Table 12.1 shows the estimated coefficients for the share of parks and public gardens in the 500 m radius circle in which the house is located. In all but one case they are of the expected sign and significant. The exception is single-family houses in Amsterdam. Also the coefficient for open space estimated for apartments is much smaller in Amsterdam than elsewhere. However, if we concentrate on the Vondelpark region, we find estimates that are comparable to those of the other two cities. These separate estimates have been listed in a second line referring to Amsterdam for apartments as well as single-family housing in Table 12.1.

These results confirm our expectation that open space has a positive value for urban residents. A household living in a single-family house in The Hague[5] is on average willing to pay €2,012 for a one per cent increase of parks and public

Table 12.1 The value of city parks and residential areas

	(1) Coefficients for share of parks	*(2) Coefficients for residential area (floor area/lot size)*	*(3) MWTP open space*[a]	*(4) MWTP lot size/floor area*
Single-family housing				
• Amsterdam	0.0001 (.001)	**0.1300** (.023)	*(not significant)*	227.84
• Amsterdam (Vondelpark)	**0.0083** (.001)	**0.1300** (.023)	1,737	227.84
• The Hague	**0.0124** (.002)	**0.3180** (.027)	2,012	288.38
• Rotterdam	**0.0095** (.001)	**0.3740** (.021)	1,548	321.24
Apartments				
• Amsterdam	**0.0030** (.001)	**0.1980** (.059)	*(not significant)*	548.22
• Amsterdam (Vondelpark)	**0.0083** (.001)	**0.1980** (.059)	2,047	548.22
• The Hague	**0.0110** (.003)	**0.5440** (.056)	1,785	937.79
• Rotterdam	**0.0063** (.002)	**0.3480** (.046)	1,020	613.25

[a] The marginal willingness to pay (MWTP) for open space refers to an increase in the share of parks and public gardens by one point (for example an increase from 0.095 to 0.105 in Amsterdam).

The first columns of the table report the coefficients for the share of open space in a 500 m radius circle around the house. The second column reports the coefficients of lot size, in the case of single-family housing, and floor area, in the case of apartments. The coefficients are estimated in a spatial error model with the dependent variable the natural logarithm of the transaction price. The equations also contained controls for the natural logarithm of volume, house type, presence of a garden and garage, number of rooms, state of maintenance, monument status, distance to city center, percentage of ethnic minorities, population density, percentage of agricultural land, percentage of industrial area, percentage of service area, percentage of open water, year of construction (before 1905, 1906–1930, 1931–1944, 1945–1959, 1960–1970, 1971–1980, 1981–1990), month of sale and for neighborhoods (14 in Amsterdam, 7 in The Hague, 9 in Rotterdam). Full estimation results are available from the authors upon request. Standard errors are given in parentheses. Statistically significant effects (at the 5 per cent level) are in bold.

gardens in the 500 m circle around the house. In Rotterdam the willingness to pay for this amenity is equal to €1,548. The willingness to pay for additional open space of households living in an apartment in The Hague and Rotterdam is lower than for households living in single-family houses. The willingness to pay is €1,785 for The Hague and €1,020 for Rotterdam.

In the first regression, that refers to the whole city, the willingness to pay for an increase in open space of Amsterdam households is €705 for households living in apartments and zero for those living in single-family housing. Closer examination of this result suggested that it is caused by heterogeneity in the valuation of parks and public gardens in Amsterdam. The open space in the outside quarters of the city – like the Bijlmer and Slotervaart – seems to be appreciated much less than that close to the center of the city, where the Vondelpark is the main amenity. This is the reason for estimating a second regression that concentrates on this amenity. The background of the difference between Amsterdam and the other two cities is that in Rotterdam and The Hague neighborhoods with a large amount of open space are generally regarded as attractive residential areas, for example the areas surrounding the Kralingse Plas in Rotterdam. In Amsterdam however, some of the neighborhoods with the highest scores of open space are widely regarded as problem areas. This suggests that the value of urban amenities is not just a matter of the characteristics of these amenities *per se*, but also of the social environment in which they are located. Parks can be attractive walking areas, but they can also be perceived as unsafe, depending on the kind of people one expects to meet there.

Optimal provision of open space

The Samuelsonian rule for optimal provision of open space can be regarded as a cost–benefit rule: on the left-hand side of (12.7) are the total benefits (*TB*) of open space, while on the right-hand side appears the opportunity cost (the amount of residential space that has to be given up to create one additional unit of open space). In the previous section we estimated the benefits of open space and its opportunity costs. Table 12.1 shows that the MWTP of open space is much smaller than the MWTP of residential land. Since open space is a public good, the total benefit it creates is equal to the sum of the MWTP of all the households who live close by. Hence, if the total benefits of open space are equal to the marginal price (MP) of residential land, equation (12.7) tells us that the provision of open space is optimal when:

$$TB_{open\ space} = MP_{residental\ land}. \tag{12.8}$$

The total benefit of open space is calculated by the multiplication of the total number of households (N) with the (individual) marginal price of open space:

$$TB_{open\ space} = N \times MP_{open\ space}. \tag{12.9}$$

Based on the estimation results in Table 12.1 the total marginal benefit of open space can be computed as:[6]

$$\begin{aligned} TB_{open\ space} &= N \times MP_{open\ space} \\ &= N \times \left(\frac{100 \times coefficient_{open\ space}}{surface\ area} \times P \right). \end{aligned} \tag{12.10}$$

The available open space exceeds its optimal level if the total benefits of open space are smaller than the marginal price of residential land. In this situation, households would like to have more private space and less open space in their neighborhoods. If the total benefits of open space are larger than the marginal price of land, the provision of open space is below its optimal level and households would like to have more open space and less private space.

The previous sub-section showed that the willingness to pay for open space differs between households living in single-family houses and apartments, and also differs between the cities. We will therefore distinguish between neighborhoods consisting only of single-family and houses and apartments and

Table 12.2 Optimal provision of open space in neighborhoods consisting of single-family houses only

	(1)[a] *Marginal price open space (m², per household)*	*(2)*[b] *Total number of households (per 500 meter circle)*	*(3) Total benefits of open space [= (1) × (2)]*	*(4) Marginal price of lot size (m², per household)*
Amsterdam	0.00 (0.03)	3,614		
Vondelpark	0.22 (0.07)	3,614	799.54 (251.91)	227.84 (40.31)
The Hague	0.26 (0.04)	3,859	988.52 (159.44)	288.38 (24.48)
Rotterdam	0.20 (0.02)	2,437	477.18 (40.23)	324.24 (18.21)

[a] Based on the marginal price of open space in a 500 m circle around the house.
[b] The corresponding number of households *per hectare* equals 46 for Amsterdam, 49 for The Hague and 31 for Rotterdam.

The table reports the results with respect to the optimal provision of open space in neighborhoods consisting of single-family houses only. The standard errors of the marginal prices are given in parentheses.

investigate whether the amount of open space is optimal in both situations in each city.[7]

Table 12.2 shows the results with regard of neighborhoods that consist of single-family houses only. The marginal prices of open space per m^2 per household are shown in the first column. Because the coefficient of open space for Amsterdam excluding the Vondelpark is insignificant we focus on the results of the Vondelpark. The coefficient of the Vondelpark is significant and the marginal price of a 1 m^2 increase of the Vondelpark is equal to €22 cents, which is in line with the marginal prices of open space in The Hague and Rotterdam. In order to obtain the total benefits of open space we multiply the marginal willingness to pay of open space with the total number of households in the corresponding surface area.[8] We assume that the willingness to pay of tenants is equal to the willingness to pay of house-owners as estimated in the hedonic price functions. The number of households includes therefore both tenants and house-owners. The results are shown in the third column.

When we compare the total benefits of open space (column (3)) with the marginal price of lot size (column (4)), it shows that the provision of open space in neighborhoods consisting of single-family houses is below its optimal level in the three cities. Households would rather like to have more open space and less residential land in their neighborhoods.

Next we look at the willingness to pay for open space and floor area of households living in apartments. For this type of housing we approximate the willingness to pay for residential land on the basis of that for floor space. The idea is that there is a relationship between the total amount of residential land-used for an apartment building and the total floor space of the apartments inside the building. We therefore use the marginal price of floor area as the basis for computing the opportunity cost for open space.

The high rise character of most apartment buildings suggests that a reduction of the residential land by 1 m^2, reduces the total available floor area inside the building by more than 1 m^2, say with k m^2 where k is an adjustment factor that could, as a first approximation be set equal to the number of floors in the building. However, we must also take into account that apartment buildings have common space (for example hall(way), walkway, elevators etc.) and that Dutch regulation states that between such buildings there must be a minimum amount of open space.[9]

By definition the total amount of floor area in a neighborhood is equal to the number of apartments in the neighborhood (N) multiplied with the average size of floor area of one apartment (h_i). Using the CBS-data on land-use type, we obtain the average size of residential area in a neighborhood (H_i). The correction factor k can now be calculated as:

$$k_i = \frac{N_i \times h_i}{H_i}. \tag{12.11}$$

Its value equals 1.41 for Amsterdam, 1.04 for The Hague and 1.10 for Rotterdam. This means that in Amsterdam 1 m^2 floor area is equal to 1.41 m^2 residential land.

Table 12.3 shows the results with respect to the optimal provision of open space in neighborhoods consisting of apartments only. The first column shows the marginal WTP of 1 m^2 open space, the second column the total number of households per 500 m circle. The total benefits of open space are shown in column (3). The WTP of floor area, as calculated in the hedonic price function is shown in column (4). In order to compare the WTP of 1 m^2 open space with the total benefits of open space we have to take into account that apartments are built on top of each other. The WTP of residential land is given in the last column.

The conclusions with respect to the optimal provision of open space in apartment neighborhoods are equal to those reached for single-family housing neighborhoods in Amsterdam (Vondelpark) and The Hague. In Amsterdam and The Hague the total benefits of open space are higher than the price of residential land (column (5)), which suggests that households living in an apartment would like to have more open space and less private space in their neighborhood. In Rotterdam the price of residential land is higher than the total benefits of open space and households would rather have more private space and less open space.

These results will certainly not provide the final verdict on the optimal provision of open space in large cities in the Netherlands. Their intention is to illustrate the usefulness of the hedonic method for policy oriented analyses that intend to provide useful information about the costs and benefits of – in this case – parks and public gardens in cities. Using good data and up-to-date methodology it is possible to provide reasonable estimates of these key variables that provide guidance for policy makers. In the next section we discuss if this could also be the case for the provision of open space outside urban areas.

Table 12.3 Optimal provision of open space in neighborhoods consisting of apartments only

	(1)[a] *Marginal price open space (m^2)*	*(2)*[b] *Total number of households (per 500 meter circle)*	*(3) Total benefits of open space [= (1) × (2)]*	*(4) Marginal price floor area (m^2)*	*(5)*[c] *Correction factor (k)*	*(6) Marginal price of residential land (m^2) [= (4)/(5)]*
Amsterdam	0.09 (0.07)	5,977				
Vondelpark	0.26 (0.11)	5,977	1,557.62 (652.03)	548.22 (163.36)	1.41	388.81 (115.79)
The Hague	0.23 (0.06)	4,684	1,064.34 (290.27)	937.79 (96.54)	1.04	901.73 (92.84)
Rotterdam	0.13 (0.05)	3,736	485.16 (169.42)	613.25 (81.06)	1.10	567.83 (73.60)

[a] Based on the marginal price of open space in a 500 meter circle around the house.
[b] The corresponding number of households *per hectare* equals 76 for Amsterdam, 60 for The Hague and 48 for Rotterdam.
[c] The correction factor (k) is equal to the exchange rate between 1 m^2 floor area and $k \times 1$ m^2 residential land.

The table reports the results with respect to the optimal provision of open space in neighborhoods consisting of apartments only. The standard errors of the marginal prices are given in parentheses.

The benefits and costs of landscape

In the previous section we showed that by using the hedonic pricing method we are able to estimate the benefits of open space in cities and its opportunity costs, namely the price of residential areas. In Amsterdam, The Hague and Rotterdam the presence of open space within the vicinity of the house increases the value of the house. This means that households are willing to pay more for a house if the house is located in a neighborhood with open space. What can these results tells us about the benefits and costs of landscape? Is the model and the estimation method still applicable or should some adjustments be made? In order to answer this question we take a closer look at one of the main characteristics of landscape, namely its openness. Thereafter we deal with a number of pros and cons associated with the application of the hedonic method to landscape.

Openness and landscape protection

Current practice in Dutch spatial planning with respect to landscapes is based on general principles that are broadly endorsed among the population, but that are difficult to make precise. For instance, in Dutch national planning practice open spaces are usually considered to be large areas with few visual obstacles that allow a free view over a relatively large area (see for example LNV 2002; VROM et al. 2004). That such landscapes have to be protected is easy to argue in general terms, but it is hard to argue on the basis of a vague definition like this that specific small scale changes in land-use should or should not be allowed. The Dutch research institute Alterra has attempted to formalize the notion of open space by using detailed geographical data sets on the amount of buildings and high-rising vegetation per grid cell of 250 × 250 m (e.g. Alphen et al. 1994; Farjon et al. 2004). Protection of open space could be based on "critical" values of this quantitative indicator. However, openness in itself does not imply that much value is attached to the area concerned. Neither does a lack of openness indicate a negligible value of the area. To illustrate this point, the visual concept of openness used by Alterra naturally produces the result that extensive woodlands without much human presence are considered to be less open than the big Dutch cities.

Mainly for this reason Dekkers (2010) prefers to define open space as "being free of buildings and other proofs of human presence" (e.g. greenhouses or infrastructure), a concept that corresponds roughly to the inverse of urbanization. Although this clearly solves some problems, it introduces new ones. A paradoxical aspect of this alternative definition is that most (if not all) landscapes in the Netherlands are the result of a particular type of human land-use. In other words: the landscape itself, even if it is a forest, is proof of human presence.

The economic approach to the valuation of natural amenities emphasizes the point that landscapes are valuable because they are appreciated by humans. According to this view, it would be pointless to try to protect a landscape if no one cared about its presence. Although we do not want to argue that the point of view used above is the only possible one, we think it has particular merits that – at least – must be understood before one dismisses it. Perhaps the main argument in favor of it is that it has the potential to lead to a reasoned policy that is based on verifiable information about preferences and trades off the various costs and benefits for all the stakeholders.

Landscape and hedonic price analysis

In this section we discuss a number of pros and cons associated with the application of the hedonic method to landscape protection.

Benefits of non-residents and the WTP

Perhaps one of the major limitations of the hedonic method is that it restricts attention to the direct beneficiaries of the landscape. This is probably less of a drawback when we analyze city parks or public gardens, which are more or less exclusively there for inhabitants of the neighborhood in which they are situated. With landscapes this could be different, at least in some cases. After all, some areas are designated as "world heritage" because the landscape and other aspects are judged to have value that far outreaches the benefit attached to it by just the local inhabitants. The hedonic method does not take such considerations into account. In what follows we first argue that proximity to an area is in general important for the value attached to it. Next, we discuss a number of extensions of the cost–benefit rule that have to be introduced when other considerations are taken into account.

The effect of distance and the WTP

Application of the hedonic method to landscape evaluation would proceed by comparing the value of houses located in a particular (scenic) landscape with otherwise comparable houses located in areas without this amenity. The difference would provide an indication of the value attached to the scenic landscape by its inhabitants. One could in principle argue that this methodology ignores the possibility that inhabitants of a much wider region attach value to the landscape. The fact that the area with the particular landscape is at a small distance could be an important amenity for all or a substantial part of the inhabitants of the wider region. Indeed, many hedonic studies indicate that proximity is very important. For instance, Dekkers (2010) estimated open space at three different spatial scales. His results indicate that the availability of local open space (a view of open space and/or the presence of local patches of open space) has a substantial positive impact on house prices. This contribution is most prominent within a relatively short distance range: the added value becomes negligible at a distance of around 50 m from the investigated houses, depending on the case study area.

Our study of open space within cities also illustrates the phenomenon. In order to investigate how the effect of the scope of open space influences the willingness to

Table 12.4 Mean marginal price for open space per m² for different specifications of open space

	100 m circle	*300 m circle*	*500 m circle*	*1000 m circle*
(1a) Amsterdam	0.95	**7.73**	0.003	−0.002
∗ apartment	4.29	**8.24**	0.09	0.004
(1b) Amsterdam	−0.69	−0.10	−0.002	−0.002
∗ apartment	0.41	0.06	0.01	−0.002
∗ Vondelpark	**11.13**	**1.80**	**0.26**	0.02
(2) The Hague	**7.73**	**1.54**	**0.26**	**0.02**
∗ apartment	**8.12**	**1.51**	**0.23**	**0.02**
(3) Rotterdam	**8.14**	**1.31**	**0.20**	**0.02**
∗ apartment	**5.54**	**0.84**	**0.13**	**0.01**

The table reports the mean marginal price of open space per m² for different specifications of the variable. Statistically significant effects (at the 5 per cent level) are in bold.

pay for open space we have estimated the hedonic equations with the percentages of open space within 100 m, 300 m, 500 m and 1000 m circles around the houses as explanatory variables.[10] The estimated coefficients for the percentage of parks and public gardens are the highest when we use the open space in a 300 m circle around the houses. Table 12.4 shows the marginal willingness to pay for open space per m² for the different scope of the circle implied by our estimates. The figures shows that the willingness to pay for 1 m² of parks or public gardens decreases considerably when we increase the radius of the circle within which we measure this amenity.

It should be realized that our measure of parks and public gardens includes large parks like the Vondelpark in Amsterdam, the Kralingse Plas and surroundings in Rotterdam and the Haagse Bos. Even for these amenities, which are presumably of city-wide significance, the marginal willingness to pay becomes negligible for a distance larger than 1 km.

It is mainly for this reason that we are a bit skeptical about arguments that are based on a very low willingness to pay of a very large population. One can easily imagine that the average willingness to pay among the inhabitants of China for doubling the present size of the Amsterdam Vondelpark would be something like €0,001, but to draw the conclusion that at least €1,000,000 (the total willingness to pay of one billion Chinese people for this extension) could be spent on such a project would in our view be unjustified.

Adding the value of non-inhabitants

As we indicated above, probably the main shortcoming of hedonic analysis is that it tends to focus exclusively on the value attached to an amenity by those who live in its immediate proximity. There are, however, various ways to extend the analysis with the values attached to the landscape by others using revealed preference methods.

If the landscape is of high scenic quality, visitors from outside the area will probably be attracted. The fact that those people are willing to incur travel costs to experience the beauty of the area is evidence of their willingness to pay. The travel cost therefore plays a similar role to the house price difference. In order to incorporate this travel cost into the cost–benefit rule, one needs to know the sensitivity of the number of visits for changes in the amount of open space. That is, one needs to estimate the marginal benefits of open space as indicated by these travel costs.

A related issue is that for some people an area with outstanding landscape quality is too far from their residential location to be able to make a short trip to it during the weekend or on a summer evening. Instead they like to stay in the area for a short or longer vacation. Indeed the value of vacation homes and the rent paid for such homes can be interpreted as a valuable indication of the willingness to pay for such landscapes that should be included in a full cost–benefit analysis.

In principle one should also add the benefits experienced by accidental visitors that come to the area for other purposes, or just pass through it. It can be argued that the large number of car drivers who pass through parts of the Green Heart of the Randstad in the Netherlands are the main beneficiaries of the open polder landscape in this area. These people are also the ones who experience the cost of the protection of this open landscape by the difficulties they have in adjusting their residential and work locations, which makes it particularly useful to investigate their willingness to pay for the landscape. However, here we reach the limits of what could be done with revealed preference methods.

Other valuation methods

The alternative family of evaluation methods is known as stated preference methods. They are based on statements of survey respondents about the value they attach to particular amenities or their willingness to pay in hypothetical situations. We mention some examples. Kline and Wichelns (1998) use survey data to examine public preferences for preserving farmland and open space. Their results show that farmland-used for fruit, vegetables, crops, and pasture is preferred over most other types of open space. Kotchen and Powers (2006) include a variable in their model indicating whether the open space to be preserved is agricultural or non-agricultural land, and find that voters tend to favor local farmland preservation over other types of land. However, in their meta-analyses Koetse and Brander (2010) find that urban parks are more highly valued than other types of open space (forests, agricultural and undeveloped land). This is not surprising given that urban parks are generally more accessible and intensively used than other types of open space. Koetse and Brander conclude on the basis of their meta-analyses that there are important regional differences in the preferences for open space, which may constrain the potential for transferring estimated values between regions.

Traditionally economists have been skeptical about the use of stated preference techniques, mainly for the reason that incentives to answer a particular question are different from those for choosing a particular type of behavior. Put differently, when answering questions about their behavior in hypothetical situations respondents answer on the basis of a utility function that differs from the one that would be relevant when they were actually in such a situation. Accumulating experience with these methods has shown that – at least in some cases and under particular circumstances – stated preference techniques can provide valuable information about people's behavior in actual choice situations, as was realized by marketing scientists much earlier. The case of travelers who pass through a particular landscape mentioned at the end of the previous subsection could be a relevant case for application of such a technique.

Conclusion

In this chapter we have discussed the main benefits of the hedonic method for the evaluation of landscapes. Its main virtue is that it is based on evidence about consumers' willingness to pay for amenities. It is especially this feature that – in our view – makes it valuable. Its limitations are also clear: we can only hope to find the willingness to pay for those who experience direct benefits from the amenity concerned. Supplementary information about the willingness to pay of visitors and others who experience the benefits of the landscape is needed to complete the picture. This additional information should preferably also be based on market based evaluation methods – like the travel cost method – but the stated preference technique can provide valuable supplementary information.

We realize that even then the last word has not been spoken. The cultural and historical value of particular landscape for mankind is not necessarily identical to the value attached to it by its current inhabitants or visitors. There may be good reasons to protect landscapes even if a cost–benefit analysis does not end up in a positive final number. However, here we reach the boundaries of economic science. This chapter is an attempt to show what can be done while staying within these boundaries. We hope to have convinced the reader that economists have a number of meaningful things to say about the evaluation of landscapes and that their voice should be heard in the policy discussions.

Notes

1 For simplicity we assume that there are only two types of land-use: agricultural and residential.
2 This is also a simplification. In particular cases it may be necessary to distinguish between various types of agricultural land-use (e.g. traditional and modern) as well.
3 The data are transaction prices and housing characteristics for houses sold in 2001 and 2002 as registered by the Dutch Association of Real Estate Agents (NVM). The data were enriched with area characteristics as provided by Statistics Netherlands (CBS) containing information about the types of land-use on grid cells of 100 by 100 m. The log linear hedonic equation was estimated by a spatial error model to deal with spatial autocorrelation. Estimation results are available on request.

4 We used information from the *CBS Bodemstatistiek* (in Dutch).
5 Strictly speaking we should note that our results refer only to households who bought a house in the year 2000.
6 See Rouwendal and Van der Straaten (2008) for details of the computation.
7 We realize, of course, that neighborhoods consisting only of single-family housing or only apartments are hard to find. Nevertheless, there exist many neighborhoods that have predominantly one of these two types of housing and the exposition is clarified by concentrating on "pure" cases.
8 The total of households equals the average number of households in a 500 m circle around a single-family house.
9 This type of open space is not registered as parks or public gardens in the CBS statistics.
10 Our data do not allow us to make a distinction between houses that have a view of the park, and houses that are located close by a park, but do not have a view of it. The smallest range our data provides is 100 meters around the house. White papers in the Netherlands use a 500 m circle around the house as a criterion. This criterion however is criticized by some researchers and they argue that a 300 m circle should be used instead. Therefore we also include this range in our analyses. Finally we include a larger scope of amenities around the house, namely 1 km.

References

Alphen, B.J. van, H. Dijkstra, H. and Roos-Klein Lankhorst, J. (1994) *De Ontwikkeling van een Methode voor Monitoring van de Maat van de Ruimte.* Wageningen: DLO-Staring Centrum, report 334, Onderzoeksreeks Nota Landschap, no. 2 (in Dutch).

Chiesura, A. (2004) 'The role of urban parks for the sustainable city', *Landscape and Urban Planning*, 68(1): 129–138.

Dekkers, J.E.C. (2010) *Externalities, Land Use Planning and Urban Expansion.* Amsterdam: Vrije Universiteit, PhD thesis.

Farjon, J.M.J., Roos-Klein Lankhorst, J. and Verweij, P.J.F.M. (2004) *KELK 2003 – Landschapsmodule; Kennismodel voor de Bepaling van Effecten van Ruimtegebruiksverandering op de Landschappelijke Kwaliteit.* Bilthoven/Wageningen/Den Haag: RIVM/Alterra/LEI, NPB-Werkdocument 2004/10 (in Dutch).

Fujita, M. (1989) *Urban Economic Theory: Land Use and City Size.* Cambridge: University Press.

Geoghegan, J. (2002) 'The value of open spaces in residential land-use', *Land Use Policy*, 19(1): 91–98.

Kline, J. and Wichelns, D. (1998) 'Measuring heterogeneous preferences for preserving farmland and open space', *Ecological Economics*, 26(2): 211–224.

Kotchen, M.J. and Powers, S.M. (2006) 'Explaining the appearance and success of voter referenda for open-space conservation', *Journal of Environmental Economics and Management*, 52(1): 373–390.

Koetse, M.J. and Brander, L. (2010) 'Meta-analyse naar de waarde van openbaar stedelijk groen'. In: M.J. Koetse and P. Rietveld (eds) *Economische Waardering van Omgevingskwaliteit.* pp. 65–82, The Hague: Sdu Uitgevers (in Dutch).

LNV (2002) *Structuurschema Groene Ruimte 2, Samen Werken aan Groen Nederland.* The Hague, Ministry of Agriculture , Nature and Food Quality (in Dutch).

Luttik, J. (2000) 'The value of trees and open space as reflected by house prices in the Netherlands', *Landscape and Urban Planning*, 48(3–4): 161–167.

Mansfield, C., Pattanayak, S.K., McDow,W., McDonald R. and Halpin, P. (2005) 'Shades of green: Measuring the value of urban forests in the housing market', *Journal of Forest Economics*, 11(3): 177–199.

Rouwendal, J. and Straaten, J.W. van der (2008) *The Costs and Benefits of Providing Open Space in Cities*. The Hague: CPB, Discussion Paper, no. 98.

VROM, LNV, VenW and EZ (2004) *Nota Ruimte: Ruimte voor Ontwikkeling*, The Hague, Ministerie van Volkshuisvesting, Ruimtelijke Ordening en Milieu, Ministerie van Landbouw, Natuurbeheer en Visserij, Ministerie van Verkeer en Waterstaat and Ministerie van Economische Zaken (in Dutch).

Visser, P. and Dam, F. van (2006) *De Prijs van de Plek: Woonomgeving en Woningprijs*. Rotterdam and Den Haag: NAi Uitgevers and Ruimtelijke Planbureau (in Dutch).

Ward Thompson, C. (2002) 'Urban open space in the 21st century', *Landscape and Urban Planning* 60(2): 59–72.

13 The monetary value of open space in urban areas

Evidence from a Dutch house price analysis

Jasper Dekkers and Eric Koomen

Introduction

In urban areas the landscape consists of a mix of built-up areas and open spaces. These open spaces surround, intrude and perforate the urban fabric and contain many societal values: agricultural areas produce food; parks and sports grounds offer opportunities for recreation; and natural areas contain ecological values. In fact, most open spaces offer several values simultaneously: natural areas, for example, can also be important for recreation or the infiltration and storage of rain water, while agricultural land can be of high ecological value and offer recreational opportunities. These many additional values of open space are not generally acknowledged, which may be partly due to the lack of a clear price. Not having an obvious monetary value makes green, open areas vulnerable to construction activities and infrastructure. Such use of open space entails the imposition of externalities of certain actors on others, but since the market value of open space does not fully reflect its societal value, these externalities are market failures that call for corrective measures by the public sector in the form of land-use interventions or pricing measures. Incorporation of the public interest in open space in planning requires quantitative valuation of this asset. This can help policy makers improve their decisions. The difficulty with this is of course that environmental and general societal values are normally not traded, and hence no market price can be observed that would reflect or approximate marginal costs or benefits.

In the past decade, economists have developed a number of procedures that, at least in the case of some externalities, provide reasonable estimates of the monetary value of part of the amenities of open space, despite the remaining uncertainty and dispersion in values produced (see, for example, Button 1993). In recent years the level of sophistication used in this process has risen considerably, for example in the work of Luttik (2000), Geoghegan (2002), Ward Thompson (2002), Chiesura (2004) and Mansfield et al. (2005) who suggest that the presence of open space improves urban living conditions and individual well-being. However, some limitations remain, partly because the valuation of severance and visual intrusion is hampered by many complications. These especially include difficulties in objective quantification, uncertainties regarding the impacts on human and ecological communities, and collinearity with other pressures on the metropolitan open space (for example noise disturbance from infrastructure). The current analysis attempts to determine a value for the non-built-up landscape, generally referred to as open space, in urban areas in a systematic and integrated way using an econometric approach.

In this chapter, we first present our methodological-technical design. Subsequently, we define the concept of open space and make it operational. Then, we attach a monetary value to it using two separate hedonic house price analyses at the local housing market level in the Randstad region, the strongly urbanised western part of the Netherlands. The studies focus specifically on the impacts of the availability of different types of open space on house values. The final section summarises the results and discusses their implications for open space preservation policies.

Methodological-technical design of the study

The research focuses on those aspects that can be related to the appreciation of individual residents of the urban landscape, the so-called "use values". We use hedonic pricing analysis, a revealed preference valuation method, to determine the effect of the availability of open space on residential property prices. Revealed preference valuation has the advantage of potentially capturing all utilities associated with open space based on observed human behaviour, rather than having to trust stated preferences that are limited to the specific utilities included in such surveys (see also other contributions to this book). The added value of open space observed in a hedonic price regression is a combination of all positive and negative characteristics that house buyers associate with open space.

Defining open space

The non-built-up landscapes in urban areas that we refer to as open space can comprise a substantial part of the urban fabric. Even within the major cities in the densely populated Netherlands only about 50 per cent of the land area can be considered built-up, while only 10 to 20 per cent is actually covered by buildings (Koomen et al. 2009). The built-up area contains all types of land-use related to residential and non-residential buildings, such as gardens, pavements etc. Open spaces in and around urban areas consist of parks, sports fields, allotment gardens, water bodies, agricultural land and natural areas.

Most types of open space are under continuous threat of further urbanisation: small inner-urban open spaces are claimed in densification processes, whereas more extensive urban expansions are found in the larger open spaces surrounding cities. These threats make the valuation of open space crucial for helping urban and regional planners to better assess its societal values. Recent economic literature has seen a steady increase in valuation studies on open space at the local level (see, for instance, Brander and Koetse 2007; Van der Straaten 2010).

An additional concern to include in the study on open space is disturbance, since this aspect is considered to significantly influence the human perception of open space. We will therefore take into account the presence of motorways and related disturbance through traffic noise as separate control variables in our regression model. Other sources of disturbance, such as stench, light and visual disturbances are excluded from the analysis because these are either considered to be less important to the general perception of openness or are difficult to quantify objectively. Accessibility of open spaces (whether or not one can physically access the open space) is indirectly considered in the analysis in the form of inclusion of different types of open space. In general, (private) agricultural land is less accessible than (public) nature and forest areas. Finally, for operational reasons the shape of the open areas is excluded from the analysis.

In order to make the aforementioned issues involving open space operational for analysis in a GIS and spatial-statistical environment, we use a land-use map relating to the year 2000 (CBS 2002). This highly detailed data set, which has a minimum mapping unit of up to about 30 m^2 (0.1 ha) for individual land-use types, allows for the distinction of small patches of different types of open space. This fine resolution is essential for this type of analysis as it is known that house prices are only influenced by open space at relatively short distances (Dekkers et al. 2009). The land-use base map was reclassified in such way that four different types of open space were distinguished: urban open space (also referred to as urban green), agricultural open space, nature/forest and water. Using this definition, we are able to distinguish between open spaces within cities (urban green) and those surrounding them (agricultural and natural areas). Water can be found both within (e.g. as canals) and outside urban areas (e.g. lakes).

Hedonic house price analysis

House prices are determined by the moment and type of transaction, the structural characteristics of the sold object and its locational or spatial characteristics. Since all these characteristics are embedded in a single house sale transaction, they only have an implicit value. The hedonic pricing method determines the implicit value of non-tradable characteristics of goods by analysing the observed value of tradable goods that incorporate all or part of those non-tradable characteristics. The method was already applied at the beginning of the last century (see Taylor 1916; Waugh 1928; Court 1939), but Lancaster (1966) first provided the theoretical justification, while the method was documented in a general framework by Rosen (1974). The method is able to provide a detailed quantitative valuation of a wide range of structural and spatial characteristics when reliable transaction data and spatial data are available. A limitation of the method is that it assumes perfect competition, fully informed actors and no transaction costs when actors choose to relocate. This is an obvious simplification of reality where, for example, zoning restrictions create artificial submarkets. For a more detailed overview of advantages and limitations of the hedonic pricing method, we refer to King and Mazotta (2005).

Model formulation

The basic regression model used in this analysis to explain house prices is formulated as follows:

$$P = \alpha + \beta S + \gamma L + \tau G + \varepsilon, \tag{13.1}$$

where P is an $(n \times 1)$ vector of housing prices, S is an $(n \times i)$ matrix of transaction-related characteristics, L is an $(n \times j)$ matrix of structural characteristics, G is an $(n \times k)$ matrix of spatial characteristics, α; β; γ; τ are the associated parameter vectors and ε is an $(n \times 1)$ vector of random error terms. For this analysis, we choose to estimate a loglinear model, since this functional form is widely used in similar studies and, thus, allows for a straightforward comparison of results.

Spatially explicit regression models always have to be tested for spatial dependence as model estimates may prove to be biased or inefficient. There are two types of spatial dependence. The first type, *lag or structural dependence*, means that the prices of parcels that are close to each other are correlated. The second type, *error dependence*, means that the error term for parcels that are close to each other are mutually dependent. A spatial error model estimates the effect of this heteroskedasticity. See Anselin (1988a, 1988b) for a more elaborate discussion on spatial dependence.

For our analysis, we can describe both types of spatial dependence by reformulating our basic model (Equation 13.1) into:

$$P = \rho WP + \alpha + \beta S + \gamma L + \tau G + \varepsilon, \tag{13.2}$$

where ε is equal to:

$$\varepsilon = \lambda W \varepsilon + \mu \quad \text{and} \quad \mu \sim N(0, \sigma^2). \tag{13.3}$$

In this model, W represents a row-standardized spatial weight matrix, while ρ and λ are spatial-econometric coefficients that describe the importance of the spatial lag and spatial error component respectively.

Selection of variables

A scan of recent hedonic pricing studies that include spatial characteristics, such as open space availability, in the explanation of residential property values pointed at a whole list of transaction and structural variables that should be included in the regression analysis (see Table 13.1). For a more extensive overview of environmental valuation studies, see Van Leeuwen (1997), Garrod and Willis (1999), Ruijgrok (2004) and Hanley and Barbier (2009).

The study area

The economically important Dutch Randstad area (see Figure 13.1) consists of an intricate mix of urban and open areas and offers an interesting case study area

Table 13.1 Overview of spatial variables impacting housing prices in selected publications

Variable	Studies
Distance to city	Bastian et al. (2002), Geoghegan (2002), and many more
Distance to local shopping centre	Mathis et al. (2003)
Historical value of neighborhood/house	Mathis et al. (2003), Ruijgrok et al. (2004)
Neighborhood quality (measured by proxies as population composition, average income)	Geoghegan et al. (1997)
Distance to public transport station	Mathis et al. (2003)
Distance to/disturbance from school/college	Powe et al. (1995)
Distance to/disturbance from major road	Geoghegan et al. (1997), Powe et al. (1995)
Noise	Bateman et al. (2002), Mathis et al. (2003), Oosterhuis and Van der Pligt (1985)
Percentage of high voltage masts/wind mills	Mathis et al. (2003)
Residential land-use (percentage)	Geoghegan et al. (1997)
Scenic view (and/or view on or gardens adjacent to public green/forest/open space/ water)	Bastian et al. (2002), Garrod and Willis (1992), Geoghegan et al. (1997), Luttik (2000), Mathis et al. (2003), Tyrväinen (1997), Tyrväinen and Miettinen (2000)
Natural amenities (and/or value increase due to the creation of new nature/recreation areas)	Geoghegan et al. (1997), Briene et al. (2001), Ruijgrok et al. (1999)
Distance to/view on forest	Mathis et al. (2003)[1], Tyrväinen (1997), Tyrväinen and Miettinen (2000)[1], Morales (1980), Powe et al. (1995)
Distance to local and/or regional park	Bervaes and Van den Berg (1995)[2], Fennema (1995)[2], Hammer et al. (1974)[3], More et al. (1988), Van Leeuwen (1997), Weicher and Zerbst (1973)
Distance to green (local and regional scale)	Sijtsma et al. (1996)
Open space (various types and distances)	Ready and Abdalla (2005), Geoghegan (2002), Powe et al. (1995)
Open space (percentage)	Geoghegan et al. (1997), Geoghegan (2002)
Fragmentation of land-uses	Geoghegan et al. (1997)

[1] Only significant at distances smaller than 600 m.
[2] Including both view on and/or distances to a park up to 400 m.
[3] Includes distances to a park up to 800 m.

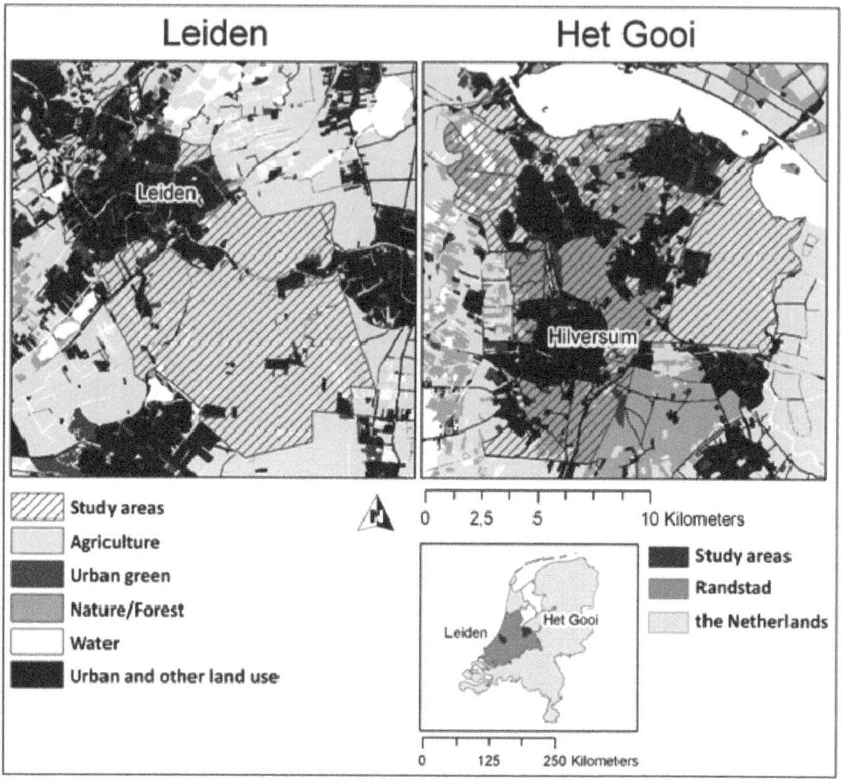

Figure 13.1 Location of the selected local housing market regions within the Dutch Randstad urban constellation.

because it is a sizeable metropolitan area with a peculiar layout. It has a total population that is comparable to the greater Paris or London areas, but it lacks the focus on a central city and has a substantial amount of open space between its constituting cities. The Randstad area is located in the western part of the Netherlands and has high population density and economic growth figures compared to the Dutch national averages. Urban growth threatens the remaining open space here and has given rise to various restrictive zoning regimes, known as Buffer Zones and the Green Heart contour. These restrictive development zones have helped preserve part of the open space in the Randstad area since their designation in the 1960s (Koomen et al. 2008; Van Rij et al. 2008).

Smaller, more homogeneous case study areas than the complete Randstad area must, however, be selected for the analysis. This is done because

> [...] the market for a hedonic analysis [...] should contain all the options available to potential buyers. If the market is defined larger than individuals

> actually choose from, then the regression results will be biased. On the other hand, by limiting the size of the market, the investigator loses information, so the estimation may become less efficient.
>
> (Geoghegan et al. 1997:258)

For the Netherlands, the Dutch Association of Real Estate Brokers (NVM) distinguished 76 regions that they consider homogeneous local housing market areas. For the estimation of our model, two of these housing market regions are selected that differ in size (e.g. of the urban area and number of inhabitants), amount and types of open space in the metropolitan area and composition of the housing market (i.e. number and types of houses available). The selected regions of Leiden and Het Gooi (Figure 13.1) have in common that they border the Green Heart restrictive development zone. Leiden, a city in the western part of the Randstad is situated near the major city of The Hague and has around 117,000 inhabitants (CBS 2010). The city has a number of different types of open spaces in its surroundings: dunes and forest in the west, lakes to the north and the grasslands of the Green Heart to the east. Het Gooi is a relatively nature-rich region that is situated between the big cities of Amsterdam, Utrecht, Amersfoort and Almere. The largest city in Het Gooi is Hilversum with over 84,000 inhabitants (CBS 2010). Most of the open spaces in Het Gooi are protected natural areas, thus strongly limiting the options for urban growth.

The summary statistics of the data included in the analysis (Table 13.2) show that the average transaction price is highest in Het Gooi. This area also has the largest share of the more expensive house types with more conveniences which is, for example, reflected by the fact that houses here, more often, have a garage, an open fireplace and a private open space in the form of a garden. Leiden offers the largest provision of local (mostly urban-type) open space; more houses have a view on open space than in Het Gooi and the average distance to open space is also smallest here. In Het Gooi, however, the average distance to nature is smallest. Accessibility measures in the form of distance to railway stations and distance to motorway ramps indicate that Het Gooi is less well connected to the main forms of infrastructure than the city of Leiden. The negative externalities associated with especially the presence of motorways do, however, also occur here. We have selected all NVM house transactions in the year 1998 for our analysis.

Estimation results

The statistical results for the estimated loglinear regression models for the two case-study areas are presented in Table 13.3. Based on the Langrange Multiplier spatial dependence test results, for Het Gooi a lag model was specified, while for Leiden an error model was specified.

The obtained λ (ρ) value for Leiden (Het Gooi) indicate that spatial error (spatial lag) is indeed substantial and statistically significant. This means for spatial error that some unobserved characteristics that are present on a local

Table 13.2 Summary statistics for the two local case studies (year: 1998)

Variable	*Leiden (1,388 obs.)*			*Het Gooi (2,142 obs.)*		
	Min.	*Max.*	*Mean*	*Min.*	*Max.*	*Mean*
Transaction price (× € 1,000)	18.15	1008.4	150.96	26.09	2631.9	211.80
Idem per m^2 (× 1,000 €/m^2)	0.01	17.47	1.36	0.01	55.30	1.24
Structural characteristics						
Surface area (m^2)	9.974	40,945	109.76	5.003	89,322	171.6
Number of rooms	1	11	4.09	1	14	4.51
Number of bathrooms	0	3	0.70	0	3	0.83
Quality of inside maintenance (1=bad to 5=excellent)	1	5	4.00	1	5	3.85
Monumental status (0/1)	0	1	0.03	0	1	0.01
Presence of gas heater instead of central heating (0/1)	0	1	0.13	0	1	0.14
Presence of open fireplace (0/1)	0	1	0.13	0	1	0.12
Presence of garage (0/1)	0	1	0.10	0	1	0.30
House type (indicating any of the 5 possible types)	–	–	–	–	–	–
Spatial characteristics						
Amount of agricultural open space within a 100 m radius (ha)	0.00	2.98	0.03	0.00	2.94	0.03
Amount of urban open space within a 100 m radius (ha)	0.00	2.05	0.10	0.00	1.68	0.08
Amount of nature within a 100 m radius (ha)	0.00	2.58	0.00	0.00	2.94	0.05
Amount of water a 100 m radius (ha)	0.00	1.25	0.12	0.00	1.10	0.02
Distance to nearest railway station (km)	0.09	5.08	1.50	0.04	7.96	2.60
Distance to nearest motorway ramp (km)	0.20	7.53	1.83	0.07	5.00	1.99
Presence of railway line within 100 m (0/1)	0	1	0.05	0	1	0.03
Presence of motorway within 100 m (0/1)	0	1	0.07	0	1	0.03
Percentage foreigners of non-western origin per neighbor (%)	1	25	9.38	0	20	6.06
Level of urban facilities on 0 to 1 scale in 500 m grid cells	0.00	0.59	0.17	0.00	0.51	0.15

Table 13.3 Estimation results for the two local case studies (year: 1998)

Variable	*Leiden (error-model)*		*Het Gooi (lag-model)*	
	Coeff.	*(St.Err) Prob.*	*Coeff.*	*(St.Err) Prob.*
Constant	9.759	(0.078)***	6.493	(0.526)***
Physical characteristics				
Ln (surface area)	0.259	(0.014)***	0.233	(0.010)***
Ln (number of rooms)	0.399	(0.020)***	0.318	(0.020)***
Number of bathrooms	0.060	(0.011)***	0.110	(0.013)***
Inside maintenance	0.095	(0.008)***	0.086	(0.009)***
Monumental status	0.151	(0.033)***	0.338	(0.083)***
Gas heater	−0.082	(0.016)***	−0.091	(0.017)***
Open fireplace	0.033	(0.016)**	0.082	(0.018)***
Garage	0.103	(0.020)***	0.152	(0.014)***
House type				
Corner house	0.001	(0.016)	−0.013	(0.017)
Semi-detached house	0.123	(0.034)***	0.199	(0.019)***
Detached house	0.215	(0.039)***	0.453	(0.027)***
Apartment	−0.149	(0.014)***	−0.005	(0.019)
Spatial characteristics				
Amount of agricultural open space within 100 m (ha)	−0.081	(0.026)***	0.045	(0.028)
Amount of urban open space within 100 m (ha)	0.033	(0.018)*	0.028	(0.026)
Amount of natural open space within 100 m (ha)	0.098	(0.072)	0.060	(0.028)**
Amount of water within 100 m (ha)	0.113	(0.023)***	0.159	(0.072)**
Non-western foreigners per neighborhood (%)	−0.010	(0.001)***	−0.018	(0.002)***
Level of urban facilities (0/1)	0.432	(0.059)***	0.021	(0.062)
Dist. railway station (km)	−0.026	(0.009)***	−0.015	(0.003)***
Dist. motorway ramp (km)	−0.009	(0.011)	0.012	(0.008)
Pres. of railway line within 100 m (0/1)	−0.048	(0.024)**	−0.086	(0.034)**
Pres. of motorway within 100 m (0/1)	−0.019	(0.020)	−0.065	(0.033)**
Lambda	0.762	(0.077)***	–	
Rho	–		0.289	(0.043)***
Number of observations	1,388		2,142	
R^2	0.81		0.80	

*** = significant at 0.01; ** = significant at 0.05; * = significant at 0.10.

Note: For the dummy variables the remaining categories (e.g. house type = terraced house) act as reference values.

(i.e. neighborhood) scale level are important. For spatial lag this means that indeed the prices of parcels that are close to each other are correlated and that this correlation is not entirely captured by the current model setup and selection of variables. The added error (lag) variables correct for these omissions.

In general we can say that the models for both case studies are similar in terms of sign and significance of the explanatory variables. Both models explain about 80 per cent of the observed variance in property values. The structural characteristics adhere to what can be expected based upon the literature review. With regard to the house types, we can see that the semi-detached and detached house types clearly are more expensive. Further, apartments are cheaper than terraced houses. The expected price difference between a corner house and a terraced house does not appear in the results.

When we turn to the spatial characteristics, we see that the accessibility variables show the expected negative signs; that is, further away from a railway station, property prices drop. In our setup, distance to a railway station in combination with the level of urban facilities serve as a combined proxy for distance to city centre. We observe a relatively high positive impact of the level of urban facilities on house prices in Leiden, which corresponds with other hedonic pricing studies in which we included this variable (see, for instance, Dekkers and Van der Straaten 2009). A possible explanation for the somewhat low coefficient value of urban facilities in Het Gooi is the fact that these are less abundant here; only few observations in the sample are within a short distance of urban facilities. Next, we also observe a negative impact on property prices as the per cent of foreigners of non-western origin increases in a neighborhood. This variable can be seen as a proxy for (perceived) neighborhood quality as discussed in Table 13.1.

Finally, when we focus on the open-space related results it is apparent that the presence of water in the vicinity of a residence has the highest positive influence on property prices, followed by nature and urban open space. This corresponds with the findings by, for instance, Luttik (2000) and Bervaes and Vreke (2004). The impact of agricultural open space on property prices is ambiguous; a negative impact for Leiden and a positive albeit not significant impact for Het Gooi. A reason for this result may lie in the fact that agricultural land in general is less accessible than nature and forest areas. This confirms the results of Brander and Koetse (2007) who found that accessibility of open space for recreational purposes adds additional value, on top of the presence of open space. Another reason can be the fact that houses with a lot of agricultural land in their surroundings are located relatively far from the actual city of Leiden and thus have a very different price setting and location value than the city itself. Also, from other stated preference studies in the Netherlands it is apparent that people generally value nature and forest areas more highly than agricultural areas (see Roos-Klein Lankhorst et al. 2002). This is different from the United States, where, according to two stated preference studies, people attach a higher value to agricultural open space then to other types of open space (Kline and Wichelns 1998; Kotchen and Powers 2006). The impact of water on property prices is higher in Het Gooi, while the impact of the other open space types is higher in Leiden. This may be explained by the fact that water in Het Gooi is not as ubiquitous as in Leiden. It tends to be more concentrated in larger water bodies that are part of attractive natural wetlands. Previous research also indicated that a connection to sizeable water bodies adds a premium to property prices (Luttik 2000).

Conclusion and discussion

This study aims to quantify the non-economic value of open spaces surrounding, intruding and perforating the urban fabric that is relevant to spatial policy, making these values easier to consider in policy-making. The hedonic price method proved to be a useful tool to assess the impact of open space on residential property values. For this study open space is interpreted as being a non-urban area that provides opportunities for rest and recreation.

The presented estimation results for the Dutch cases show a mixed picture. Agricultural open space has a negative impact on property prices for Leiden and a positive albeit not significant impact for Het Gooi. For urban open space, a positive impact is found in Leiden, while the impact is positive but not significant in Het Gooi. For natural areas, the impact is positive but not significant for Leiden, and in this case positive for Het Gooi. Finally, the impact of water on property prices is positive in both cases. In summary, we observe one significantly negative impact, three not significant positive impacts and three significantly positive impacts. The price impact thus differs between different types of open space indicating that in these Dutch cases, residents seem to prefer having water in the direct vicinity of their house and, in some cases, nature or urban open space as well.

In a preceding analysis of the value of open space that partially focussed on the same case study areas no distinction was made between different types of open space (see Dekkers and Koomen 2008; Dekkers et al. 2009). Instead, we focussed on the presence of patches of local open space (within a distance of 100 m from sold property) and the presence of larger areas of regional open space at larger distances. The estimation results of that study showed that the presence of local open space has a substantial positive impact on property prices: houses directly overlooking open space have a 4–8 per cent higher price than identical properties located more than 50 m from (local) open space. At the same time we found inconclusive evidence for a price impact of regional open space. These results may indicate that local house prices are not well suited to explain this essentially regional characteristic. The availability of regional open space may, in fact, be more important in determining regional house prices at the aggregated level of housing market regions than at the local level of individual houses.

These earlier results indicate that in general the presence of open space has a positive impact on property prices. The current study makes clear that this impact depends on the type of open space and other local conditions. The observed, generally positive effect of the proximity to local open space on house prices corresponds with findings in similar national and international hedonic pricing studies (e.g. Powe et al. 1995; Bolitzer and Netusil 2000; Visser and van Dam 2006; Brander and Koetse 2007).

Locally diverging impacts are also found for other specific characteristics, for example in Het Gooi where the level of urban facilities has an impact that differs significantly from Leiden. This, again, points at the existence of considerable differences between housing-market regions and thus signals the importance of performing this type of analysis at this specific scale level.

The positive impact of various types of open space in the immediate vicinity of houses provides policy makers with additional arguments for the preservation of green spaces in new residential areas. The results can, in fact, be used in negotiations about a (partial) recovery of the construction and maintenance costs for the green spaces that increase the property values. An approach to cover the costs associated with preserving and maintaining landscapes that have a positive impact on property values that has recently received considerable attention is the establishment of a landscape fund. Such funds can be organised at the national level (VNC 2005) or the local level (Kloen et al. 2005). At the local level, house owners that directly benefit from, for example, a view of open space can contribute to such a fund. In return they could receive a guarantee that the landscape remains open and maintained (Wing 2008).

The observed impact of agricultural open spaces differs too much per region so that it does not offer unequivocal arguments for new ready-made intervention strategies aimed at preserving rural open space. In the studied areas open space outside the existing cities also offers the possibility to internalise its positive impact on residential property in (sub-)urban development. However, it is important to note that such open spaces may comprise many other societal values that are not necessarily expressed in house prices. For instance, the intrinsic biodiversity value related to the occurrence of many different animal and plant species is not included in our analysis as its value to residents is something that is very difficult to monetarise. Also the agricultural production value and, for forest, the wood production value, is not included in our analysis. Furthermore, we know from previous studies (Brander and Koetse 2007; Dekkers et al. 2009) that people only pay an additional amount of money for the presence of open space when it is very close to their intended home. These studies observed an impact of open space availability up to 100 m from houses. Apparently people are indifferent to being located at 200 m or more from open space. They are, however, willing to spend money and time to visit attractive landscapes in their spare time, pay entrance fees for entering designated natural areas, or pay a membership fee for a nature-conservation organisation. This clearly underpins the need for alternative valuation studies with different objectives and techniques.

Acknowledgements

This research is funded by the research programme 'Environment, Surroundings and Nature' (GaMON) of the Netherlands organisation for scientific research (NWO). We would furthermore like to thank the Netherlands Environmental Assessment Agency (PBL) for providing the necessary spatial data for our analysis. Finally, we thank the Dutch Association of Real Estate Brokers (NVM) for making available their data on house transactions for the case studies.

References

Anselin, L. (1988a) *Spatial Econometrics: Methods and Models*. Boston: Kluwer Academic.

Anselin, L. (1988b) 'Lagrange Multiplier test diagnostics for spatial dependence and spatial heterogeneity', *Geographical Analysis*, 20: 1–17.

Bastian, C.T., McLeod, D.M., Germino, M.J., Reiners, W.A. and Blasko, B.J. (2002) 'Environmental amenities and agricultural land values: a hedonic model using geographic information systems data', *Ecological Economics*, 40(3): 337–349.

Bateman, I.J., Jones, A.P., Lovett, A.A., Lake, I.R. and Day, B.H. (2002) 'Applying geographical information systems (GIS) to environmental and resource economics', *Environmental and Resource Economics*, 22: 219–269.

Bervaes, J.C.A.M. and Van den Berg, L.M. (1995) 'De compacte stad, het groen aan snee?', Wageningen: Instituut voor Bos- en Natuuronderzoek (IBN-DLO), IBN rapport 130 (in Dutch).

Bervaes, J.C.A.M. and Vreke, J. (2004) 'De invloed van groen en water op de transactieprijzen van woningen', Wageningen: Alterra, rapport 959 (in Dutch).

Bolitzer, B. and Netusil, N.R. (2000) 'The impact of open spaces on property values in Portland, Oregon', *Journal of Environmental Management*, 59: 185–193.

Brander, L.M. and Koetse, M.J. (2007) 'The value of urban open space: Meta-analyses of contingent valuation and hedonic pricing results', Amsterdam: Institute for Environmental Studies (IVM), VU University, IVM Working Paper, I. 07/03.

Briene, M.F.M., Boeckhout, I.J., Verschuren, A.F.M., Verster, A.C.P. and Annema, J.A. (2001) 'Kosten en baten 750 ha natuur- en recreatiegebied regio Rotterdam', Rotterdam/Bilthoven: NEI/RIVM (in Dutch).

Button, K. (1993) *Transport, The Environment and Economic Policy*. Aldershot: Edward Elgar.

CBS (2002) 'Productbeschrijving bestand Bodemgebruik', Voorburg/Heerlen, Central Bureau of Statistics (in Dutch).

CBS (2010) *Statline*, Voorburg/Heerlen: Central Bureau of Statistics. http://www.statline.nl (accessed 22 December 2010).

Chiesura, A. (2004) 'The role of urban parks for the sustainable city', *Landscape and Urban Planning*, 68: 129–138.

Court, A.T. (1939) 'Hedonic price indexes with automotive examples'. In: American Statistics Association, *The Dynamics of Automobile Demand*. pp. 99–117, New York: General Motors Corporation.

Dekkers, J.E.C. and Koomen, E. (2008) 'Valuation of open space; hedonic house price analyses in the Dutch Randstad region', Amsterdam, Vrije Universiteit, FEWEB Research Memorandum 2008-24.

Dekkers, J.E.C., Koomen, E., Koetse. M. and Brander, L. (2009) 'Does proximity to open space increase the values of dwellings?: evidence from three Dutch case studies'. In: A. van der Valk and T. van Dijk (eds), *Regional Planning for Open Space; The RTPI Library Series vol. 18*. pp. 107–124, London and New York: Routledge.

Dekkers, J.E.C and Van der Straaten, W. (2009) 'Monetary valuation of aircraft noise; a hedonic analysis around Amsterdam airport', *Ecological Economics*, 68(11): 2850–2858.

Fennema, A.T. (1995) 'Wonen in het groen; de invloed van groen op de prijs van een woning', Wageningen: DLO-Staring Centrum (in Dutch).

Garrod, G. and Willis, K.G. (1992) 'The environmental economic impact of woodland: A 2-Stage hedonic price model of the amenity value of forestry in Britain', *Applied Economics*, 24(7): 715–728.

Garrod, G. and Willis, K.G. (1999) *Economic Valuation of the Environment; Methods and Case Studies*. Cheltenham, UK and Northampton, MA, USA: Edward Elgar.

Geoghegan, J., Wainger, L.A. and Bockstael, N.E. (1997) 'Spatial landscape indices in a hedonic framework: an ecological economics analysis using GIS', *Ecological Economics*, 23(3): 251–264.

Geoghegan, J. (2002) 'The value of open spaces in residential land-use', *Land Use Policy*, 19: 91–98.

Hammer, T.R., Coughlin, R.E. and Horn, E.T. (1974) 'The effect of a large urban park on real estate', *Journal of the American Institute of Planning*, 40(4): 274–277.

Hanley, N. and Barbier, E.B. (2009) *Pricing Nature; Cost–Benefit Analysis and Environmental Policy*. Cheltenham, UK and Northampton, MA, USA: Edward Elgar.

King, D.M. and Mazotta, M. (2005) *Ecosystem Valuation*. Universities of Maryland/University of Rhode Island. http://www.ecosystemvaluation.org (accessed 1 July 2011).

Kline, J. and Wichelns, D. (1998) 'Measuring heterogeneous preferences for preserving farmland and open space', *Ecological Economics*, 26(2): 211–224.

Kloen, H., Padt, F., Verschuur, G., Joldersma, R., Lobry, E. and De Graaff, R. (2005) 'Lokaal Landschapsfonds voor natuur en landschap; Handleiding voor het organiseren van een Landschapsfonds', Platform Nederland Mooi (in Dutch).

Koomen, E., Dekkers, J. and Van Dijk, T. (2008) 'Open space preservation in the Netherlands: planning, practice and prospects', *Land Use Policy*, 25(3): 361–377.

Koomen, E., Rietveld, P. and Bacao, F. (2009) 'The third dimension in urban geography; the urban-volume approach', *Environment and Planning B: Planning and Design*, 36(6): 1008–1025.

Kotchen, M.J. and Powers, S.M. (2006) 'Explaining the appearance and success of voter referenda for open-space conservation', *Journal of Environmental Economics and Management*, 52(1): 373–390.

Lancaster, K.J. (1966) 'A new approach to consumer theory', *Journal of Political Economy*, 74: 132–157.

Luttik, J. (2000) 'The value of trees, water and open space as reflected by house prices in the Netherlands', *Landscape and Urban Planning*, 48: 161–167.

Mansfield, C., Pattanayak, S.K., McDow, W., McDonald, R. and Halpin, P. (2005) 'Shades of green: Measuring the value of urban forests in the housing market', *Journal of Forest Economics*, 11(3): 177–199.

Mathis, M.L., Fawcett, A.A. and Konda, L.S. (2003) 'Valuing nature: A survey of the non-market valuation literature', Houston, TX: Houston Advanced Research Center, Discussion paper VNT03-01.

Morales, D.J. (1980) 'The contribution of trees to residential property value', *Journal of Arboriculture*, 6: 305–308.

More, T.A., Stevens, T. and Allen, P. (1988) 'Valuation of urban parks, USDA Forest Service', *Landscape and Urban Planning*, 15: 139–152.

Oosterhuis, F.H. and Van der Pligt, J. (1985) 'Kosten en baten van de wet geluidshinder', Commissie Evaluatie Wet Geluidshinder, Rapport CW-AS-06 (in Dutch).

Powe, N.A., Garrod, G.D. and Willis, K.G. (1995) 'Valuation of urban amenities using an hedonic price model', *Journal of Property Research*, 12: 137–147.

Ready, R.C. and Abdalla, C.W. (2005) 'The amenity and disamenity impacts of agriculture: Estimates from a Hedonic Pricing Model', *American Journal of Agricultural Economics*, 87(2): 314–326.

Roos-Klein Lankhorst, J., Buijs, A., Van den Berg, A., Bloemmen, M., De Vries, S., Schuiling, R. and Griffioen, A. (2002) 'BelevingsGIS versie februari 2002', Wageningen: Natuurplanbureau, NPB Werkdocument 2002/08.

Rosen, S. (1974) 'Hedonic prices and implicit markets: product differentiation in pure competition', *Journal of Political Economy*, 82(1): 34–55.

Ruijgrok, E.C.M. (2004) 'Valuation of nature and environment. A historic overview of Dutch socio-economic valuation studies', Rotterdam: Platform for Economic Valuation of Nature.

Ruijgrok, E.C.M., Nillesen, E.E.M. and Atman, R.E. (2004) 'Economische waardering van cultuurhistorie. Een case-studie in de Tieler- en Culemborgerwaard', Utrecht: Projectbureau Belvedere (in Dutch).

Ruijgrok, E.C.M., Goosen, H. and Vonk, S. (1999) 'Meervoudig ruimtegebruik en natuurwaarden; Een studie naar de ecologische- en belevingswaarden van multifunctionele natuurtypen', Gouda: Stichting LWI (in Dutch).

Sijtsma, F.J., Stelder, T.M., Elhorst, J.P., Oosterhaven, J. and Strijker, D. (1996) 'Ruimte over, ruimte tekort', Groningen: Stichting Ruimtelijke Economie (in Dutch).

Taylor, F. (1916) 'Relation between Primary Market Prices and Qualities of Cotton', U.S. Department of Agriculture, *Bulletin no. 457*, Nov. 24.

Tyrväinen, L. (1997) 'The amenity value of the urban forest: an application of the hedonic pricing method', *Landscape and Urban Planning*, 37: 211–222.

Tyrväinen, L. and Miettinen, A. (2000) 'Property prices and urban forest amenities', *Journal of Environmental Economics and Management*, 39: 205–223.

Van der Straaten, J.W. (2010) 'Essays on urban amenities and location choice', Vrije Universiteit Amsterdam/Tinbergen Institute, Amsterdam: PhD Thesis.

Van Leeuwen, M.G.A. (1997) 'De meerwaarde van groen voor wonen; een regionale analyse', The Hague: Landbouw-Economisch Instituut (LEI-DLO), report no. 576, (in Dutch).

Van Rij, E., Dekkers, J. and Koomen, E. (2008) 'Analysing the success of open space preservation in the Netherlands: the Midden-Delfland case', *Tijdschrift voor Economische en Sociale Geografie (Journal of Economic and Social Geography)*, 99(1): 115–124.

Visser, P. and Van Dam, F. (2006) 'De prijs van de plek. Woonomgeving en woningprijs', Rotterdam/The Hague, NAi Uitgevers/Ruimtelijk Planbureau (in Dutch).

VNC (2005) 'Nederland weer mooi, Deltaplan voor het landschap', Beek-Ubbergen: Vereniging Nederlands Cultuurlandschap (in Dutch).

Ward Thompson, C. (2002) 'Urban open space in the 21st century', *Landscape and Urban Planning*, 60: 59–72.

Waugh, F. (1928) 'Quality factors influencing vegetable prices', *Journal of Farm Economics*, 10: 185–196.

Weicher, J.C. and Zerbst, R.H. (1973) 'The externalities of neighborhood parks: an empirical investigation', *Land Economics*, 49: 99–105.

Wing (2008) 'Binnenveld pilotgebied LNV voor financiering landschap', Wageningen. http://www.wing.nl/projecten/gebied/landschapsfonds_binnenveld (accessed 1 July 2011) (in Dutch).

Part IV

Outlook for landscape economists – burgeoning perspectives on recreation, agriculture and urban agglomeration

14 The importance of landscapes for recreational firms

Nico B.P. Polman, Arianne T. de Blaeij, Louis H.G. Slangen and Stijn J. Reinhard

Introduction

Since the 1960s there has been growing social demand for nature conservation in the Netherlands. This demand led to a turning point; landscape is no longer changing because of the conversion of nature areas into agricultural land, but because of conversion in the opposite direction, from agricultural land to nature areas. Since the 1980s the Dutch government has gradually developed a new nature zoning policy, consisting of purchasing agricultural land and converting it to different types of nature. In 1990 the Dutch government introduced the Nature Policy Plan (*Natuurbeleidsplan*) of the Dutch Ministry of Agriculture, with Nature and Food Quality as a "policy concept". An important component of this Plan was the National Ecological Network (NEN): a network of nature areas including nationally and internationally important nature and landscape areas and biodiversity "hotspots". The objective of the NEN is to protect and enhance nature areas and landscape structures.

Developing the NEN is based on the island theory of McArthur and Wilson (1963). According to this theory the number of species increases when different populations of the same species that were geographically separated from one another are connected again. Based on this theory the scattered nature areas or landscapes should be expanded and connected in a network of patches where flora and fauna will have priority. This network of areas, which is planned to be completed by 2018, would cover about 15 per cent of rural areas in the Netherlands. It will consist of existing nature areas, nature development areas, connecting zones, and agricultural areas with potential for agri-environmental management. The NEN contributes to the coherence of the larger European Natura 2000 network. The nature areas to be developed are integral parts of man-made landscapes.

A landscape is defined by the European Landscape Convention (Council of Europe 2000) as an area, as perceived by people, whose character is the result of the action and interaction of natural and/or human factors. Landscapes, such as most of the nature areas, are able to provide many different landscape or ecosystem services, which are defined as the capacity of a landscape to provide goods and services to society (Millennium Ecosystem Assessment 2005). In this chapter we will focus on a specific part of the preamble of the convention, namely that landscape constitutes a resource favourable to economic activity and whose protection, management and planning can contribute to job creation. In ecosystem assessment terms, we focus on one landscape service, namely the attractiveness of the landscape for recreational purposes. Tourists are attracted to landscapes which provide recreational benefits. The attractiveness of natural areas in the Netherlands is reflected by the number of tourists who visit the National Parks (belonging to the NEN) each year.

Developing an NEN implies a large public investment in the Netherlands (see, for example, Jongeneel et al. 2005). It is therefore important to know what the social benefits of nature areas are. Within economic theory, it is well known that location is an important factor for growth of employment by firms (Hoogstra and van Dijk 2004). It is even one of the foundations of spatial economics (Boschma et al. 2002). We may expect that location as classified by distance to an NEN (and its size) has a significant effect on the net value added (NVA) of recreational firms. This effect cannot be isolated without considering other determinants of NVA such as the type of recreational businesses in question, i.e. restaurants, camp sites or bungalow parks.

In this chapter we will analyse to what extent different location factors determine firm employment growth based on individual firm data. The location factors consist of landscape factors concerning the type of nature, the location of the firm in relation to the surrounding NEN and economic factors such as the growth of the population and employment in the neighborhood of the firms. Because we focus in this chapter on the NVA of recreational firms, we do not consider the non-market benefits of the NEN.

This chapter proceeds as follows. In the following section, we elaborate on the relationship between social recreational benefits and private profit earned by recreational firms. Subsequently, the Landscape Reilly-Index, a measure that can take into account the density of landscape in the neighborhood of the recreational firm, is described. The section thereafter presents the theoretical background and the employment data for the different types of recreational firms. This is followed by a section that discusses the method and empirical model. Then, in two sections, we present estimation results and give further insight into the implications of these results. The chapter finishes with some conclusions.

Landscape and recreational benefits

One of ecosystem services provided by the NEN is the recreation service. Different stakeholder groups will benefit from this specific ecosystem service. Due to the fact that normally there is no market for ecosystem services like recreation, the benefits have to be measured with non-market valuation studies. There is a lot of stated preference monetary valuation literature that estimates the citizen/consumer value of changes in nature areas and their identified landscapes. Examples for the Netherlands are Brouwer and Slangen (1998) who analyse the non-market benefits of preserving Dutch nature areas, Van der Heide et al. (2008) who focus on the value of defragmentation of the Veluwe, and Nunes et al. (2009) who estimate the value of reducing cockle fishing in order to improve the ecological quality of the Waddenzee. These studies focus on non-marketable benefits of landscapes.

Another kind of valuation method, investigating the influence (of the value) of landscape elements on market prices, is hedonic pricing. Landscape characteristics can be included as location characteristics, which is an explanatory variable for restaurant, hotel room and package deal prices. Landscape characteristics that appear to be important are distance to the city centre (Monty and Skidmore 2003), distance to the beach (Espinet et al. 2003), and dikes and open coasts (Hamilton 2007).

In this chapter, we will focus on another group of beneficiaries, namely recreational firms located in proximity to the NEN. The performance of recreational firms is the result of a complex combination of (i) firm-specific factors including entrepreneurial skills of the owner; (ii) surrounding man-made and natural amenities, including climate (weather); and (iii) ease of access to these amenities, including restrictions for example due to spatial policy. An obvious example of the importance of landscape for recreational firms concerns firms which directly depend on the nature of the landscape. For example, "active" activities such as scenic flights and mountain biking or more passive activities such as painting or environmentally-friendly lodging. Recreation can also be classed as consumptive (i.e. fishing and hunting) or non-consumptive (i.e. walking, bird-watching and photographic tours) (Marcouiller and Prey 2005; Bell et al. 2007; Marcouiller et al. 2009). This study focuses on surrounding man-made and natural amenities indicated as location specific factors. Entrepreneurial skills and access to amenities are not taken into account because of limited data.

Location factors which qualify the spatial environment traditionally play a crucial role in the field of regional science, in which the economic performance of regions is the main concern of the study. Among economists, there is a widespread belief that 'space matters' (Krugman 1991: 8). However, in these fields of research the units of observation are spatial entities and not individual firms (cf. Hoogstra and Van Dijk 2004).

According to Hoogstra and Van Dijk (2004), it is possible to formulate the hypothesis that location factors are significant in determining regional employment growth. It is reasonable to assume that environmental factors determining regional growth are also important determinants for the employment growth of individual firms because regional growth is the aggregated of individual firms growth. It is, however, important to take into consideration that employment growth is often not regarded as the major business target by the firm itself. This is based on neoclassical economic theory in which a firm focuses on profit maximization, not number of employees per se. In any case, firms often follow a strategy of increasing turnover in order to increase total profit. This may well require increasing the number of employees. Furthermore, this strategy fits with the idea of exploiting economics of scale and of firm survival as a major business objective (Hoogstra and Van Dijk 2004).

The Landscape Reilly Index; a landscape density index

To investigate whether the surrounding landscape is one of the factors influencing the performance of a recreational firm, the level of nature-landscape in the surroundings of the firm has to be measured. The Landscape Reilly Index (or in short Reilly-index) can be used to measure the density of landscape in proximity to a recreational firm. This index (as distinguished by Geohegan et al. 1997) provides a means to quantify for the individual recreational firm the importance of distance to and the size of the NEN, which is relevant from the human perspective of management and valuation. This index captures, in one number, the size of the NEN-area in proximity to the recreational firm and the distance from the firm to certain landscape areas (Cotteleer 2008). Strong points of the Reilly-index are the combination of distance with size, and the fact that natural areas located further away or which are smaller in size are weighted less. As such, the Reilly-index is a measure for the share of land-used for a certain land-use function in the surroundings of a specific location. Other examples of land-uses that can be captured by a Reilly-index are agriculture and glasshouse horticulture.

Equation (14.1) gives the formula for the calculation of the Reilly-index. The calculation of the Reilly-index starts at the point where the firm is located. After that, the size of the NEN-area within a certain radius (i.e. 5 km) is determined. Based on the sum of all the distances of the firm to the NEN-areas located within the chosen radius, and on the size of the NEN-areas the Reilly-index can be calculated. Distance is measured in m, size in m^2

$$\text{Landscape Reilly-index for firm}_i = \sum_{j=1}^{J} \frac{\text{Size of the NEN } j \text{ (within radius)}}{(\text{distance of firm } i \text{ to NEN})_{ij}^2}. \tag{14.1}$$

We applied the Reilly-index to the impact of the NEN on recreational firms in the Netherlands. The Netherlands has about 3.5 million ha of land. In 1990, 60 per cent was agricultural landscape and 13 per cent was nature-landscape. The target of the Nature Policy Plan (LNV 1990) was to develop the NEN, for which about 275,000 ha of new nature have to be develop before 2018. This implies an increase of about 60 per cent, partly realised by purchasing agricultural land and converting it into nature.

The Netherlands Environmental Assessment Agency (NEAA) defined 18 different types of NEN types of nature-landscape (Koeijer et al. 2008). Based on personal communication with the NEAA, we have taken into account 12 types of nature, clustered in four main groups, namely: forests, marshes, grasslands and coastal areas (see Table 14.1).

We illustrate the calculation of the Reilly-index in Table 14.2 and Figure 14.1 for the location of two restaurants: A and B. The two restaurants (A and B) are situated in the proximity of four different areas of the NEN, numbered from 1 to 4. Table 14.2 shows the distance between the restaurants and the four NEN-areas.

Table 14.1 Types of nature and clusters of nature

Type of nature	*Cluster of nature*
Cultural historic forests	Forest
Dry heath lands	Forest
Dry natural forests	Forest
Dry poor grasslands	Forest
Salt marshes	Coastal areas
Marshland	Marshes
Open dunes	Coastal areas
Wet natural forests	Forest
Wet poor grasslands	Grasslands
Poor fen pools and moorlands	Forest
Nutrient rich grasslands and fields	Grasslands
Grassland specific for birds	Grasslands

Source: The clustering is based on personal communication with Tanja de Koeijer and Rogier Pouwels (Netherlands Environmental Assessment Agency).

Table 14.2 Reilly-index for two restaurants given the size of and the distance to the NEN

NEN area	*Size of the NEN in ha*	*Distance to restaurant A (in 1,000 m)*	*Size/ (distance)^2*	*Distance to restaurant B (in 1,000 m)*	*Size/ (distance)^2*
1	100	1	1	1.4	0.51
2	50	2.1	0.11	0.4	3.13
3	20	0.6	0.56	0.7	0.41
4	90	1.2	0.63	0.9	1.11
Reilly-index		2.29			5.15

Source: Adapted from Cottelleer (2008: 101).

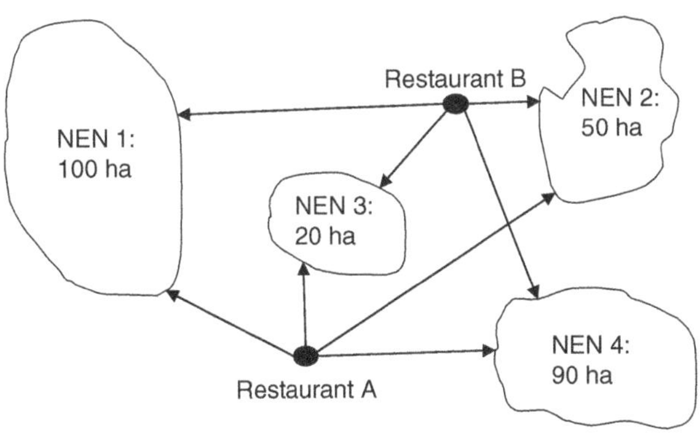

Figure 14.1 Graphical representation of the Reilly-index.

Table 14.3 The Reilly-index for seven restaurants with 10 workers in six cities

City	*Reilly-index*				
	NEN	*Forests*	*Marshes*	*Grasslands*	*Coastal areas*
Amsterdam	0	0	0	0	0
Amsterdam	0.039	0.022	0.012	0.0050	0
Den Haag	1.12	0.71	0	0.057	0.36
Apeldoorn	1.47	1.44	0.0067	0.018	0
Beekbergen	19.50	19.35	0.011	0.14	0
Winterswijk	8.70	3.78	0.54	4.37	0
Zandvoort	9.64	0.32	0	0.028	9.29

Source: Own calculations.

The size of the four NEN-areas is also given. Figure 14.1 shows the two restaurants and their location in relation to the four NEN-areas. The arrows in Figure 14.1 give the distance to the four NEN-areas. The size and distance correspond with Table 14.2.

From Table 14.2 and Figure 14.1, it is apparent that for restaurant B the Reilly-index is much larger than for restaurant A, because restaurant B is located closer to one of the NEN-areas. Although location 2 of the NEN is not the largest NEN-area, the shorter distance from restaurant B to this area is largely responsible for the larger Reilly-score for this restaurant.

From this theoretical example, it can be seen that a restaurant located near or in an NEN-area has a higher Reilly-index than a restaurant that is located further away Also larger NEN-areas result in a higher Reilly-index when compared with smaller NEN-areas. Table 14.3 gives an illustration of Reilly-indexes for seven randomly chosen restaurants with 10 workers in six cities in the Netherlands. For each restaurant, the Reilly-index for all the different NEN-areas together, and for the four types of nature-landscape (Table 14.1) are given (radius 5 km). These restaurants are examples taken from the LISA database with information about all locations in the Netherlands where people have paid work It includes the addresses of recreational firms in the Netherlands.

In Amsterdam, there is a restaurant with a Reilly-index of 0, i.e. as low as possible (no NEN-landscape within 5 km of the restaurant). Notwithstanding, there is another recreational firm in Amsterdam with a low but a positive Reilly-index for the entire NEN, for forests, marshes and for grasslands. However, for the coastal areas the Reilly-index of this restaurant is 0. The two restaurants in the cities of Den Haag and Apeldoorn have a comparable Reilly-index, indicating that the landscape density of nature within a 5 km radius is comparable. Restaurants in Beekbergen and Zandvoort have a relatively large Reilly-index, mostly affected by one type of nature, forests in Beekbergen and coastal areas in Zandvoort. In Winterswijk, the two NEN types grasslands and forests contribute to the relatively large Reilly-index.

Data and empirical method

As said above, in this study we would like to analyse to what extent different location factors affect firm employment based on data which use individuals firms as the units of observation. The firm employment is taken as a proxy for the NVA of recreational firms, because we do not have data to represent the NVA of individual recreational firms directly.

The LISA database distinguishes full-time and part-time workers (LISA is a Dutch acronym for *Landelijk InformatieSysteem voor Arbeidsplaatsen*, which can be translated as National Information System for Jobs). Full-timers are all the workers who are working more than 12 hours a week and part-time workers less than 12 hours. The number of workers includes all the people who are working for the firm, including the owner, manager, members of the family household and employees. Table 14.4 gives an overview of the numbers of workers in the recreational sector.

From the employment point of view, restaurants constitute the most important sub-sector. It has almost one third of the full-time jobs and about 38 per cent of the part-time jobs. Due to the fact that landscape characteristics can have a different impact on different types of recreational firms (Shaw and Ozog 1999; Yeh et al. 2006), we included single day recreational firms and overnight recreation. The most important of the first category are restaurants and cafés, representing respectively 21 and 35 per cent of the total number of people employed. Overnight recreational firms include hotels, youth hostels, camping and bungalow parks, together representing nine per cent of the people employed. Not included are sub-sectors that fit both categories and heterogeneous sub-sectors.

In this section we will discuss the method for estimating the relationship between firm employment and firm-specific variables. The firm-specific variables refer to the type of the recreational firm and whether the firm has relocated. The location-specific variables can be split up into (i) economic location factors such as the local growth of population and growth of employment (see Hoogstra and Van

Table 14.4 Overview of the recreational sector and its employment

	Number of working people		*In per cent*	
Sectors	*Full-time*	*Part-time*	*Full-time*	*Part-time*
Cafés	37,309	29,773	19.0	23.6
Hotels	7,656	2,551	3.9	2.0
Youth hostels	474	427	0.2	0.3
Camping sites	5,942	2,669	3.0	2.1
Restaurants	64,155	47,827	32.7	37.9
Bungalow parks	6,293	3,071	3.2	2.4
Other	74,404	39,818	37.9	31.6
Total	196,233	126,136	100.0	100.0

Source: LISA database, 2003 and 2007.

Dijk 2004); (ii) a variable for the location of recreational firms in rural or non-rural areas; and (iii) Reilly-indices for each recreational firm for different types of nature belonging to the NEN. For testing the relationship we use the following econometric model:

$$\ln(WP) = \alpha + \beta_1 BEDR + \beta_2 LOC + \varepsilon, \quad (14.2)$$

where ln(*WP*) is the number of working people on a recreational firm; *BEDR* is a set of firm-specific factors; *LOC* is a set of location-specific factors; ε is error; and α, β_1 and β_2 are coefficients.

Table 14.5 presents the variables used in the econometric model. The firm-specific variables consists of 29,026 observations; originating from firms existing in 2003 and 2007. These variables refer to the numbers of employees per firm and whether the firm has relocated during the last five years.

On an average five people are working on a recreational firm. Looking to the minimum and the maximum the dispersion is very large. In the regression analysis, this variable is used as the dependent variable. It is a proxy variable for the NVA.

Concerning the variable "relocation during the last 5 years", the theoretical viewpoint could be that a firm chooses a feasible location that maximizes profits. Due to the dynamics in the spatial economic environment and/or firm, the optimal location may change over time rather often. This does not imply that firms also move frequently, because relocation itself is rather costly especially when large investments in building and equipment are required (cf. Hoogstra and Van Dijk 2004). Table 14.5 shows that less than one per cent of the firms have moved in the last five years.

Table 14.5 Firm-specific and location-specific variables

Variable	*Average*	*Median*	*Maximum*	*Minimum*
Location-specific variables:				
Growth of population between 2004 and 2007 in post code area (ln)	0.0041	−0.0018	2.01	−1.39
Growth employment between 2003 and 2007 in post code area (ln)	0.0068	0.0027	1.79	−2.15
Dummy for location in rural area	0.43	0.00	1.00	0.00
Reilly-index NEN-marshes	0.11	0.021	28.11	0.00
Reilly-index NEN-forests	1.69	0.61	49.57	0.00
Reilly-index NEN-coastal areas	0.24	0.00	44.99	0.00
Reilly-index NEN-grasslands	0.72	0.25	44.09	0.00
Firm-specific variables:				
Relocation during the last 5 years	0.0064	0.00	1.00	0.00
Dummy is 1 if restaurants	0.42	0.00	1.00	0.00
Dummy is 1 if cafés	0.42	0.00	1.00	0.00
Dummy is 1 if hotels	0.049	0.00	1.00	0.00
Dummy is 1 if youth hostel	0.0041	0.00	1.00	0.00
Dummy is 1 if bungalow park	0.040	0.00	1.00	0.00

The variables "growth of population" and "growth of employment in the post code area" of the recreational firms belong to economic location variables. These variables represent the demand side of the goods and services of the recreational firms. Hoogstra and Van Dijk (2004) consider the regional growth of employment as an indicator for the regional economic development. The variables "growth of population" and "growth of employment" are presented as a logarithm. We assume that both variables will have a positive relationship with the size (measured in employment) of the recreational firms.

Another variable is the location of the firms in the spectrum from urban to rural area. Areas with less than 1,000 addresses per km^2 are considered to be rural and areas with more than 1,000 addresses per km^2 as urban. Table 14.5 shows that about 40 per cent of the firms are located in the rural areas. Given the fact that the NEN is located in rural areas, we could expect a positive effect of the dummy variable on the size of the recreational firms.

The Reilly-indices for the four types of nature are taken as independent variables. Table 14.5 shows that the dispersion of these variables is relatively large. If the NEN has a positive effect on the number of employees in recreational firms we will find a positive effect of the variables "Reilly-indices NEN" on the size of the recreational firms.

The last group of firm-specific variables refers to the type of the recreational firms, presented by a dummy variable. However, the term firm-specific is not completely right, because the dummy variable indicates the type of recreational firm (i.e. the sector) and not characteristics of the firm itself. Based on Table 14.4 we know that restaurants and cafés are the most important sectors for "day recreation" with respect to the number of people employed. They also represent about 50 per cent of the total NVA of all the recreational firms. For the "overnight recreational" firms we take hotels, youth hostels, camping and bungalow parks. Together, these recreational firms produce about €700 million NVA or about 11 per cent of the total NVA. It is difficult to make an assumption about the impact of these dummy variables on the size of the recreational firm in terms of employment.

Results

Table 14.6 gives the results of the estimations. The independent variable is the number of people working per recreational firm. The unit of this variable is taken as a logarithm in the estimation procedure. A Breusch–Pagan test on heteroskedasiticity showed that the assumptions of Ordinary Least Squares (OLS) are rejected. Therefore, we leave the OLS out of consideration and present in the four columns of Table 14.6 the so-called quantiles estimations (for more information, see Koenker and Hallock 2001). The pseudo R^2 is included as the bottom row of Table 14.6. The rather low R^2 is of less importance because the main focus of this study is on finding significant relations between NEN and recreation firm employment and not on maximising explanatory power (see also Hoogstra and Van Dijk 2004).

As mentioned earlier, the dependent variables "growth of population" and "growth of employment in the post code area" of the recreational firms are presented as logarithms. This means that the estimated coefficients of these variables are elasticities. An elasticity is the percentage change in the dependent variable when the explanatory variable changes by one per cent. The other dependent variables are linear. Because of that, the estimated coefficients of these variables are semi-elasticities. A semi-elasticity is the percentage change of the dependent variable when the explanatory variable changes by one unit (see Woolridge 2009).

Table 14.6 suggests that being involved in a process of moving or having recently moved has a positive effect on the size of the firms for two out of the four quantiles estimations. Apparently, the investments in building and equipment as result of a moving are not an impediment to employment growth (cf. Hoogstra and Van Dijk 2004).

The coefficients for the variable "growth of population in postcode area" are positive and statistically significant for all the estimations. Given the size of the coefficient, it is an important variable for the explanation of the amount of the employment per recreational firm. Because the variable "growth of population" is presented as a logarithm, the estimated coefficients are also elasticities. The estimated coefficients are positive and about one. This means that an increase

Table 14.6 The relationship between number of people working per recreational firm and a set of explanatory variables

Variable	*Quantile estimate*			
	0.25	*0.5 (median)*	*0.75*	*0.90*
Moving during the last five years	−0.24***	−0.15***	−0.098	−0.080
Growth of population in the postcode area (ln)	0.72***	0.78***	1.093***	1.06***
Growth of employment in the postcode area (ln)	−0.0061	0.061	0.089	0.037
Dummy for location in rural area	−0.0077	−0.0078	−0.038**	0.014
Reilly-index NEN marshes	0.0039	0.0029	−0.0015	0.0041
Reilly-index NEN forests	0.015***	0.018***	0.032***	0.033***
Reilly-index NEN coastal areas	0.022***	0.017***	0.0091*	0.0040
Reilly-index NEN grasslands	−0.0012	−0.0041	−0.0050	−0.013*
Dummy is 1 if restaurants	0.84***	0.76***	0.90***	0.81***
Dummy is 1 if cafés	0.27***	0.19***	0.36***	0.34***
Dummy is 1 if hotels	−0.0088	0.021	0.34***	0.67***
Dummy is 1 if bungalow park	−0.087***	−0.40***	−0.27***	−0.028
Dummy is 1 if youth hostel	−0.028	−0.0067	0.30***	0.57***
Constant	0.0024	0.64***	1.06***	1.70***
Pseudo R^2	0.081	0.078	0.062	0.037

* Significant at the 10 per cent level.
** Significant at the 5 per cent level.
*** Significant at the 1 per cent level.

of the population by one per cent leads to an increase of the employment per recreational firm by one per cent also.

The coefficients for the variable "growth of employment in the post code area" of the recreational firms are not statistically significant for the four quantiles estimations. This means that this variable does not influence the size of the firms. We have already noted that Hoogstra and Van Dijk (2004) consider the growth of employment as an indicator for the regional economic development. This means and implies that the regional economic development has no influence on the size of recreational firms.

For all of the four quantiles estimations, the coefficients for the dummy variable "location in the rural areas" are not statistically significant. These results of the quantiles estimations indicate no influence of this variable. We can say that a location in a rural area does not have an effect on the size recreational firms.

The coefficients for the "Reilly-index for marshes" for the four estimations are small and not statistically significant. This means that there is no statistical relationship between marshes and the size of recreational firms. The coefficients for the "Reilly-index for forests" for the four estimations are statistically significant and positive. As said earlier, these coefficients are semi-elasticities. Given the size of the coefficients, it means that if the Reilly-index increases by one unit, the increase of the employment per recreational firm in the neighborhood of the forest in terms of percentage will be: (i) 1.5 per cent for 0.25 quartiles estimation; (ii) 1.8 per cent for the 0.5 quartiles estimation (median); (iii) 3.3 per cent for the 0.75 quartiles estimation; and 3.5 per cent for the 0.90 quartiles estimation.

Consequently, if the Reilly-index increases by one unit, the increase of the employment per recreational firm in the neighborhood of forest in terms of percentage will, for example be 1.8 per cent for the 0.5 quantiles estimation.

The coefficients for the "Reilly-index for coastal areas" are statistically significant and positive for three quantiles estimations. For the 90 per cent quantiles, the estimation is not statistically significant. The size of the statistically significant coefficients is, in general, smaller than those for forests. Therefore, an increase of the Reilly-index by one unit has a smaller positive effect on the employment of recreational firms in the neighborhood of the coastal areas than of firms that are located near or into forests.

The coefficients for the "Reilly-indices for grasslands" are negative for quantiles estimations, but not statistically significant. This means that grassland has no positive effect on the size and scope of recreational firms.

The type of the recreational firms is presented by a dummy variable. The coefficients for the dummy variables "restaurants and cafés" are statistically significant and positive for all estimations. This means that they are larger (in employment) than the other types of recreational firms in Table 14.6. Based on Table 14.4 we know that restaurants and cafés are the most important sectors for recreation. They represent about 50 per cent of the total NVA. The coefficients for the dummy variable "hotels" are positive and statistically significant for two quantiles estimations. However, for two quantiles estimations the results are not statistically significant. This means hotels as a type of business do not always have a positive effect on the size of the firm.

The coefficients for the dummy variable "bungalow parks" are statistically significant and negative for three quantiles estimations. Only the 90 per cent quantiles estimation is not statistically significant. This means that in general, they are smaller (in employment) than the other types of recreational firms in Table 14.6.

The coefficients for the dummy variable "youth hostel" are positively signed and statistically significant for two quantiles estimations. This means that we can only safely say that for these two quantiles they are larger in terms of employment than the other types of recreational firms in Table 14.6. A general statement concerning the size is not possible.

So all in all, only the variables for the Reilly-indices for forests and coastal areas lead to positive and statistically significant effects on the size of recreational firms. The effect of the variable "Reilly-index NEN forests" is for all four estimations positive and statistically significant and so is the variable "Reilly-index coastal areas" for three estimations. However, the size of the coefficients of the Reilly-indices varies between the different estimations. This implies that the influence of the NEN on the size of recreational firms is heterogeneous. It can be a result of the size of the areas of the NEN or of the distance of the NEN-areas to the recreational firms.

Landscape benefits: an application to developing natural areas

In the Netherlands, some private parties have indicated interest in developing natural areas to generate income by getting paid for ecosystem service provision. Developing natural areas implies, however, the change of a landscape. In this chapter we focus in particular on the recreational services provided by these areas. To that end, the estimation results presented above will be discussed from two perspectives: the effect of a landscape change from the viewpoint of recreational firm, and from the viewpoint of a so-called "entrepreneur of a natural area". Finally, both perspectives will be integrated and discussed from an economic perspective.

For entrepreneurs of natural areas, it is interesting to get hold of the benefits that accrue to them through the provision of the ecosystem services. This makes it interesting to interpret the results from a landscape change perspective. Extending natural areas in the neighborhood of recreational firms or developing new natural areas close to these firms will increase the Reilly-index, due to a change in size of a natural area or a shorter distance between the firm and an area. But a high Reilly-index does not automatically guarantee a boost in performance benefits for firms. To put it more specifically, increasing nature has, on average, a very small effect on the employment and NVA of recreational firms. Nevertheless, and as expected, investments to protect or develop nature-areas close to the location of the firm have a larger effect on the NVA of a firm than investments in nature that are situated further away.

From the viewpoint of a "nature entrepreneur", developing new natural areas implies producing ecosystem services. It can be shown that different stakeholders benefit (sometimes hugely) from these services – but without realising it or paying for them. One of the reasons is that many ecosystem services have public good characteristics. This research suggests that recreational firms can reap benefits from newly developed natural areas. However, the effects on employment are small when analysing individual cases.

In specific situations, it can be expected that certain firms benefit (also) from location factors other than mentioned earlier, such as nearby urban areas, good infrastructure or the neighborhood of other firms. Renski (2009) concludes that for the United States, firms in rural areas have lower growth compared to firms in urban areas.

To show the impact of the result of regression analysis, we take Reilly-index coefficients for forests. Given the size of these coefficients, it means that if the Reilly-index increases by one unit, the increase of the employment per recreational firm in the neighborhood of the forest will be 2.5 per cent. According to our calculations the median of the Reilly-indices for forest is 0.62. An increase of the Reilly-index with by unit means a rise of the Reilly-index from 0.62 to 1.62, which is an increase of about 160 per cent. The meaning of this can be made clearer by the following example. Suppose restaurant A is located at a distance of 500 m from a forest with a size of 25 ha. The Reilly-index in this case is 1. By increasing the Reilly-index from 1 to 1.60 and keeping the same distance (500 m), the size of the forest should increase from 25 to 40 ha. Given the results of our regression analysis, this will lead to an increase of the employment for restaurant A of 1.8 per cent for the median firm.

The average NVA per worker in our data set is estimated at about €28,000. The median of the number of workers per recreational firm is about three. This means that the NVA for the median recreational firm is about €84,000. So, an increase of the Reilly-index by one point means an increase of NVA per recreational firm of about €1,500 (0.018 × 84.000).

We focused in our example only on one type of nature, namely "NEN forest". However, as the regression analysis shows, the results are not the same for each type of (NEN) nature. This is reflected in different values of the Reilly-index for marshes, forests, coastal areas and grasslands. Nonetheless, taking all the arguments into account, our main conclusion is that increasing nature has only a very small effect on the employment and NVA of the recreational firms.

All in all, paying particular attention to the link between developing landscapes and benefits for recreational firms can be useful – not only for nature policy makers, but also for spatial planners and economic developers. Calculating a Reilly-index could contribute to this analysis as a means of quantifying benefits. Our analysis was limited to four types of nature. Further research is needed to analyse the effects of, for instance, other types of natural areas, such as multifunctional commercial wetlands, in the neighborhood of recreational firms.

Conclusion

The objective of this chapter is to analyse the economic benefits of the Dutch National Ecological Network (NEN). Due to the lack of data for the net value added (NVA) per recreational firm, the employment per firm is used as a proxy for NVA. To explain the employment by recreational firms, we made use of firm-specific and location-specific variables. The firm-specific variables include factors such as whether the firm has relocated in the last five years. The location-specific variables consist of (i) economic location factors such as the growth of population and growth of employment in the post code area of the recreational firms; (ii) a dummy variable for location of recreational firms in a rural or non-rural area; (iii) the Reilly-indices for four different types of nature belonging to the NEN; and (iv) dummy variables for five types of recreational firms.

The empirical analysis shows that recreational firms located in the neighborhood of marshes, forests, coastal areas or grasslands have more workers than those that do not have such types of nature in their proximity. However, the effect is small. The size of the effect depends on the size of the NEN and the distance to the recreational firm, and on the specific type of nature. A smaller size and/or a larger distance has a decreasing effect on the employment of recreational firms.

Concerning the type of nature, only forests and coastal areas contribute in a statistically significant and positive way to the employment of recreational firms. However, increasing the employment level of recreational firms in the neighborhood of forests or coastal areas with a small percentage, say 2.5 per cent, requires a very strong increase in the area of such a type of nature.

It is worthwhile to consider that we did not pay attention to: (i) the quality of the entrepreneur as an explanatory variable for the size of recreational firms; (ii) characteristics, quality and more detailed location aspects (such as the number of passers-by or visibility) of the business premises of the recreational firms; and (iii) the quality of the service and provisions of the recreational firms.

Acknowledgements

This research is partly funded by the former Dutch Ministry of Agriculture, Nature Conservation and Food Quality (KB-01-003-011) and based on research for the Netherlands Environmental Assessment Agency (WOT-04-002 2008-2.5).

References

Bell, S., Tyrvainen, L., Sievanen, T., Probstl, U. and Simpson, M. (2007) 'Outdoor recreation and nature tourism: a European perspective', *Living Reviews. Landscape*, 1(2), http://www.livingreviews.org/lrlr-2007-2 (accessed 18 February 2010).

Boschma, R.A., Frenken, K. and Lambooy, J.G. (2002) *Evolutionaire Economie.* Bussum: Coutinho.

Brouwer, R. and Slangen, L. (1998) 'Contingent valuation of the public benefits of agricultural wildlife management: The case of Dutch peat meadow land', *European Review of Agricultural Economics*, 25: 53–72.

Cotteleer, G. (2008) *Valuation of Land Use in the Netherlands and British Columbia; A Spatial Hedonic GIS-based Approach*. Wageningen University: PhD thesis.

Council of Europe (2000) *European Landscape Convention,* http://www.coe.int/t/dg4/cultureheritage/heritage/Landscape/default_en.asp (accessed 23 February 2010).

Espinet, J.M., Saez, M., Coenders, G. and Fluvià, M. (2003) 'Effect on prices of the attributes of holiday hotels: a hedonic prices approach', *Tourism Economics*, 9(2): 165–178.

Geoghegan, J., Wainger, L.A. and Bockstael, N. (1997) 'Spatial landscape indices in a hedonic framework: an ecological economics analysis using GIS', *Ecological Economics*, 23(3): 251–264.

Hamilton, J. (2007) 'Coastal landscape and the hedonic price of accommodation', *Ecological Economics*, 62(3–4): 594–602.

Heide, C.M. van der, Bergh, J.C.J.M. van den, Ierland, E.C. van and Nunes, P.A.L.D. (2008) 'Economic valuation of habitat defragmentation: A study of the Veluwe, the Netherlands', *Ecological Economics*, 67(2): 205–216.

Hoogstra, J.G. and Dijk, J. van (2004) 'Explaining firm employment growth: Does location matter?' *Small Business Economics*, 22: 179–192.

Jongeneel, R.A., Slangen, L.H.G., Bos, E., Koning, M., Ponsioen, T. and Vader, J. (2005) *De Effecten van Natuurprojecten op de Economie: Financiële en Economische Analyse van Kosten en Baten,* Wageningen University and LEI: Leerstoelgroep Agrarische Economie en Plattelandsbeleid (in Dutch).

Koeijer, T.J. de, Bommel, K.H.M. van, Clement, J., Groeneveld, R.A., Jong, J.J. de, Oltmer, K., Reijnen, M.J.S.M. and Wijk, M.N. van (2008) *Kosteneffectiviteit Terrestrische Ecologische Hoofdstructuur; Een Eerste Verkenning van Mogelijke Toepassingen*, Wageningen, Wettelijke Onderzoekstaken Natuur & Milieu: WOt-report no. 73 (in Dutch).

Koenker, R. and Hallock, K.F. (2001) 'Quantile regression', *Journal of Economic Perspectives*, 15(4): 143–156.

Krugman, P. (1991) *Geography and Trade.* Cambridge: MIT Press.

LNV (1990) *Natuurbeleidsplan: Regeringsbeslissing,* Tweede Kamer, vergaderjaar 1989–1990, 21 149, nrs.2–3,'s-Gravenhage: Sdu (in Dutch).

MacArthur, R.H. and Wilson, E.O. (1963) 'An equilibrium theory of insular zoogeography', *Evolution*, 17: 373–387.

Marcouiller, D.W. and Prey, J. (2005) 'The tourism supply linkage: Recreational sites and their related natural amenities', *Journal of Regional Analysis and Policy*, 35(1): 29–39.

Marcouiller, D.W., Prey, J. and Scott, I. (2009) 'The regional supply of outdoor recreation resources: Demonstrating the use of location quotients as a management tool', *Journal of Park and Recreation Administration*, 27(4): 92–107.

Millennium Ecosystem Assessment (2005) *Ecosystems and Human Well-being: A Framework for Assessment*, Washington, D.C.: Island Press, http://www.millenniumassessment.org/en/Framework.aspx (accessed 23 February 2010).

Monty, B. and Skidmore, M. (2003) 'Hedonic pricing and willingness to pay for bed and breakfast amenities in Southeast Wisconsin', *Journal of Travel Research*, 42: 195–199.

Nunes, P.A.L.D., Blaeij, A.T. de and Bergh, J.C.J.M. van den (2009) 'Decomposition of warm glow for multiple stakeholders: Stated choice valuation of shellfishery policy', *Land Economics*, 85(3): 485–499.

Renski, H. (2009) 'New firm entry, survival and growth in the United States; a comparison of urban, suburban and rural areas', *Journal of the American Planning Association*, 75: 60–77.

Shaw, W.D. and Ozog, M.T. (1999) 'Modeling overnight recreation trip choice: application of a repeated nested multinomial logit model', *Environmental and Resource Economics*, 13(4): 397–414.
Woolridge, J.M. (2009) *Introductory Econometrics; A Modern Approach*, 4th edition. Singapore: South-Western.
Yeh, C-Y, Haab, T.C. and Sohngen, B.L. (2006) 'Modeling multiple-objective recreation trips with choices over trip duration and alternative sites', *Environmental and Resource Economics*, 34(2): 189–209.

15 Agricultural policies and rural landscape

Some insights from theoretical and empirical literature

Sylvie Ferrari, Christian Lippert and Olivier Aznar

Introduction

In the field of the European Landscape Convention (2000), landscape means 'an area, as perceived by people, whose character is the result of the action and interaction of natural and/or human factors' (article 1). Such a definition implies that landscape dynamics follow from various actions taken by private and public stakeholders, which are led by different targets and subject to heterogeneous constraints. It also brings into play elaborate interactions between human action and natural dynamics. Rural landscapes and agricultural activities provide a relevant example of such interactions.

Over the past century, rural landscapes have been influenced in different ways by a succession of public policy orientations. At the same time, many of the policies that impacted on landscapes did not relate to them *per se* but targeted other finalities instead, including infrastructure and economic development based on territorial reconstruction activity. What this means is that observed changes in rural landscapes have been the *indirect* consequence of economic and social developments and not the result of a public policy aimed at improving or preserving rural landscapes. Similarly, landscapes that are often idealised as being part of the "good old days" usually date from the 19th century and are no more than the temporary outcomes of economic and social changes that began in the aftermath of the Napoleonic wars (Küster 1996: 321).

Given the progressively greater significance ascribed over the second half of the 20th century to landscape issues within overall agricultural policies, the developments that the present chapter covers mainly pertain to agriculture-related policies and not to all policies impacting on rural landscapes. Major regional initiatives like regional development, irrigation, wetlands management or reforestation had been undertaken in an effort to control the conditions of agricultural production by directly acting upon its natural settings (INRA 2008). Globally, until the 1980s–1990s the main purpose of the Common Agricultural Policy (CAP) in Europe was to improve production conditions by land consolidation, intensifying agricultural practices and increasing farms' production capacities.[1] The first attempts within the European Economic Community (EEC) to prevent changes in agricultural production structures and to preserve an extensive agricultural sector dated from 1975, with the introduction of a payments system targeting areas with natural handicaps, mainly mountain areas (Henrichsmeyer and Witzke 1994).

More recently, rural landscapes have been altered significantly due to changes in land-uses resulting from new forms of public intervention in the agricultural domain. More specifically, rural landscapes in many European countries have been transformed by the modification of agricultural practices in the wake of new public incentive mechanisms. Changes in the Common Agricultural Policy (CAP) have played a decisive role in a succession of reforms centered around issues that were initially purely economic in nature before gradually evolving to integrate agricultural activities' environmental and social dimensions.

During the course of the 1990s, the concept of multifunctionality entered the agricultural sector at the European and international levels in a context defined by the liberalisation of economic policies and a greater focus on environmental issues (Garzon 2005). This concept introduces the idea that an agricultural activity generates market products (food, raw materials for industry) as well as non-market products (environmental amenities, pollutions). It has propelled a number of new debates in the agricultural domain, like the sector's expected role in the dynamics of territories or in the protection of the environment. Questions about the relations between traded agricultural goods and non-market goods that derived, as by-product, from agricultural activity (like landscapes, cultural heritage, biodiversity or water pollution) therefore appear key to any discussion of agricultural policy reform (Abler 2001).

In this context, three points should be highlighted:

- The implementation of agri-environmental measures related to compensatory mechanisms associated with the 1992 CAP reform. These measures were intended to promote complementarity between farms' productive and territorial management activities. The main goals were the fight against abandonment of agricultural land, wetlands management and fire prevention. Today, the concept of agri-environmental measures applies to all measures implemented within the European Union under the CAP framework, attempting to support the farmers who voluntarily adhere to environmentally beneficial farming practices.
- The 1999 reform associated with the Agenda 2000 which introduced a European agricultural model incorporating non-market functions that match the goals pursued by public policies. Examples include animal welfare, biodiversity, rural development and food security.
- The 2003 reform specifying fixed payments subject to cross compliance in order to ensure the multifunctionality of the European agricultural system – a move that clearly demonstrates the relative priority attributed to the preservation of rural landscapes (European Commission 2005).

By the way, highlighting agriculture's multifunctional nature within a broader environmental policy is also a direct response to the existence of a social demand for non-market services that are environmental in nature and which then become a public policy goal (Bonnieux and Desaigues 1998; Randall, 2002). At the European level, the 2003 CAP reform can be seen in this respect as an attempt to give incentives to encourage specific non-market functions, like environmental protection, spatial planning and viable rural zones.

The purpose of the present chapter is twofold: to determine the conditions that ensure the provision of environmental services by rural landscapes that have been shaped in part by agricultural activities; and to analyse the economic values associated with these services and thereby account for the existence of specific social demand. Within this context, the conditions of public intervention capable of ensuring rural landscapes' provision of environmental services – and of deriving value from their non-market functions – will be studied via an approach both theoretical and applied.

The ambivalent economic status of the rural landscape

In economic terms, landscapes can be perceived in different ways, and hence imbued with variable contents. A given landscape might be apprehended as a rural amenity; a positive externality associated with an agricultural activity; a joint production; or a localised public good. Depending on the approach, emphasis may be placed on the supply of or demand for rural landscapes, highlighting certain specificities that must be incorporated into public environmental policy. The purpose of this first section is to apply concepts from recent economic literature to the analysis of rural landscapes.

Rural landscapes as an amenity

Rural landscapes are part of a wider category of "rural amenities" that, where agricultural activities are concerned, entail all of the environment services resulting from the presence of natural settings associated with a particular use (i.e. hydrological processes activated in wetlands) or non-use (i.e., desire to preserve irreplaceable species, habitats and ecological functions for future generations). In general, these are wilderness, cultivated landscapes, historical monuments, traditional cultures or amenities enabling recreational activities like hunting and fishing. Rural amenities are defined as 'natural or synthetic goods that the public appreciates independently of the role they play in production processes. They originate from strictly defined territories presenting specific physical and cultural characteristics' (OECD 1999).

Additionally, by their very nature, rural amenities resulting from agricultural activity are endowed with a number of economic and societal values (Bonnieux and Rainelli 2000). Such economic values are either use or non-use values (existence value, bequest value or option value). Given that a landscape will be embedded in a certain space, its use value for local residents is first and foremost an object of local consumption, even if non-resident visitors sometimes invest in the space under concern. A distinction can be made between the use value associated with a rural landscape's utilisation and its option value, the latter being linked to the idea that this landscape must be preserved today so that people can benefit from its use in the future. As for the existence value, this is associated with the idea that a landscape simply exists (without anyone intending to benefit from it at present or in the future). Lastly, the concept of bequest value implies that a landscape should be preserved so that future generations can use it. Rural landscapes comprise a multitude of use values, including their natural characteristics (physical, biotic and abiotic components), the effects of agricultural activities (transformation, maintenance, degradation, etc.) and other values enabling the satisfaction of final needs (final uses). Thus, a cultivated area sustains agricultural use but also constitutes a specific living framework for its residents (expressing its recreational, aesthetic and cognitive values).

As for non-use values, these vary depending on potential consumers' locations, whether or not they want to preserve the landscapes because they exist or for the benefit of future generations. The value of a landscape derives in part from some of its attributes or qualities, which are therefore crucial to its assessment and to the evaluation of rural amenities in general. Aesthetic, cultural, cognitive, recreational and social attributes are all elements that possess a highly structuring effect (Guérin and Michalland 2000).

As a rural amenity with ties to a particular territory, a rural landscape possesses a value that is largely connected to the hedonic elements it creates. The linkage between land and landscape implicitly introduces space as an attribute. From an economic perspective, space is both an area that sustains economic activities and a scarce resource (Paelinck and Sallez 1983). According to Jayet (1996), space, through the coexistence that it establishes between public and private contents, introduces a public dimension into activities that were originally private in nature. A prime example of this is agricultural land, by definition a private good characterised by rival and exclusive uses. Yet agricultural land is also a spatialised good and therefore has a landscape dimension whose uses will be only partially rival in nature. A perfect exclusion from use is often too costly to be feasible (ibid.), even if it is technically possible.

This means that a connection exists between the dynamic process underlying the attribution of value to land and the production of a particular rural landscape. Where an agricultural activity is involved, changes in the use of the space will have a strong impact on the linkages between the natural and man-made objects contained within this landscape. Such transformations will necessarily affect the landscape's hedonistic content. Production from cultivated landscapes involves different acts of production (OECD 1999). Some acts are non-specific (e.g. a landscape produced without any cost to be borne by the farmer) whereas others are specific and have direct associated costs (creation of a hedge) or are intended to protect a particular landscape (refusal to eliminate a hedge) and therefore entail an opportunity cost for farmers.

Lastly, classifying amenities according to their nature depends on the market value ascribed to them which finally concerns amenities that were originally agricultural in nature,[2] notably involving tourist or recreational activities.

Table 15.1 Classification of rural amenities

Type of value	*Nature of the amenity*	*Economic valuation*[3]
Use value		
1. direct use	Countryside	Product labelling
2. recreational use	Fishing, hunting	Hunting or fishing permits
3. aesthetic use	Quality of landscapes	Green tourism
Indirect use value	Prevention of flooding, subsidence or avalanches	No market valuation (need for indirect assessment)
Ecological value	Ecological functions sustaining life, assimilative capacity of environment	
Non-use values		
4. existence	Biodiversity, rare species (flora and fauna)	No market valuation (need for indirect assessment)
5. shared use		
6. bequest/option	Habitats with significant ecological value	

Derived from Bonnieux and Rainelli (2000: 86).

Rural amenities like hunting, for instance, depend on the state of the natural setting (cf. Table 15.1).

Amenities that have a use value will generally lack an exchange value, meaning that no market exists for them as rural amenities. Thereby, the higher an amenity's use value, the greater the likelihood that it will be "marketable", whether or not the amenity derives from a production externality. Thus, the actor holding ownership rights over the amenity will have cause to seek benefits from it. Conversely, the greater an amenity's non-use value, the less feasible it becomes to derive a benefit from it via a market transaction, and the greater the need for public intervention to ensure its preservation.

Rural landscape as an externality and a localised public good

In general, one complementary way of apprehending rural amenities is to view them as positive agriculture-related environmental externalities, or as agricultural production-related public goods (OECD 2001). Public policies seeking to derive value from a rural amenity will be different depending on whether the amenity in question can be characterised as an externality or a public good (OECD 1999).

An externality is generally defined as a situation where one actor's economic activity (production or consumption) has an indirect effect on another actor in the absence of any market transaction between the two. It will be positive (negative) if value is transferred to (from) the actor enjoying the gains (incurring the losses), even in the absence of direct monetary compensation. Thus, farmers who implement an intensive mode of production will have no incentive to pay for the damages caused by agricultural pollution (an example of negative environmental externalities). Conversely, the environmental services that a farmer provides a community when maintaining grasslands (preservation of biodiversity, preventing soil erosion, etc.) are not paid for by the parties enjoying these services (an example of positive environmental externalities) (Le Goffe 2003). This is a situation where public intervention can be justified (notably in the case where social demand exceeds the supply of the positive externality under concern). Indeed, by using suitable instruments like incentives or regulations, it should be possible to ensure a sufficient quantity of rural amenities (which involves somehow revealing the existence of a social demand).

In short, landscape is largely the outcome of the productive actions of actors who, without getting anything in return, have modified the utility of other actors located in their immediate vicinity, which is in line with the canonical definition of externalities (Meade 1973; Baumol and Oates 1988). This is because an externality is a non-intentional, market-external direct interdependence between an issuing actor and a receiving actor.

Farmers are often unaware of how deeply they modify a landscape's qualities. For instance, a relatively extensive dairy farm can have positive secondary effects insofar as it simultaneously offers leisure possibilities and accommodation. As highlighted by Beuret and Mouchet (2000), the sources of landscape externalities are first and foremost non-specific productive acts. The supply of landscapes is usually a by-product with deep connections to the process leading to the production of a primary good. However, this primary good is not directly linked to the demand for the by-product.

Viewed in this light, landscapes can be construed as localised externalities. Depending on users' preferences (which are likely to evolve over time), the effect can be positive or negative.

Alongside the non-intentional aspect of an output that is private in origin (and which can be assessed using a production externality construct), analysing a rural landscape's economic status by viewing it as an object shaped by agricultural activity means highlighting its identity as a localised public good. These two aspects might be analysed in light of the fact that externalities per se are not necessarily a cause of market failure. Economic theory teaches us that only those externalities possessing the characteristics of public goods require public regulation to ensure their provision to the community.

In economic theory, a public good is generally one that is shared by several actors, a quality that distinguishes it from a private good. A public good is considered "pure" if it satisfies two properties: non-excludability and non-rivalry in consumption. A good is non-excludable if it is physically or institutionally impossible or very expensive to keep certain individuals from consuming it. A good is non-rivalrous when an individual can consume one unit without restricting other people's opportunity to consume the same unit. Thus, no one should be excluded from the good's consumption since a new consumer's arrival will have zero marginal cost. Conversely, by definition a private good is something subject to rivalry that can be excluded from consumption. For economists, there are four main types of goods, depending on the rivalry/excludability criteria that each features (cf. Table 15.2).

Table 15.2 The economic status of different goods

	Excludability	*Non-excludability*
Rivalry	Private good *(family garden)*	Mixed public good *(new houses built in a natural setting not protected by law)*
Non-rivalry	Club good *(golf course)*	Public good *(mountain vista)*

Since landscapes usually satisfy a principle of non-rivalry, several actors can simultaneously enjoy one and the same landscape, which can then be considered a localised public good. According to Cheshire (1989), the composite good represented by rural amenities is in fact a quasi-public good for which no real market currently exists. Thus, landscapes are not marketable objects per se but are associated with the exchange of commodities and services (landed property, tourist accommodations etc.). Under certain conditions, this kind of public good will be subject to congestion effects caused by a relative lack of privacy. Consequently, rivalry does exist, in which case the landscape must be considered a mixed public good. It remains that rural landscapes usually fulfil the two properties of non-excludability and non-rivalry. Normally, the characteristics of exclusive use and congestion are not given. One example would be when people contemplate a landscape from a particular viewing platform (that can only be accessed upon payment of an entrance fee) or when access to a particular landscape suffers from congestion and overcrowding. The end result is that shared rural landscapes are tantamount to pure public goods.

Rural landscapes as a provider of environmental services

Acknowledging the multifunctional nature of agriculture given the many environmental services it provides ultimately means attributing a crucial role to the rural landscapes associated with agricultural activities (Ferrari and Rambonilaza 2008). The function that agriculture fulfils in the production of environmental services has recently been treated in a number of interesting analyses (Heal and Small 2002; Millennium Ecosystem Assessment 2005; FAO 2007; World Bank 2008). Studies of ecosystems' characteristics and dynamics generally emphasise the benefits that populations can derive directly or indirectly from these ecosystems (Fisher et al. 2009). In general, ecosystem services[4] resulting from the interaction between living organisms and their natural settings are comprised of services like climate or water regulation, pollination or the preservation of flora and fauna habitats. With respect to climate change mitigation and preservation of biological diversity, the services that have an agricultural origin appear to be playing an increasingly fundamental role (FAO 2007).

Classifying such services involves several categories of functions or services[5] associated with ecosystems: production services; regulation services; and cultural services (Hein et al. 2006). The former refers to those goods and services produced within an ecosystem (genetic resources, fibers, etc.) and are services that help to sustain life on earth. The latter results from an ecosystem's ability to regulate the climate, hydrological cycles and biological processes. The third relates directly to the (recreational, cultural or cognitive) benefits that society enjoys when ecosystems perform very well.

In general, the approaches found in the literature on this topic adopt a utilitarian perspective and consider that one precondition for the specification of ecosystem services has to be fulfilled: namely, the existence of a social demand and/or of needs to be satisfied (De Groot et al. 2002; Fisher et al. 2009; Hein et al. 2006). Assessing ecosystems' values for society depends directly on the goods and services that these ecosystems produce. In other terms, the economic value of an ecosystem is tied to the market and non-market products. For the former, a market price is available. For the latter, we need to rest upon social demand so as to infer the economic value that then includes a use value and non-use value (see above).

This orientation must be put into perspective where rural landscapes are concerned since consideration must be given to the way in which agricultural activity encourages or hampers a particular ecosystem's ability to produce ecosystem services. It is crucial to comprehend agricultural practices' and production systems' general effects on the quality of the natural processes at work, checking how they might lead, after a certain period, to a diminution in ecosystem services – a diminution that is directly attributable to human actions. For example, land degradation or soil erosion can lower agricultural productivity and undermine environmental services like habitat availability or biodiversity (World Bank, 2008). Thus, the supply of a rural landscape depends not only on agricultural practices but also, over a longer period, on natural processes' ability to enable the joint production of non-market goods – with the characteristics of public goods – and traded goods. To sum up, even if it may not be possible to assess the environmental services associated with a rural landscape without knowledge of the relevant social demand – thus without evaluating the associated public good – it is crucial to note the role that agricultural practices play in a territory's ability to produce many ecosystem services.

Rural landscape as a joint product of agricultural activity

As a non-intentional result of agricultural activity, a rural landscape might be seen as a joint non-commodity production of agriculture that often shows the character of a public good. Agricultural activity can be analysed as a process of joint production under whose aegis both market and non-market goods[6] will be produced (Boisvert 2001). If we consider that the multifunctional nature of agricultural activity might constitute a specific property of production processes, this means then analysing the links between multiple productions, on the one hand, and productions and inputs, on the other, will reveal the relations of complementarity/substitution that exist between various productions, and the way production linkages influence production levels for non-market goods (Ferrari and Rambonilaza 2008).

In the case of rural landscapes, it is important to show how agricultural practices and cultivation choices affect the linkage between market and non-market goods. More generally, the properties of an input that is part of a production process will have a crucial impact on the conditions enabling the acquisition of non-market goods (Ferrari 2004; Marcouiller and Clendenning 2005).

By identifying the origins of production linkages, three separate cases may be highlighted (Boisvert 2001).

1. A case where the existence of technical interdependencies entails physical links between the level of the externality being created and the level of agricultural output (e.g. nitrogen pollution following the use of fertiliser inputs). When applied to agriculture, it may be shown that, depending on the agronomic practices, this particular link gives rise to the production of negative externalities like pollutants, soil erosion, or positive ones like an improvement of soil fertility. In the former case, there is a link between the negative externality and the agricultural production level. Furthermore, the commodity production depends both on the allocated factors and on the production level of the other product. These interdependencies have an important implication: an increase in the production of the commodity good which follows the increase in the amount of one production factor can lead to a change in the marginal productivity of this factor with respect to the non-commodity output. Thus, the marginal productivity of a productive factor with respect to the production of one output depends on the level of the production of another output.
2. A case where the production process involves the use of one or several factors of production whose allocation cannot be determined. This situation is observed when different products are obtained from a single input. Examples may be the production of wheat and straw, or of ovine meat and wool from sheep breeding. In this case, the amounts of the produced goods depend on each other (e.g. using grasslands to produce milk, meat and biodiversity).
3. A case where there is an agricultural production factor that can be allocated but which is only available in fixed quantities at the farm level (examples include land or water for irrigation). In other terms, the products compete for a production factor which is available in a fixed quantity at the farming level. Any increase of the production of one output reduces the quantity of the fixed input available for the production of the other product. Thus, there is interdependence within the products and within the inputs (for at least two production factors).

The two latter instances apply in analyses of the production of rural landscapes resulting from agricultural activities. Where the production linkage stems from the fact that the production factors in question cannot be allocated, it is impossible to separate different inputs' contributions to different kinds of production.

In addition, this linkage implies significant complementarity between traded and non-market goods. For example, the case of joint productions of milk and landscape amenities – together with grazing activities – exemplifies the extent to which quantities of these two highly complementary productions are dependent in nature. Furthermore, it is impossible to determine which parts of the "land" input will be used, respectively, to produce a traded or a non-market good. If productions are in competition for an input that is available in fixed quantities (at the farm level), an increase in the production of one output will lead to a reduced quantity of the production factor being made available for the production of another output. Where rural landscapes are concerned, this kind of linkage will be valid as long as land constitutes a production factor that can be allocated to different growing activities occupying spaces defined within a total area.

Generally, at the level of a given production process, the proportions of market and non-market goods are considered to be fixed. However, this hypothesis seems difficult to sustain in reality (Blandford and Boisvert 2002). For instance, the same quantity of a traded good can be obtained using different combinations of land and labor, whereas the production level of non-market goods can be modified by changing their productive combination (in other words, by modifying the degree of linkage, thus the connections between traded and non-market goods). As an illustration, government support of milk production that seeks to preserve bucolic landscapes by encouraging natural grazing methods might undermine the quality of the landscape (notably damaging biodiversity) if farmers shift towards a milk production system based on intensive foraging because of insufficient public incentives (OECD 2003).

Public support for rural landscapes: what is at stake?

The basic justification for public intervention lies in the public goods nature of environmental externalities like rural amenities in general, and rural landscapes in particular. Due to the existence of positive, agricultural-activity-related externalities, it is sometimes put forward that agriculture should be broadly subsidised to ensure the preservation of rural landscapes. This raises two questions, *viz.*, whether such public payments are justified from the perspective of the World Trade Organisation (WTO) (Glebe 2007) and, if so, which remunerative and institutional arrangement modalities should be implemented? After clarifying the justifications for public intervention, the section below will mainly deal with the latter question. Accounting for transaction costs and economies of scale, it will try to analyse and classify the institutional arrangements that are used in Europe in order to ensure the preservation and even improvement of certain environmental landscape qualities.

Is public regulation required?

Depending on which positive externalities are being considered, beneficiaries will represent different social groups linked to different spatial levels (i.e. people consuming local drinking water; residents living near forests; regional populations; tourists; and even global society, which is for instance susceptible to the beneficial effects of a landscape's ability to retain CO_2). In general, fixed payments for agricultural practices like grazing (or simply for the maintenance of agricultural activities leading to positive externalities) are only justified in the case of real market failure and if there are hardly any non-governmental possibilities for providing consumers (OECD 2001; Vanslembrouck and Van Huylenbroeck 2005).

In the absence of a market for rural landscapes, however, the question becomes how social demand might be revealed for such environmental externalities. This involves answering two questions relating to price and quantity. What is the value of a rural landscape for the beneficiaries/consumers of such amenities? And what quantity of rural landscapes might they wish to "consume"? Knowing about these two points helps to determine the quantity that should be produced, as well as the incentives that farmers should be paid.

Figure 15.1 illustrates a situation where a rural landscape is a by-product of a non-compulsory agricultural activity (like grazing). The marginal opportunity cost for farmers (MC) will depend on the net revenues derived from the agricultural activity (the higher these profits, the lower the costs MC). If the quantity of the amenity, X, is directly related to the area under cultivation, a fixed payment for each hectare corresponding to an amount P would be optimal, leading to total expenses of $a + b$ for beneficiaries. Payments reflecting opportunity costs for farmers would correspond to the area b in Figure 15.1. The higher the marginal

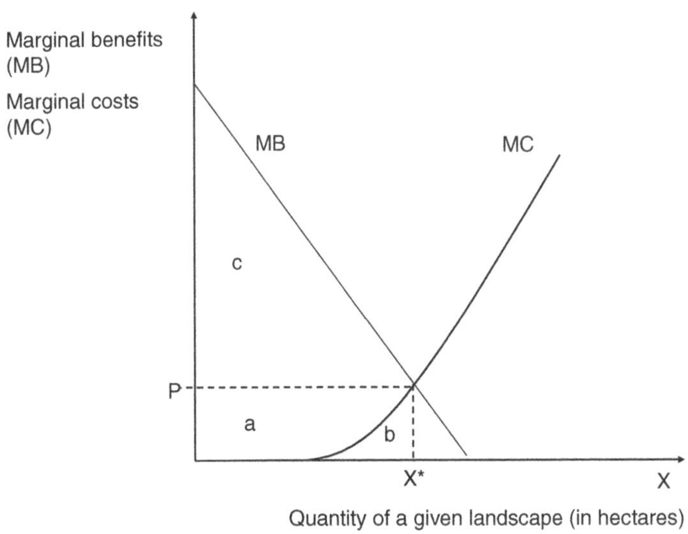

Figure 15.1 Supply of and demand for a landscape of which the amenity relates to a particular agricultural practice (i.e. grazing).

Source: Scheele and Isermeyer (1989, modified).

social benefits (MB) or marginal costs (MC), the higher the optimal payment needed to maintain the agricultural activities. In practice, the MC and MB curves are, like the optimum (P; X*) difficult to estimate. In addition, they are entirely unstable due to the frequent changes in costs or consumer preferences.

Different econometric approaches help to reveal actors' preferences regarding the environmental attributes that should be associated with a given agricultural activity. For instance, the study by Madureira et al. (2007) highlights the relevant role of attributes like biodiversity preservation, landscape cultivation, food security or recreation. Analysis shows that an economic assessment of agricultural landscapes' non-market benefits, in conjunction with social demand, suffers from limitations largely caused by the difficulty of coming up with a precise definition of which goods and services should be assessed.

Current studies generally find that the European population is willing to financially support the provision of agricultural amenities, notably the preservation of agricultural landscapes, even if some uncertainty remains with respect to the extent of this financial commitment (Ferrari and Rambonilaza 2008). The section below analyses the ways to deliberately organise the provision for landscape qualities that feature the characteristics of a public good (a case of true market failure).

Funding and economic instruments supporting the production of landscape externalities

The role of public intervention consists of regulating the production of externalities by calling upon incentive mechanisms. Prime examples in Europe include the agri-environmental measures implemented during the 1990s. These measures were intended to encourage the production of rural amenities by supporting biodiversity and landscape preservation. Without over-emphasising the difficulty of estimating real preferences for the externalities associated with certain landscapes (thereby justifying these measures), it remains to be seen how the beneficiaries would contribute to financing such measures so as to ensure its effects over time.

Generally, this kind of social dilemma can be resolved in different ways:

(i) public funds used by state authorities acting in lieu (and in the name) of beneficiaries;
(ii) calls to citizens' sense of civic duty (moral suasion);
(iii) combined purchase of private and public goods (with consumers who are buying a high quality private good simultaneously contributing to the funding of a public good).

Alongside this, it is possible to distinguish four different ways of stimulating supply (see Table 15.3). Remuneration can occur directly or indirectly, with or without a monetary exchange. Direct payment signifies that the farmer generating the positive externality receives a premium (example A, with a monetary exchange) or certain property rights, like the right to use a pasture so that his/her herd can graze there (example B, non-monetary exchange). Regarding funding,

Table 15.3 Four ways of stimulating the supply of positive agricultural externalities

Remuneration	*Monetary*	*Non-monetary*
Directly	A Direct payments (premiums)	B Property rights
Indirectly	C High prices for agricultural products	D Intangible reward (social prestige, altruism, etc.)

both arrangements will generally correspond to solution (i) above due to the need to raise necessary funds. Indirect remuneration (example C) occurs if the actor provoking the externalities sells other goods being produced at a higher price (internalisation involving the payment of a bonus on top of the agricultural production).

To illustrate this point, note several regional marketing initiatives in Germany (cf. www.reginet.de) where members are required to adhere to certain supplementary rules ensuring the sustainability of their agricultural activities. This can be exemplified by farmers participating in the "Freisinger Land" initiative (running in the Upper Bavarian province of Freising, cf. http://www.freisinger-land.de) where they engage in environmental protection measures across 5% of their land area to ensure the preservation of local landscapes. In exchange, they received marketing assistance for their agricultural output, the end effect of which is to allow them to sell at above-market prices. In a similar vein, see the case of certified French regional cheeses like Comté. Note also current efforts in Reutlingen (Württemberg, Germany) to preserve traditional mixed orchards through the marketing of apple juice (cf. http://www.reutlinger-bio-apfelsaft.de). Lastly, in France, a number of national regional parks are pushing a similar initiative by labelling certain products and services with a bespoke acronym (PNR).

Another example of indirect, market-based remuneration entails attracting tourists to pretty areas around farms offering accommodation. Remuneration of this kind supposes that beneficiaries behave as disinterested parties and not as "free riders", or that they have the conviction that the private good they are acquiring (tourist accommodation) is better than the standard alternatives. In a similar vein, since nobody can be required to acquire local foods produced according to sustainable principles, this kind of funding modality corresponds to categories (ii) or (iii) above. Example D (indirect remuneration, without monetary exchange) includes all of the arrangements where the "producers" of a public good expect no material rewards. These are situations where actors can be motivated by the search for social prestige, or by other factors like local patriotism or altruism.

Where potential "producers" of an external effect hold the corresponding property rights and starting with the example of direct remuneration, there are broadly two categories of arrangements enabling the internalisation of an externality:

Arrangement 1. According to Coase (1960) one solution is to negotiate a price for the property rights to the resource in question. Thus, "consumers" of an externality will negotiate with its potential "producers" and ultimately reward them to get them to modify their agricultural activities or relinquish some other property right (e.g. the right to use a certain quantity of nitrogen fertilisers or to not let their herds out to graze). These rights will be exchanged in a way that will generate the desired landscape qualities. The remuneration should at the very least cover the agricultural benefits that will be lost or more broadly, the relevant opportunity costs (corresponding to area *b* in Figure 15.1). On the other hand, compensation for "producers" cannot exceed an amount corresponding to area $a + b + c$ in Figure 15.1 (total benefits for "consumers" of the externality). A clear definition and unequivocal distribution of property rights are necessary conditions for this kind of solution. In addition, according to Coase, it can only happen if all participants in a negotiation have perfect information and if the exchange and the enforcement of property rights have negligible transaction costs.

Arrangement 2. Applying a Pigovian approach, public authorities who act in favor of the parties benefiting from a landscape's quality might pay compensation to potential "producers". At best, the corresponding premium P would be chosen in such a way as to achieve the social optimum (P, X*) indicated in Figure 15.1. In this case, only those actors whose opportunity costs are equal to or less than this premium would help to produce the quality of the landscape in question (i.e. by continuing to practice extensive agriculture). Those whose costs are lower would obtain a rent that would, in total, correspond to area *a* in Figure 15.1.

The best solution for internalising an externality actually depends on the type and number of actors on both the demand and the supply sides (private parties, local authorities, farmers, etc.). Where externalities (like a landscape's ability to retain CO_2) spill over into neighboring regions or even countries, they tend to be remunerated with Pigou-type subsidies to reduce the transaction costs. Conversely, a Coase-type solution is preferable when a small number of actors and very specific landscape qualities are concerned. These two conditions tend to be satisfied if beneficiaries and "producers" all operate within one and the same region, a condition that also enables the use of indirect forms of remuneration without any monetary exchange (B, C and D in Table 15.3). Similarly, when the quality of a local landscape creates benefits for the whole of a country or for all of Europe, specific solutions will have to be negotiated with local or regional actors. It is self-evident that internalising at the local level implies costs for the communities involved. Notably when related benefits are broadly shared by individuals who live elsewhere, some of the expenses will have to be paid by a higher administrative level. Ideally, the financial contribution that a community has to make to preserve a landscape corresponds exactly to the benefits that the local population derives from it. Note that this kind of co-funding leads to an effective use of budgetary resources insofar as the community in question will only commit to such spending if real local demand exists for the landscape in question.

Finally, the nature of the public intervention depends on the kind of public good which is of concern. Non-excludability prevents providers from charging user fees while a voluntary provision tends to lead to under-provision of goods. However, even if a government decides to provide pure public goods, it is often difficult to estimate people's true willingness to pay for the good (i.e. the marginal value of the good to them). There is therefore a substantial risk of policy failure associated with the over- or underestimation of the willingness to contribute to the provision of a pure public good. For excludable but non-rival goods, private provision can be sustained by user fees. However, they may be some efficiency losses due to decisions that tend to under-provisioning. Providers would take into account only those who can pay the prices that have been determined by the capacity for providing the public good and, consequently they would eliminate all other users whose willingness to pay is positive but inferior to the price. Nevertheless, the impact of market failures may be smaller than that caused by policy failures associated with government provision. Market provision could at least force users to reveal their true willingness to pay, which is often difficult in the case of government provision.

Should remuneration be related to agricultural outcomes or practices?

As long as the sum of opportunity and transaction costs associated with the preservation or improvement of a certain landscape quality is less than its value as rewarded by the market, remuneration will reflect producers' efforts. Good examples of this include zoos or golf courses. However, where other landscape qualities like biodiversity are involved, the public good attribute will impede this kind of market solution (see above). In this case, a public agency can promote actions aimed at increasing the availability of the qualities under concern, as long as the ensuing social benefits are considered to be superior to the opportunity and transaction costs that society has to pay. In principle, there are two kinds of financial support for the provision of a particular landscape quality: one where the level of remuneration reflects a visible result; or one where it reflects a particular agricultural activity or practice (this is the case of most agri-environmental schemes conducted within the European Union). The second modality involves a close relationship between the agricultural practice being subsidised and the desired landscape effects (i.e. a joint production, see above), one example being France's *Prime Herbagère Agri-Environnementale* (PHAE) paid for certain grasslands.

The first point is that payments reflecting the outcomes of efforts to improve the quality of a landscape (i.e. result related payments) will only be feasible if the value of this quality is easy to measure. Such is the case, for instance, with the grassland plant diversity premium awarded by the German Federal State of Baden-Württemberg (Ministerium für Ernährung und Ländlichen Raum Baden-Württemberg 2000), where farmers receive premiums for plots of land containing at least four relatively rare plant species (out of the 28 herbaceous species listed in a catalogue). This kind of arrangement has the general advantage of stimulating a permanent search for processes which fulfill certain ecological and aesthetic aims at lower opportunity costs. One disadvantage is that in the short run, risk-averse farmers will ask for higher premiums compared with situations where the government pays a fixed sum for the relevant agricultural practice. This is because they face the risk that the desired plants might not grow, despite extensive farming efforts. Here, the risk is that farmers will be reluctant to participate in the scheme unless the government pays allowances higher than the opportunity costs to offset the "risk premium" that farmers expect on top of their costs whenever returns are uncertain.

Generally, most agri-environmental payments based on European Council article 37 relating to support for rural development are payments remunerating an agricultural practice that does better than "Usual Good Farming Practice". Such payments are efficient when the scope of the activity being subsidised covers a number of landscape qualities all at once. This is often the case with support for organic farms in zones characterised by their high ecological value. In this instance and beyond the significant economies of scope that are achieved with the production of public goods derived from a single agricultural activity, transaction costs for programmes' administration and surveillance will be lower since they will be distributed to several ecological objectives (Dabbert et al. 2004).

A relevant analysis here involves envisioning rural landscapes as joint products of agricultural activity (Ferrari and Rambonilaza 2008). What seems essential is that production linkages are studied in a way that defines the nature of those public policies that are most likely to encourage the provision of the environmental services associated with a given rural landscape. The greater or lesser degree of this linkage is what will drive the public policy being implemented to regulate the production of externalities. In general, it is targeted or decoupled public policies that should be highlighted (independently of agricultural policies) where a linkage is weak (i.e. if productions are not very complementary). This is an instance where dissociation can be envisaged between traded and non-market products. Conversely, where linkages are strong, policies could be coupled with specific products (Le Goffe 2003). Regarding rural landscapes, in case we consider that the qualities of a landscape mainly relate to a process of agricultural production, decoupling does not appear to be relevant. However, even if it is technically possible to separate the production of traded and non-market goods, it can be more expensive to produce them separately due to the existence of economies of scope (Casini et al. 2004). Thus, farmers who let their herds out to graze will face lower costs than companies specialising in landscape protection whose main concern is to stave off agricultural abandonment (i.e. to prevent a meadow from reverting to wilderness) because the milk and meat that the herd provides constitute precious "secondary products" that will help to preserve the landscape.

In short, measures concerning prices for private goods can play an important role in terms of the supply of certain desired public goods. Due to the existence of transaction costs, it may be useful to implement policies that will benefit from the joint production of private and public goods (Vatn 2002). It remains that given the shortcomings of price support actions (inefficient income transfers, intensification incentives, burdens on consumers, conflicts within the WTO), it seems more appropriate to mobilise fixed per-hectare payments, as the European

Union does with the natural handicap payments that it grants in mountain areas and other areas with handicap (Council of the European Union 2005).

Specialisation and scale effects leading to the need for a "nature agent"

An efficient allocation of property rights relating to the attributes of a particular asset like agricultural land depends on the actors' ability to influence the asset's various attributes. The party who is most likely to impact an attribute is supposed to become the residual claimant of the ensuing gains and to work even harder to further improve the attribute (Barzel 1989; North 1990), that is, as long as the opportunity costs of its provision are lower than its remunerated value. Due to specialisation and spatial scale effects, this can lead to "divided ownership" over an asset's different attributes (North 1981: 23–26). As a result, property rights relating to the agricultural and ecological attributes of a landscape are often owned by different actors, each of whom will try to improve "his/her" attribute. For instance, in Germany or France, property rights (and obligations) relating to hunting activities in some zones are usually owned by people other than those who have the right to affect local agricultural attributes like soil structure, humus content in soil, etc. Similarly, there are many examples where landscape or ecological attributes of land cultivated by various farmers (e.g. its groundwater table quality or its aptitude to act as an animal habitat) are governed and controlled by actors operating on a much broader spatial scale than the farmers themselves.

Therefore, if there is the possibility of a payment related to an improvement of a landscape attribute – and if major scale effects exist (for instance when trying to optimise particular species' habitat systems), then an intermediary actor might be needed because farmers operating on a smaller spatial scale (working their own fields instead of the whole of a region) do not have the means to influence the corresponding attribute efficiently. This intermediary actor, who can also be called a "nature agent" (Lippert 1999), will purchase the property rights to "inputs" (like extensive practices) enabling the desired landscape quality. Payments will be made to farmers rewarded on the basis of their opportunity costs. Such agents are akin to the "Conservation, Amenity and Recreation Trusts (CARTs)" described by Hodge (1991: 191f.); South German "beaver consultants" (Schwab 2001); or Austrian "bear advocates" (Coordination Board for Bear Management in Austria 2005). They resemble ideal-types of nature agents who, in a pluralistic society, act as advocates resolving conflicts caused by the presence of "their" animals.

Note that since the 1990s, European Union agricultural policy has focused on the preservation of rural landscapes' environmental qualities (see chapter introduction). State authorities along with other political actors are paying greater attention to certain landscape qualities or attributes, like the aptitude to serve as a habitat or the overall appearance reflecting aesthetic qualities, panoramic views, etc. In these cases, where the landscape structure and derived benefits become explicit and even priority policy objectives, the presence of a planning and coordinating agent may be required for three reasons:

1. Some landscape elements possess non-additive effects: a single motorway bridge can ruin the whole of a panoramic view and not just part of it (technical argument);
2. Presence of the aforementioned scale effects (economic argument);
3. Need for different actors possessing divergent interests in terms of landscape appearance and use to negotiate and reconcile their viewpoints (social argument).

A precondition for the work of a "nature agent" is a clear definition and dissemination of the property rights to natural resources and the main elements comprising the landscape. The aforementioned Coase approach already deemed this condition necessary for the internalisation of externalities. Where resources are free and belong to no one, corresponding rights will first have to be established and allocated via a political process that will clearly be contentious. Negotiations between the "nature agent" and different stakeholders (farmers, environmental protection associations, public authorities, etc.) will have to respect the initial distribution of rights. This means that each transfer of property rights must receive compensation. Thus, rights held by other actors must be purchased by the "nature agent" who will be reimbursed by public monies because of the landscape's public good aspect.

An alternative to this market-based solution might be an expropriation with or without state compensation, something that would be tantamount to a redistribution of property rights. Any redistribution that is not indemnified, as per the "public trust doctrine" (Brewer and Libecap 2009), might appear more efficient at first glance, since public authorities can quickly take whatever measures are needed to improve the landscape. It remains that an approach of this kind can lead to very high costs if the actors being disadvantaged decide to sue. Furthermore, given that the state has less information than the different actors, a market-based solution based on negotiations between the interested parties will reveal greater information and ultimately lead to an optimal use of natural resources and landscapes. Lastly, an expropriation without compensation may be contrary to legal principles. For instance, in the Federal Republic of Germany, article 14 of the Constitution states that any expropriation for the common welfare must be rooted in a law establishing compensation for the expropriated parties, who can still go to court if they are unhappy with the compensation offered.

Conclusion: towards a sustainable management of rural landscapes?

Until now, public policy based analyses of rural landscapes have taken place within a static framework. Yet research into the dynamics underlying rural landscapes is a crucial stage in the development of long-term policies aimed at landscape preservation and a regulated provision of environmental services. Long-term management of rural landscapes cannot be organised without accounting for the interactions between agricultural and natural systems, no matter which spatial scale is considered.

Pragmatically, this means considering rural landscapes as assets that are specific, i.e. that will be strongly devalued if used for ends other than those for which they were designed (Williamson 1983). According to Hagedorn et al. (2002), three categories of specific assets can be considered in the agri-environmental field:

1. site specificity (e.g. a precious biotope cultivated by a farmer);
2. capital specificity leading to sunk costs (e.g. hedged farmland planted by a farmer seeking promised annual payments, and where the planting costs would be lost if the government suddenly decided to no longer pay the initially promised yearly premium);
3. specific knowledge that land-users progressively acquire.

These specificities are important when farmers are forced to completely change their agricultural practices. This is the case, for instance, when a farmer starts to cultivate extensive grazing lands for milk cows. For the person concerned, this means significant investments that will take around 20 years to write off. If the State only offers contracts lasting five years, it is possible that the risk-averse farmer will choose to eschew this agri-environment scheme, even if the annual payments seem largely sufficient. What is crucial here is that the public authorities be able to guarantee long-term contracts when specific assets are involved.

Furthermore, preserving rural landscapes implies major public spending. The budgetary constraint can be quite significant, and may be tightened at any moment. One way of saving budgetary funds for public payments for rural landscape improvements is to implement auctions (Latacz-Lohmann and van der Hamsvoort 1997; Müller and Weikard 2002). At best, if after an auction farmers are only compensated for their marginal opportunity costs (see area b in Figure 15.1), the ensuing budgetary savings could equal the total producer surplus (area a). Compared to a unitary premium of P leading to budgetary expenditures of (P times $X^* = a + b$), an auction is a promising solution if the opportunity costs vary significantly among the bidding farmers, and when these cost differences are not obvious to the bidders (i.e. if they are unfamiliar with the other participants' costs). In addition, the landscape qualities being remunerated should not be too specific. If, for example, the main objective of a reforestation program is to recover CO_2, an auction could help to lower total costs insofar as the areas involved are interchangeable, with relatively unproductive farms ultimately being the only ones to participate in the program.

In addition, preferences that constitute social demand can change rapidly whereas landscapes evolve slowly. This means that the equilibrium outlined in Figure 15.1 may be impossible to achieve in the short run. Similarly, a public intervention that only integrates short-term considerations may be incompatible with the dynamics underlying rural landscapes in the long run. Although a few landscape attributes (like farmland hedges) may be possible to govern, a full control of landscapes appears illusory and might occasionally create paradoxical situations (e.g. protected beavers felling protected trees). Similarly, due to the indivisibility of any given landscape, a marginal allocation by small units – the underlying assumption in Figure 15.1 – has got its obvious limits.

Lastly, the analysis conducted here has revealed the impossibility of unequivocally establishing modalities for a public intervention aimed at the governance of rural landscapes. Public support policies can be broken down into different spatial levels (local, regional or global) and according to the interventions have a more or less decentralised nature. Furthermore, since each rural landscape is the product of social interactions (involving a multitude of production and consumption actors) with the natural environment, it is essential that different actors get a chance to express their interests and their landscape preferences. At the same time, defining long-term landscape policies requires that consideration be given to the co-evolution of agricultural processes and natural environments (Dalgaard et al. 2006). This is a major hurdle for the development of public agricultural policies insofar as it implies that agricultural activities' environmental sustainability be scrutinised *a priori* to enable as precise an understanding as possible of the long-term interactions between a given production activity and the relevant ecosystems. This then is the path towards a more sustainable management of rural landscapes in the future.

Notes

1 For an overview of landscape environmental changes during the period in question, see Küster (1996: 354), Chaléard and Charvet (2004: 157) and Renard (2005: 126).
2 The concept of amenity is akin to a positive externality of production.
3 On the approaches and methods for environmental economic valuation of agricultural non-commodity outputs, see Madureira et al. (2007).
4 Ecosystem services are environmental services produced and defined on the basis of one specific ecosystem. At the biosphere (global ecosystem) level, for instance, the "climate regulation" service is achieved via the biogeochemical cycles.
5 Natural functions and processes at work within ecosystems become services if populations can directly or indirectly take advantage of them (Fisher et al. 2009).
6 Where a negative (positive) externality is involved, the good would be described as negative (positive).

References

Abler D. (2001) *A synthesis of country reports on jointness between commodity and non-commodity outputs in OECD agriculture*. Workshop on Multifunctionality, OECD, Paris.

Baumol W.J. and Oates W.E. (1988) *The Theory of Environmental Policy*. Cambridge University Press, New York.

Barzel, Y. (1989) *Economic Analysis of Property Rights*. Cambridge University Press.

Beuret J.-E. and Mouchet C. (2000) 'Pratiques agricoles, système de production et espace rural: quelles causes pour quels effets?', *Cahiers d'études et de recherches francophones/Agricultures*, 9(1): 29–37.

Blandford, D. and Boisvert, R.N. (2002) *Non-trade concerns and domestic/international policy choice*, Working Paper #02-1, January 2002, International Agricultural Trade Research Consortium.

Boisvert, R.N. (2001) 'A note on the concept of jointness in production'. In: *Multifunctionality: Toward an Analytical Framework.* OECD, Paris.

Bonnieux, F. and Rainelli, P. (2000) 'Aménités agricoles et tourisme rural'. *Revue d'Economie Régionale et Urbaine*, 5: 803–820.

Bonnieux, F. and Desaigues, B. (1998) *Économie et politiques de l'environnement.* Dalloz, Paris.

Brewer, J. and Libecap, G.D. (2009) 'Property rights and the public trust doctrine in environmental protection and natural resource conservation'. *The Australian Journal of Agricultural and Resource Economics*, 53: 1–17.

Casini, L., Ferrari, S., Lombardi, G., Rambonilaza, M., Sattler, C. and Waarts, Y. (2004) *Research Report on the Analytic Multifunctionality Framework.* MEAScope Research Project, Müncheberg.

Chaléard, J.-L. and Charvet, J.-P. (2004) *Géographie agricole et rurale.* Éditions Belin, Paris.

Cheshire, P. (1989) 'L'application de la rationalité économique aux problèmes d'aménagement de l'espace rural: une perspective européenne'. In: OCDE (ed.) *Ressources Naturelles Renouvelables: Incitations Économiques pour une Meilleure Gestion.* Paris, pp. 154–165.

Coase, R. (1960) 'The problem of social cost'. *Journal of Law and Economics*, 3: 1–44.

Commission Européenne (2005) *Agri-environment measures, Overview on general principles, types of measures and applications*, Directorate General for Agriculture and Rural Development, March.

Coordination Board for Bear Management in Austria (ed.) (2005) *Bears in Austria – A Management Plan.* Reviewed version 2005. WWF Autriche, Vienne, 2005.

Council of the European Union (2005) Council Regulation (EC) No.1698/2005 on support for rural development by the European Agricultural Fund for Rural Development. Bruxelles.

Dabbert, S., Häring, A.M. and Zanoli, R. (2004) *Organic Farming – Policies and Prospects.* ZedBooks, London.

Dalgaard, T., Ferrari, S. and Rambonilaza, M. (2006) 'Features of environmental sustainability in agriculture: some conceptual and operational issues', *International Journal of Agricultural Resources, Governance and Ecology*, 5(2/3): 107–115.

De Groot, R.S., Wilson, M.A. and Boumans, R.M.J. (2002) 'A typology for the classification, description and valuation of ecosystem functions, goods and services', *Ecological Economics*, 41: 393–408.

FAO (2007) *La situation mondiale de l'alimentation et de l'agriculture: payer les agriculteurs pour les services environnementaux.* FAO, Rome.

Ferrari, S. (2004) 'Multifunctionality of agriculture and joint production', contribution to the 90th EAAE Seminar: *Multifunctional agriculture, policies and markets/Understanding the critical linkage*, October 27–29 2004, Rennes.

Ferrari, S. and Rambonilaza, M. (2008) 'Agricultural multifunctionality promoting policies and the safeguarding of rural landscapes: How to evaluate the link?', *Landscape Research*, 33(3), June 2008: 297–309.

Fisher B., Turner R.K. and Morling P. (2009) 'Defining and classifying ecosystem services for decision making', *Ecological Economics*, 68: 643–653.

Garzon I. (2005) 'Multifunctionality of agriculture in the European Union: Is there substance behind the discourse's smoke?' Working Paper, 2005-36, Institute of Governmental Studies, University of California, Berkeley.

Glebe, T.W. (2007) 'The environmental impact of European farming: How legitimate are agri-environmental payments?' *Review of Agricultural Economics*, 29: 87–102.

Guérin, M. and Michalland, B. (2000) *Le paysage: quelles caractéristiques économiques?* Séminaire 'Le paysage a-t-il une valeur économique?' Clermont Ferrand, Engref-Cemagref.

Hagedorn, K., Arzt, K. and Peters, U. (2002) 'Institutional arrangements for environmental co-operatives: A conceptual framework'. In: Hagedorn, K. (ed.) *Environmental Co-operation and Institutional Change. Theories and Policies for European Agriculture.* Edward Elgar, Cheltenham (UK) and Northampton (USA), pp. 3–25.

Heal G.M. and Small A.A. (2001) 'Agriculture and ecosystems services'. In: *Handbook of Agricultural Economics*, vol. 2A, Elsevier, pp. 1341–1369.

Henrichsmeyer, W. and Witzke, H.P. (1994) *Agrarpolitik. Band 2. Bewertung und Willensbildung.* Verlag Eugen Ulmer, Stuttgart.

Hein, L., Van Koppen, K., De Groot, R.S. and Van Ierland, E.C. (2006) 'Spatial scales, stakeholders and the valuation of ecosystem services', *Ecological Economics*, 57: 209–228.

Hodge, I.D. (1991) 'The provision of public goods in the countryside: How should it be arranged?'. In: Hanley, N. (ed.) *Farming and the Countryside: An Economic Analysis of External Costs and Benefits.* CAB International, Oxford, pp. 179–96.

INRA (2008) *Agriculture et biodiversité, valoriser les synergies*, Rapport d'expertise, Juin.

Jayet, H. (1996) 'Peut-on caractériser les zones rurales par l'abondance d'espace?' *Revue d'Economie Régionale et Urbaine*, 2: 201–210.

Küster, H. (1996) *Geschichte der Landschaft in Mitteleuropa: Von der Eiszeit bis zur Gegenwart.* Verlag C. H. Beck, Munich.

Latacz-Lohmann, U. and van der Hamsvoort, C. (1997) 'Auctioning conservation contracts: A theoretical analysis and an application'. *American Journal of Agricultural Economics*, 79: 407–418.

Le Goffe P. (2003) 'Multifonctionnalité des prairies: comment articuler marché et politiques publiques?' INRA *Productions Animales*, 16(3): 175–182.

Lippert, C. (1999) 'Institutionenökonomische Überlegungen zur optimalen Bereitstellung und Entlohnung von Umweltattributen in Agrarlandschaften'. *Agrarwirtschaft*, 48: 417–430.

Madureira L., Rambonilaza M. and Kaprinski I. (2007) 'Review of methods and evidence for economic valuation of agricultural non-commodity outputs and suggestions to facilitate its application to broader decisional contexts', *Agriculture, Ecosystems and Environment*, 120(1): 5–20.

Marcouiller, D.W. and Clendenning, G. (2005) 'The supply of natural amenities: moving from empirical anecdotes to a theoretical basis'. In: G.P. Green et al. (eds) *Amenities and Rural Development.* Edward Elgar, Aldershot, pp. 6–32.

Meade J.E. (1973) *The Theory of Economic Externalities: The Control of Environmental Pollution and Similar Social Costs.* Leiden, A.W. Sijthoff, Genève.

Millennium Ecosystem Assessment (2005) *Ecosystems and Human Well-being: Biodiversity Synthesis.* World Resource Institute, Washington, D.C.

Ministerium für Ernährung und Ländlichen Raum Baden-Württemberg (2000). Richtlinie des Ministeriums Ländlicher Raum zur Förderung der Erhaltung und Pflege der Kulturlandschaft und von Erzeugungspraktiken, die der Marktentlastung dienen (Marktentlastungs- und Kulturlandschaftsausgleich – MEKA II -) vom 12.09.2000 – Az. 65-8872.53 -. Stuttgart, 2000.

Müller, K. and Weikard, H.-P. (2002) 'Auction mechanisms for soil and habitat protection programmes'. In: Hagedorn, K. (ed.) *Environmental Co-operation and Institutional Change*. Edward Elgar, Cheltenham (UK) and Northampton (USA), pp. 202–213.

North, D. (1981) *Structure and Change in Economic History*. Norton, New York/London.

North, D. (1990) *Institutions, Institutional Change and Economic Performance*. Cambridge University Press.

OECD (2003) *Multifunctionality: The Policy Implications*. Paris, France.

OECD (2001) *Multifunctionality. Towards an Analytical Framework*. Paris, France.

OECD (1999) *Cultivating Rural Amenities*. Paris, France.

Paelinck, J.H.P. and Sallez, A. (eds) (1983) *Espace et localisation*. Economica, Paris.

Randall, A. (2002) 'Valuing the outputs of multifunctional agriculture'. *European Review of Agricultural Economics*, 29(3): 289–307.

Renard, J. (2005) *Les mutations des campagnes. Paysages et structures agraires dans le monde*. Armand Colin, Paris.

Scheele, M. and Isermeyer, F. (1989) 'Umweltschutz und Landschaftspflege im Bereich der Landwirtschaft – Kostenwirksame Verpflichtung oder neue Einkommensquelle', *Berichte über Landwirtschaft*, 67: 86–110.

Schwab, G. (2001) 'Der Biber in Bayern'. In: *Haus-im-Moos-Tagungs- und Seminarbeiträge* 1, vol. 1, 23–26.

Vanslembrouck, I. and van Huylenbroeck, G. (2005) *Landscape Amenities. Economic Assessment of Agricultural Landscapes. Landscape Series*, vol. 2. Springer, Dordrecht.

Vatn, A. (2002) 'Multifunctional agriculture: some consequences for international trade regimes'. *European Review of Agricultural Economics*, 29: 309–327.

Williamson, O.E. (1983) 'Credible commitments: Using hostages to support exchange'. *The American Economic Review*, 73: 519–540.

World Bank (2008) *World Development Report: Agriculture for Development*. Washington, DC.

16 Amenity-driven migration and the spatial distribution of economic activity

Wenchao Xu and JunJie Wu

Introduction

Americans are known to be one of the most mobile peoples in the world. Their migration patterns have been always changing: these changes are reflected not only in interregional migration, as Americans were seen moving around the country throughout history, but also in intra-urban migration, as they move from the center of the city to the suburban areas in recent decades (Frey 2009). These changing patterns not only created the unique landscape and culture of the United States, but also created many issues that await research in regional science and urban economics.

In recent years, amenity-driven migration has attracted much attention. Many studies show that residential developments are expanding rapidly into amenity-rich areas. Amenities or locational amenities in this chapter refer to the desirable features of a place and are location-specific in nature, the index of which is frequently computed and compared on a city-by-city basis (Rosen 1979). The concept of amenities is broad and may include attributes evaluated from environmental, cultural, political, and socioeconomic perspectives. Landscape attributes play an important role in determining locational amenities within urban and suburban areas: per cent of the landscape covered by forests or water, closeness to green space, distance to a river, and contiguity with nearby natural attributes. These landscape features make amenities more tangible and sometimes measurable.

The effects of amenities on intra-urban and interregional migration have long been recognized and systematically studied in several disciplines from various perspectives.[1] Rosen (1979) and Roback (1982) provided a theoretical framework to understand the relationship between amenities and migration decisions. In their frameworks, individuals are mobile enough to eliminate utility differences across space in pursuit of higher quality of life in a long-run equilibrium. Applying this framework to the housing market and non-traded goods, Roback (1982) provided empirical evidence that regional wage differences can be explained by local attributes, such as temperature, total snow fall, population density, and other local urban attributes. Linneman and Graves (1983) used a multinomial logit model to reveal that both housing demand and more traditional job search motivation significantly influence migration decisions. Graves and Waldman (1991) showed that the general multimarket amenity compensation model presents a more accurate picture of the spatial equilibrium mechanism than the competing hypothesis that amenities are priced separately into either the land or the labor market. Treyz et al. (1993) estimated a net migration model by using time series data for 51 regions over the period 1971–1988. Their results revealed that the dynamic response of net migration is significantly related to stock equilibrium changes induced by amenity differentials, relative employment opportunities, relative real wages, and industry composition.

The complicated and pervasive nature of the impacts of amenities on migration and the spatial distribution of economic activity has been well recognized in the regional science and urban studies literatures. First, locational amenities can take many different forms. For example, forests, lakes, beaches, and other landscape features are widely used to evaluate amenities in intra-urban migration (e.g. Johnson and Beale 1994; Rasker and Hanson 2000; Radeloff et al. 2001, 2005, among others). Climatic factors are also frequently used to evaluate the effect of locational amenities on interregional migration. For example, Glaeser and Shapiro (2003) and Glaeser and Gyourko (2005) found that weather is significantly associated with city population growth.

Second, amenities may affect the migration decisions of various age-, skill-, and race-groups differently, and thus influence regional productivity and firm location choices. Adamson et al. (2004) suggested that urban amenities affect skilled workers' location choices. Chen and Rosenthal (2008) analyzed individual migration decisions and found that young, highly educated households tend to move towards places with higher quality business environments, whereas couples near retirement tend to move towards places with highly valued consumer amenities.

Third, by influencing residential location choices, amenities also affect public services, and thus shape urban and suburban spatial profiles and development patterns (Wu 2006). Brueckner et al. (1999) presented an amenity-based model of household location decisions and showed that the relative location of different income groups depends on the spatial pattern of amenities in a city. Wu (2006) developed an economic model to explore the causes of fragmentation in land development and found that spatial heterogeneity of amenities is a major determinant of development patterns and community profiles. Wu et al. (2012) investigated locational amenities from open space conservation and its impacts on community characteristics.

Fourth, locational amenities are found to be closely associated with economic growth and regional development. Black and Henderson (1999) found that non-coastal cities with poorer climate grow more slowly. They also suggested that agglomeration is promoted by coastal location, good climate, and high market potential. Rappaport and Sachs (2003) studied US economic activity and suggested that the investigation of coastal concentration should focus on factors from both the productivity side and the quality of life aspect. Glaeser et al. (2005) showed that housing supply plays a part in mediating urban growth, and that differences in the regulatory environment across space affect how cities respond to increases in productivity.

Fifth, the study of the impact of amenities on regional migration is complicated by the interactions between location decisions of firms and households, as reflected in a prolonged debate over "people follow jobs" or "jobs follow people". Greenwood and Hunt (1989) argued that jobs and wages are considerably more important than location-specific amenities in explaining net metropolitan migration of employed persons. Greenwood et al. (1991) suggest that disequilibrium forces are merely a reflection of income and wage differentials; Hunt (1993) questioned the efficacy of including amenities in interregional migration studies, and claimed that it does little to correct wrong signs or increase the significance of correctly signed coefficients on the economic opportunities.

Sixth, the form and nature of amenities matter. A large and growing literature estimates the effects of different types of amenities on residential property values, including the value of living close to lakes or rivers (Brown and Pollakowski 1977; Lansford and Jones 1995; Leggett and Bockstael 2000), urban parks and open space (Weicher and Zerbst 1973; Irwin and Bockstael 2001; Irwin 2002; Anderson and West 2006; Klaiber and Phaneuf 2010), urban forests (Tyrväinen and Miettinen 2000), urban wetlands (Doss and Taff 1996; Mahan et al. 2000), and general amenities (Polinsky 1977; Smith 1978; Pogodzinski 1988; Palmquist 1992; Sivitanidou 1995; Kopits, McConnell and Walls 2007). Several recent studies measure the distance to different types of open areas and find that home value increases with proximity and that the effect varies by type (Polinsky and Shavell 1976; Bolitzer and Netusil 2000; Smith et al. 2002; Kopits et al. 2007). Others distinguish between protected open space, such as public parks and land under conservation easement, and developable open space, such as privately owned agricultural land (Irwin and Bockstael 2001; Geoghegan et al. 2003). Cheshire and Sheppard (1998, 2002) distinguish between publicly accessible and inaccessible open space in two English cities. They find that these areas have roughly equal value in one city, while only accessible open space increases home value in the other.

Finally, contemporary literatures that emphasize the effects of amenities on migration cannot explain the heterogeneity in regional development. For example, regions with a high level of disamenities such as high crime rates, growing urban congestion, and heavy pollution continue to thrive, whereas amenity-rich places are left completely undeveloped. And the recession in the early 21st century made it even harder for people to readily accept the importance of amenities in interregional migration. Florida, which was once the top draw for Americans in search of work and warmer climates, lost more than 31,179 residents as of July 1, 2009 according to a Brookings report (Frey 2009). As people leave this state in search of jobs elsewhere, evidence grows that job opportunities are more important than amenities in interregional migrations. However, previous studies tend to omit the difference between intra-urban migration and interregional migration. Also, previous studies often fail to characterize various economic activities both within regions and across regions. Accordingly, there is a great need to develop a multi-market-equilibrium framework that incorporates factors influencing both the quality of life and job opportunities on household migration decision.

The objective of this study is to develop a model to analyze the roles of locational amenities in interregional migration by extending the framework of Helpman (1998). We are interested in the following questions: (i) What factors determine the equilibrium distribution of population across regions? (ii) Do amenities affect interregional migration? The answers to these questions will contribute to a better understanding of the effects of locational amenities in interregional migrations.

The chapter is organized as follows. In the next section, we present the model to study amenity-driven migration and the spatial distribution of population. In the section thereafter, we conduct simulation analyses, and discuss results under different scenarios. In the last section, we summarize the main results and suggest avenues for future research.

The model

In 1991, Krugman (1991) published a path-breaking paper to investigate the heterogeneity in regional structure and economic development. The model is a variant on the monopolistic competition framework initially proposed by Dixit and Stiglitz (1977). Krugman's general equilibrium framework has two groups of participating agents: individual households and firms. Individual households maximize their utility by way of consumption subject to their income. Firms hire labor, produce goods, and maximize their profits. This general equilibrium framework also incorporates two sectors: agricultural and manufacturing. The two sectors differ in the way of how products are traded and how demands are generated. When regional economies reach equilibriums in multi-markets (e.g. the supply and demand of tradable goods, the supply and demand of labor, and the income and expenditure of representative households) simultaneously, the equilibrium distribution of population in space is determined.

By contrast, Helpman (1998) proposed an alternative model. In Helpman's framework, the agricultural sector is replaced by an immobile housing sector. It is also assumed that the gross income generated from the housing sector, the land rent, is distributed among regions in proportion to the number of residents. These changes in setups not only emphasize the role of housing consumptions in interregional migration, but produce results different from Krugman's. This study builds upon Helpman's framework. In what follows, we explain in detail how we modify Helpman's framework to incorporate the effects of amenities, and how equilibrium conditions in a multimarket scenario can be derived therefrom.

The micro level decisions

Consider an economy with two regions. Region k $(k = 1, 2)$ has N_k households and n_k monopolistic competitive firms. Household i $(i = 1, ..., N_k)$ earns incomes $(E_{i,k})$ and chooses a consumption bundle, containing tradable commodities and housing, to maximize utility. Each firm j ($j = 1, \ldots, n_k$) hires labor and produces one unique variety. The commodity can be traded across regions at an iceberg transport cost τ – meaning that a fraction of any shipment is lost in transit. Labor cost on the production side is transformed into household's income source on the consumption side.

From the consumption side, households maximize utility subject to their earned income. Their utility comes from two main sources: tradable goods (manufactured varieties) and non-tradable goods (housing). Housing consumption is comprised of three major elements: land, structure, and locational amenities (hereafter, amenities). Housing structure is made up of materials that can be traded across regions, whereas amenities are locally exclusive and not tradable at all. The total amount of developable lands is fixed in each region. All workers live in the region where they work.

Formally, household i's utility maximization problem can be expressed as

$$\max u_{i,k} \equiv \left(A_k^{\theta} l_{i,k}^{1-\theta}\right)^{\gamma} \left\{ \left[\sum_{k=1}^{n_1} x_{i,k}^{\alpha} + \sum_{-k=1}^{n_{-k}} x_{i,-k}^{\alpha} \right]^{1/\alpha} \right\}^{1-\gamma}$$
$$s.t.\ \ p_l l_{i,k} + \sum_{k=1}^{n_k} p_{j,k} x_{i,k} + \sum_{-k=1}^{n_{-k}} (\tau p_{-k}) x_{i,-k} \le E_{i,k} \tag{16.1}$$

where A_k is the level of amenities in region k, and can be freely enjoyed by all residents living in the region; $l_{i,k}$ is the amount of land consumed by household i in region k, at a price p_l per unit; $l_{i,k}$ is assumed to be the same for all households living in a region (e.g., $l_{i,k} \equiv L_k/N_k$); $x_{i,k}$ is the demand for tradable good k produced in region k, and $x_{i,-k}$ is the demand for imported goods; γ represents households' preference for housing relative to the consumption bundles of tradable goods, and θ represents households' preference for amenities relative to housing. This setup implies a constant elasticity of substitution (ε) among the tradable goods with $\varepsilon = 1/(1-\alpha)$ and $0 < \alpha < 1$.[2] Migration between different regions is costless.

Two results can be obtained by solving this utility maximization problem. First, this maximization problem determines the optimal allocation of households' income between the tradable and non-tradable goods:[3]

$$p_l l_k = (1-\theta)\gamma \frac{E_k}{(1-\theta\gamma)},$$
$$\sum_{1}^{n_k} p_{j,k} x_{j,k} + \sum_{1}^{n_{-k}} \left(\tau p_{j,-k}\right) x_{j,-k} = (1-\gamma)\frac{E_k}{(1-\theta\gamma)}, \tag{16.2}$$

where $x_{j,k} = \sum_i x_{i,j,k}$ and $E_k = \sum_i E_{i,k}$.

Second, from the maximization problem, we can derive the demand for locally-produced goods ($x_{j,k}$) and imported goods ($x_{j,-k}$) as functions of prices

($p_{j,k}$ and $p_{j,-k}$), iceberg transport cost (τ), and the total expenditures in region k (E_k):

$$x_{j,k} = \frac{(1-\gamma)p_{j,k}{}^{\frac{1}{\alpha-1}} E_k}{(1-\theta\gamma)\left[\sum_{1}^{n_k} p_{j,k}{}^{\frac{\alpha}{\alpha-1}} + \sum_{1}^{n_{-k}} \left(\tau p_{j,-k}\right)^{\frac{\alpha}{\alpha-1}}\right]}$$
$$x_{j,-k} = \frac{(1-\gamma)\left(\tau p_{j,-k}\right)^{\frac{1}{\alpha-1}} E_k}{(1-\theta\gamma)\left[\sum_{1}^{n_k} p_{j,k}{}^{\frac{\alpha}{\alpha-1}} + \sum_{1}^{n_{-k}} \left(\tau p_{j,-k}\right)^{\frac{\alpha}{\alpha-1}}\right]} \tag{16.3}$$

Firms' production technologies stipulate that it takes $a + x_{j,k}$ units of labor to produce $x_{j,k}$ units of output. Thus, the labor demand of each firm is $lr_{j,k} = a + x_{j,k}$, where a is a fixed cost, which is assumed to be the same for all regions. The wage rate, w_k, or the marginal cost of producing the tradable good $x_{j,k}$, is assumed to be the same within each region.

The solution of profit maximization of each monopolistic competitive firm j gives the price set by the firm: $p_{j,k} = \frac{1}{\alpha} w_k$. In addition, free entry in the production market for individual firm implies

$$p_{j,k}\, x_{j,k} = \left(a + x_{j,k}\right) w_k \Rightarrow x_{j,k} = \frac{\alpha a}{1-\alpha} = x_k. \tag{16.4}$$

This suggests that all firms in region k produce the same amount of output and thus set the same price for all products:

$$p_{j,k} = \left(\frac{a + x_k}{x_k}\right) w_k = p_k \tag{16.5}$$

Labor market equilibrium and the number of varieties

The total labor demand in region k equals $(a+x_k)n_k$ with n_k the number of varieties produced in this region. The total labor supply equals the number of households living in this region (N_k). When the labor market is cleared, the number of varieties (n_k) is determined as

$$n_k = \frac{1-\alpha}{a} N_k. \tag{16.6}$$

The labor market equilibrium can be used to evaluate labor demand (job opportunities), which is an essential component in Helpman's framework.

Households' income and expenditure

An individual household has two sources of income: the wage income and the income from land rent. By assumption, collected land rent is distributed equally among households living in the region. The total amount of land rent collected in the two regions equals the total amount of household income spent on land $\frac{(1-\theta)\gamma}{(1-\gamma)}(w_k N_k + w_{-k}N_{-k})$. Thus, region k's total income I_k equals

$$I_k = w_k N_k + \frac{N_k}{N}\frac{(1-\theta)\gamma}{(1-\gamma)}(w_k N_k + w_{-k}N_{-k}), k = 1, 2. \tag{16.7}$$

The relative prices between the two regions

For each tradable good produced in region 1, the supply is $x_{1,s} = \frac{\alpha a}{1-\alpha}$. The total demand for a tradable good produced in region 1 equals:

$$x_{1,d} = \frac{p_1^{1-\varepsilon}}{n_1 p_1^{1-\varepsilon} + n_2(\tau p_2)^{1-\varepsilon}}\frac{(1-\gamma)E_1}{(1-\theta\gamma)p_1} + \frac{(\tau p_1)^{1-\varepsilon}}{n_1(\tau p_1)^{1-\varepsilon} + n_2 p_2^{1-\varepsilon}}\frac{(1-\gamma)E_2}{(1-\theta\gamma)p_1}, \tag{16.8}$$

where the first term on the right-hand side represents the total amount of the good consumed in region 1, and the second term represents the total amount of the good exported to region 2.

Let q be the ratio of prices of the tradable good in the two regions (i.e. $q \equiv p_1/p_2 = w_1/w_2$), and let f be the share of total population in the economy living in region 1 (i.e. $f \equiv N_1/N$). The equilibrium condition for the commodity market (i.e. $x_{1,d} = x_{1,s}$) implies the following condition

$$\begin{aligned} 1-\theta\gamma = {} & \frac{fq^{-\varepsilon}}{fq^{1-\varepsilon} + (1-f)\tau^{1-\varepsilon}}\{(1-\gamma)q + \gamma(1-\theta)[qf + (1-f)]\} \\ & + \frac{(1-f)\tau^{1-\varepsilon}q^{-\varepsilon}}{f\tau^{1-\varepsilon}q^{1-\varepsilon} + (1-f)}\{(1-\gamma) + \gamma(1-\theta)[qf + (1-f)]\} \end{aligned} \tag{16.9}$$

This result is axiomatically analogous to Helpman's (1998: 37). Two major implications can be drawn from this result. First, given the household's preferences, the relative price of tradable goods (q) is, as in Helpman's, simultaneously determined with the distribution of the population between the two regions. This relationship is highly non-linear, which makes simulation analysis more relevant by deriving more explicit outcomes under possible scenarios. The second major implication, which is unique to this study, is that the preference for locational amenities, as reflected by parameter θ, can significantly influence the equilibrium distribution of population and thus economic activities between the two regions; changes in amenities or preferences for amenities can lead to adjustment in both the relative price and the distribution of total population. Thus, amenity-driven migration can be analyzed using the model.

Relative utilities and the condition for spatial equilibrium

Using the price index of tradable products, the utility level of each household in region k can be derived as

$$u_k = \left[A_k^{\theta} \left(\frac{H_k}{N_k} \right)^{1-\theta} \right]^{\gamma} \left[\frac{(1-\gamma)E_k}{(1-\gamma\theta)N_k P_{dk}} \right]^{1-\gamma}, \tag{16.10}$$

where $P_{dk} \equiv \left[n_k p_k^{1-\varepsilon} + n_{-k}(\tau p_{-k})^{1-\varepsilon} \right]^{1/(1-\varepsilon)}$ is the price index in region k. Substituting E_k and P_{dk} into equation (16.10) gives

$$u_k = C_0 A_k^{\theta\gamma} H_k^{(1-\theta)\gamma} f_k^{-(1-\theta)\gamma} \left[\frac{(1-\gamma)q_k + \gamma(qf_k + f_{-k})}{\left[f_k q_k^{1-\varepsilon} + f_{-k}\tau^{1-\varepsilon} \right]^{1/(1-\varepsilon)}} \right]^{1-\gamma} N^{\frac{(1-\gamma\varepsilon)}{(\varepsilon-1)}+\theta\gamma}, \tag{16.11}$$

with $C_0 = \left[\alpha \left(\frac{a}{1-\alpha} \right)^{1/(1-\varepsilon)} \frac{1}{(1-\gamma\theta)} \right]^{1-\gamma}$.

Using equation (16.11), the ratio of utility for households living in regions 1 and 2 ($v_{1,2}$) can be derived as

$$v_{1,2} = \left(\frac{A_1}{A_2} \right)^{\theta\gamma} \left(\frac{H_1}{H_2} \right)^{(1-\theta)\gamma} \left(\frac{1-f}{f} \right)^{(1-\theta)\gamma} \left[\frac{(1-\gamma)q + \gamma(qf + 1 - f)}{(1-\gamma) + \gamma(qf + 1 - f)} \right]^{1-\gamma} \tag{16.12}$$

$$\left[\frac{(1-f) + f(\tau q)^{1-\varepsilon}}{fq^{1-\varepsilon} + (1-f)\tau^{1-\varepsilon}} \right]^{\frac{(1-\gamma)}{(1-\varepsilon)}}$$

A distribution of population between the two regions is a spatial equilibrium if no one has incentives to move. If $v_{1,2} > 1$, some households in region 2 would move to region 1; and vice versa. So, in equilibrium, $v_{1,2} = 1$. Equation (16.12) also implies that agglomeration configurations where all households living in one region cannot be a spatial equilibrium. This is because if $f = 1$, some households in region 1 would move to region 2 because $v_{1,2} = 0$; likewise, if $f = 0$, some households in region 2 would move to region 1 because $v_{1,2} \to \infty$.

Equilibrium distribution of population

Equation (16.12) indicates that transport cost influences the equilibrium distribution of population and economic activity between the two regions. Transport cost changes the relative price of locally versus non-locally manufactured products, which in turn affects households' consumption bundles. The adjustment in consumption bundles has a ripple effect on firms' operation, including adjustments in the production of manufactured products and demand for labor. The iceberg transport cost can be viewed not only as a trade barrier that determines the accessibility of local markets, but also as an indicator to evaluate how regions are connected through economic activities, such as consumption, production, and trade. In this section, we examine how transport costs affect the equilibrium distribution of population between the two regions.

No trade barrier to access local markets

When there is no transport cost ($\tau \rightarrow 1$), competition causes the price of tradable goods to be equal across regions (i.e., $q \rightarrow 1$). Substituting these conditions into equation (16.12), we get

$$v_{1,2} = \left(\frac{A_1}{A_2}\right)^{\theta\gamma} \left(\frac{H_1}{H_2}\right)^{(1-\theta)\gamma} \left(\frac{1}{f} - 1\right)^{(1-\theta)\gamma} \tag{16.13}$$

Equation (16.13) indicates that the level of utility in region 1 relative to that in region 2 decreases as more people move to region 1 (i.e. as f increases). Setting $v_{1,2} = 1$, the equilibrium distribution of population between the two regions can be derived as

$$f^* = \frac{1}{1 + \left(\frac{A_2}{A_1}\right)^{\frac{\theta}{(1-\theta)}} \left(\frac{H_2}{H_1}\right)}. \tag{16.14}$$

$f = f^*$ is a stable equilibrium because if an external shock causes the population distribution to deviate from it, the system will bring the distribution back to it. To see this, suppose a disturbance cause f to be greater than f^*. This would decrease the relative utility level in region 1, which, in turn, causes out-migration. The equilibrium is regained when $f = f^*$. From equation (16.14), it is easy to verify that if the two regions have the same level of developable lands and the same level of amenities (e.g. $H_1 = H_2$ and $A_1 = A_2$), $f^* = 0.5$, that is, the symmetric distribution of population between the two regions is the equilibrium. If the two regions have the same amount of developable lands but different levels of amenities (e.g. $H_1 = H_2$ and $A_1 \neq A_2$), the region with a higher level of amenities has a larger population.

Graphically, the relationship between $v_{1,2}$ and f for the two-region case is shown in Figure 16.1. The vertical axis stands for the relative utility level between the two regions ($v_{1,2}$). The horizontal line represents the share of population living in region 1 (f). The solid line shows the relative utility curve for $(A_1/A_2) = 1.2$, and the dotted line shows that for $(A_1/A_2) = 1.5$. With higher locational amenities in region 1, the relative utility curve shifts upward, which results in a larger f^*_{high} compared to f^*_{low}. As expected, the region with a higher level of amenities attracts more households, produces more varieties of tradable goods, and creates more job opportunities.

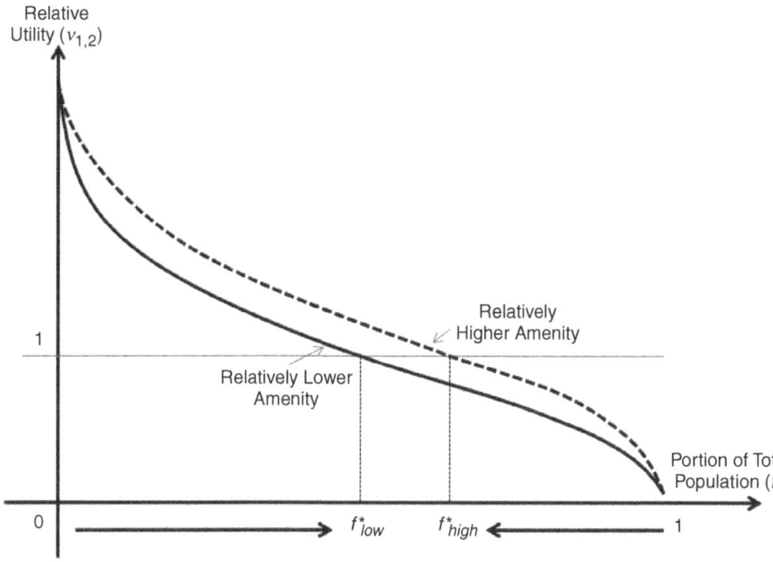

Figure 16.1 Amenities and the spatial distribution of population between the two regions. ($A_1 = 1.2/1.5, A_2 = 1, H_1 = H_2 = 1, \theta = 0.3, \varepsilon = 2, \gamma = 0.4, \tau = 1.0$, and $(1-\theta)\gamma > 0$).

Trade barrier to access local markets

When τ is greater than 1, equation (16.9) exhibits a highly non-linear relationship between the relative price (q) and the share of population living in region 1 (f). Equation (16.11) indicates that the utility level in each region is proportional to $N^{\frac{(1-\gamma\varepsilon)}{(\varepsilon-1)}+\theta\gamma}$ and welfare rises with population size if and only if $\frac{(1-\gamma\varepsilon)}{(\varepsilon-1)} + \theta\gamma > 0$. Substituting $\varepsilon = 1/(1-\alpha)$ into $R(\gamma, \alpha, \theta) \equiv \frac{(1-\gamma\varepsilon)}{(\varepsilon-1)} + \theta\gamma$ gives:

$$R(\gamma, \alpha, \theta) = (1-\gamma)(\frac{1}{\alpha} - 1) - (1-\theta)\gamma. \tag{16.15}$$

Note that $\frac{\partial R}{\partial \gamma} < 0$, $\frac{\partial R}{\partial \alpha} < 0$, and $\frac{\partial R}{\partial \theta} > 0$. These results suggest that condition $\frac{(1-\gamma\varepsilon)}{(\varepsilon-1)} + \theta\gamma > 0$ is more likely to hold if the preference for housing is low and the preferences for product varieties and locational amenities are high. Under these conditions, welfare rises with population size and a large city is likely to emerge. On the contrary, the condition $\frac{(1-\gamma\varepsilon)}{(\varepsilon-1)} + \theta\gamma < 0$ is more likely to hold if the preference for housing is high and the preferences for varieties and amenities are low. Under these circumstances, welfare falls with population increases and a dispersed distribution pattern of population will likely emerge.

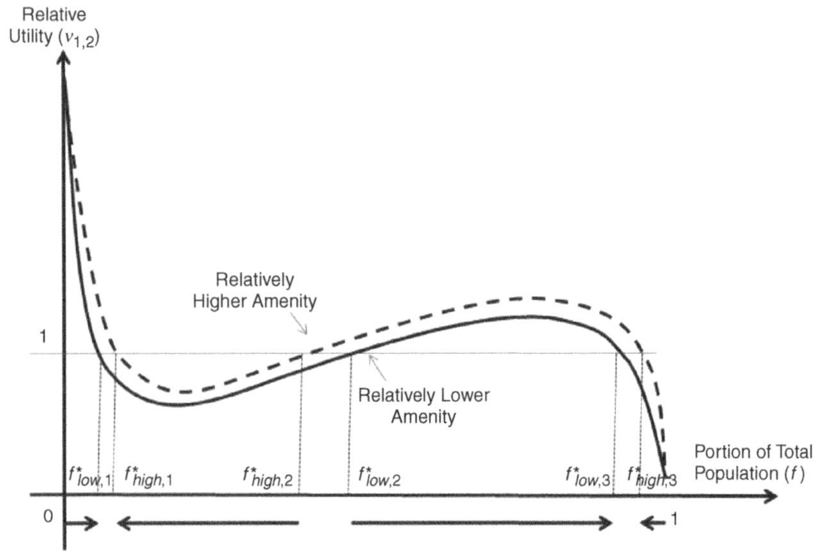

Figure 16.2a Amenities and the spatial distribution of population between the two regions. $\left(A_1 = 1.2/1.5, A_2 = 1, H_1 = H_2 = 1, \theta = 0.3, \varepsilon = 2, \gamma = 0.4, \tau = 6.0, \text{ and } \frac{(1-\gamma\varepsilon)}{(\varepsilon-1)} + \theta\gamma > 0\right)$

Our simulation results are consistent with the arguments above. Figure 16.2a shows one case for $\frac{(1-\gamma\varepsilon)}{(\varepsilon-1)} + \theta\gamma > 0$. The parameter values used in this simulation are set as $\theta = 0.3$, $\varepsilon = 2$, $\gamma = 0.4$, and $\tau = 6.0$. We assume that both regions have the same housing stock ($H_1 = H_2 = 1$), but different levels of amenities ($A_1 \neq A_2$). In Figure 16.2a, there are three equilibrium solutions for f that could be derived from $v_{1,2} = 1$. But the one in the middle, f_2^*, is unstable, whereas both f_1^* and f_3^* are stable. This can be verified by the fact that the system will bring the distribution back to f_1^* and f_3^* if an external shock causes some households to move to the other region. Suppose the initial equilibrium is $f = f_1^*$ and an external shock causes some households to move from region 1 to region 2. This would increase the relative utility level in region 1, which, in turn, would cause some households to move from region 2 to region 1. The equilibrium is regained at $f = f_1^*$.

Figure 16.2a also shows the effects of amenities on interregional distribution of population. As the dotted line indicates, an increase in the amenity level in region 1 leads to more households living in this region in a stable equilibrium. However, the effects of amenities on the distribution of population could be overshadowed by the agglomeration effects in the sense that the region with lower amenities could have a larger population (e.g. at $f = f_1^*$).

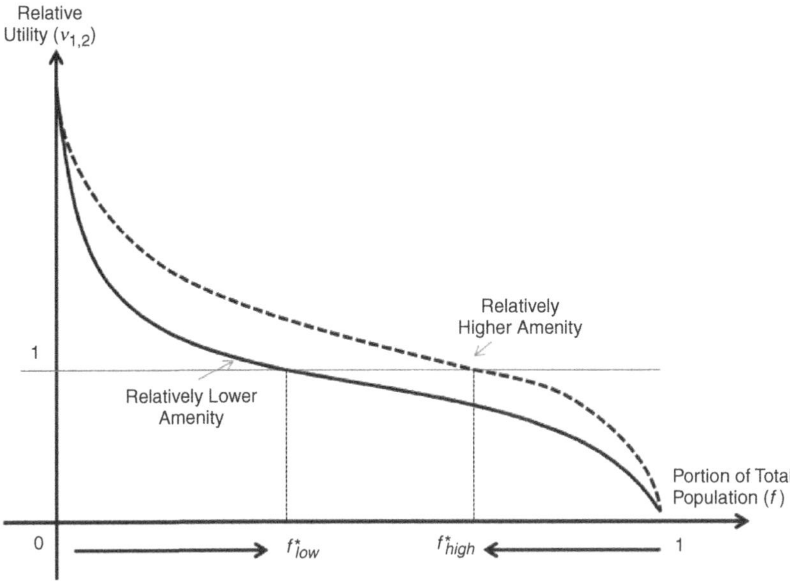

Figure 16.2b Amenities and the spatial distribution of population between the two regions. $\left(A_1 = 1.2/1.5, A_2 = 1, H_1 = H_2 = 1, \theta = 0.3, \varepsilon = 2, \gamma = 0.7, \tau = 6.0, \text{ and } \frac{(1-\gamma\varepsilon)}{(\varepsilon-1)} + \theta\gamma < 0\right)$

Figure 16.2b shows one case of $\frac{(1-\gamma\varepsilon)}{(\varepsilon-1)} + \theta\gamma < 0$. The parameter values are set at $\theta = 0.3$, $\varepsilon = 2$, $\gamma = 0.7$, and $\tau = 6.0$. We also assume that both regions have the same housing stock ($H_1 = H_2 = 1$), but different levels of amenities ($A_1 \neq A_2$). In Figure 16.2b, there is one equilibrium solution for f that could be derived by $v_{1,2} = 1$. Households' preference for housing is high, so that $\frac{(1-\gamma\varepsilon)}{(\varepsilon-1)} + \theta\gamma < 0$. In this case, the utility level declines with increases in population in each region. In contrast to Figure 16.2a, there is only one stable equilibrium ($f = f^*$), and amenities play a more decisive role in shaping the spatial distribution of population and economic activity, that is, the region with a higher level of amenities will have a larger population in equilibrium.

Finally, consider the case when the transport cost is prohibitive (i.e. $\tau \to \infty$). This simply means that trade between two regions may be impossible. Take the limit of equation (16.9) on both sides as τ approaches ∞ and we can get $q \to 1$. Substitute $q \to 1$ and $\tau \to \infty$ to equation (16.12), and the relative utility function is reduced to

$$v_{1,2} = \left(\frac{A_1}{A_2}\right)^{\gamma\theta} \left(\frac{H_1}{H_2}\right)^{\gamma(1-\theta)} \left(\frac{f}{1-f}\right)^{\frac{(1-\gamma\varepsilon)}{(\varepsilon-1)}+\gamma\theta} \tag{16.16}$$

In equilibrium, $v_{1,2} = 1$. Using equation (16.13), the equilibrium distribution of population between the two regions can be derived as

$$f^* = \frac{1}{1 + \left(\frac{A_1}{A_2}\right)^{\frac{\gamma\theta}{R(\cdot)}} \left(\frac{H_1}{H_2}\right)^{\frac{\gamma(1-\theta)}{R(\cdot)}}} \qquad (16.17)$$

Equation (16.17) suggests that if $\frac{(1-\gamma\varepsilon)}{(\varepsilon-1)} + \gamma\theta > 0$, then the relative utility, $v_{1,2}$, improves with an increase in the share of population in region 1 and a larger city will emerge. Other things being equal, the region with higher amenities has a smaller population in equilibrium and enjoys fewer varieties of manufactured goods. This equilibrium is unstable, however. A small derivation from this equilibrium would lead to an agglomerated configuration, that is, all households live in one region. The agglomerated configuration where all households live in the high-amenity region is socially optimal, although the other agglomerated configuration where all households live in the low-amenity region is also stable. On the other hand, if $\frac{(1-\gamma\varepsilon)}{(\varepsilon-1)} + \gamma\theta < 0$, then the relative welfare, $v_{1,2}$, declines as the share of population in region 1 increases and the equilibrium given by (16.17) is stable.

Figure 16.3a and Figure 16.3b confirm the results above. Figure 16.3a illustrates the case of $\frac{(1-\gamma\varepsilon)}{(\varepsilon-1)} + \theta\gamma > 0$. The parameter values are set as $\theta = 0.3$, $\varepsilon = 2$,

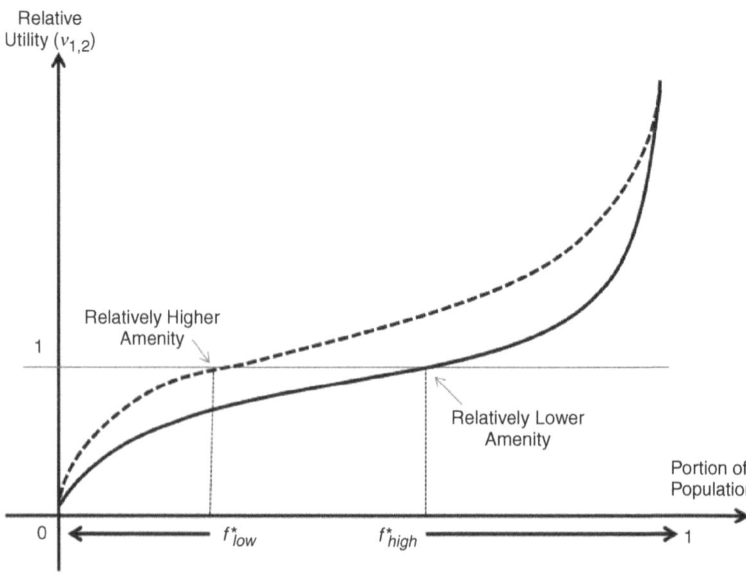

Figure 16.3a Amenities and the spatial distribution of population between the two regions. $\left(A_1 = 1.2/1.5,\ A_2 = 1,\ H_1 = H_2 = 1,\ \theta = 0.3,\ \varepsilon = 2,\ \gamma = 0.4,\ \tau \to +\infty,\right.$ and $\left.\frac{(1-\gamma\varepsilon)}{(\varepsilon-1)} + \theta\gamma > 0\right)$

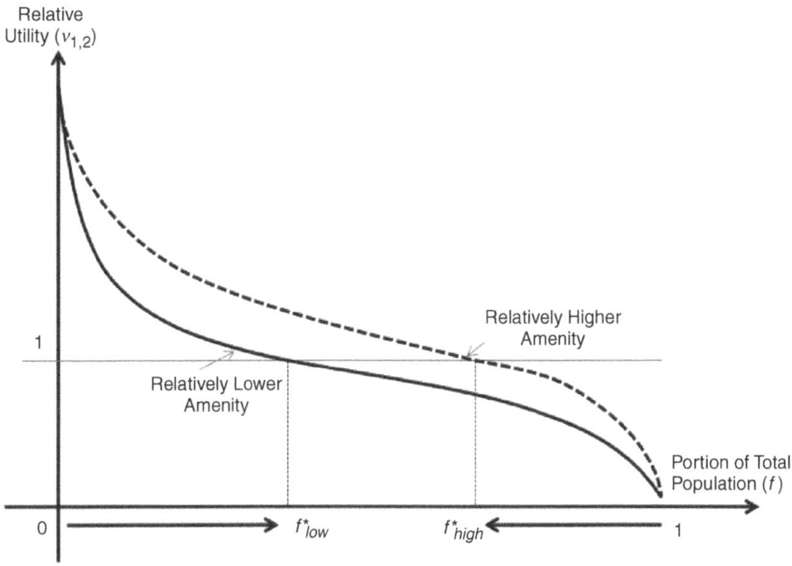

Figure 16.3b Amenities and the spatial distribution of population between the two regions. $\left(A_1 = 1.2/1.5, A_2 = 1, H_1 = H_2 = 1, \theta = 0.3, \varepsilon = 2, \gamma = 0.7, \tau \to \infty, \text{ and } \frac{(1-\gamma\varepsilon)}{(\varepsilon-1)} + \theta\gamma < 0\right)$

$\gamma = 0.4$, and $\tau \to \infty$. In this situation, $f = f^*$ is unstable. As in Figure 16.2a, the agglomeration effect from producing more varieties is the main factor influencing the spatial distribution of population and economic activities, amenities only play a limited role in interregional migrations. Figure 16.3b illustrates the case of $\frac{(1-\gamma\varepsilon)}{(\varepsilon-1)} + \gamma\theta < 0$ for $\theta = 0.3$, $\varepsilon = 2$, $\gamma = 0.7$, and $\tau \to \infty$. In this case, there is one stable equilibrium solution $f = f^*$. Other things being equal, the region with the higher level of amenities has a larger population, and migration is highly responsive to changes in amenities.

In sum, the effect of amenities on the spatial distribution of population and economic activities depends on regional economic characteristics, household preferences and trade barriers. When there is no trade barrier to access local markets, a more dispersed spatial pattern may arise, and the region with a higher level of amenities will attract more households. When there is a trade barrier to access local markets, the role of locational amenities on interregional migration depends on household preferences. When the preference for residential space is relatively low compared with the preference for amenities and varieties of manufactured goods such that $\frac{(1-\gamma\varepsilon)}{(\varepsilon-1)} + \theta\gamma > 0$, an agglomerated configuration of population will emerge. In this case, agglomeration to produce more varieties is the main factor shaping the spatial distribution of population and economic activities, whereas the amenities play only a secondary role. In contrast, if the preference for residential space is relatively high compared with the preference for varieties and amenities such that $\frac{(1-\gamma\varepsilon)}{(\varepsilon-1)}+\theta\gamma<0$, a dispersed pattern of population will emerge. In this case, amenities play an important role in shaping the spatial distribution of population and economic activities.

Discussion and conclusion

The new economic geography model developed in Krugman (1991) provides a framework to study the spatial distribution of economic activity and interregional migration. It generates renewed interest in regional studies. Building on Krugman's framework, Helpman (1998) analyzes the importance of non-tradable goods (housing stock) in shaping the economic geography. He replaces the agricultural sector in Krugman's model by a non-tradable good to highlight the role of market access and regional characteristics in shaping the spatial distribution of population (Ottaviano and Puga 1998; Puga 1999,2002; Fujita and Krugman 2004; Hanson 2005; Redding and Sturm 2008). In Helpman's model, increasing returns to scale and trade barriers provide a force for agglomeration, while a regional immobile non-tradable resource (housing stock) constitutes a congestion force and favors dispersion (Ottaviano and Puga 1998; Redding and Sturm 2008). The introduction of the immobile housing stock in Helpman's model echoes the theory that wage difference may be one reason for globalization bringing convergence in income levels, but not the only reason (Ottaviano and Puga 1998). The departure from Krugman's assumptions leads to some strikingly different results: in Krugman's model low transport costs lead to agglomeration and high transport costs lead to dispersion, whereas in Helpman's model, low transport costs lead to dispersion and high transport costs lead to agglomeration (Helpman 1998; Ottaviano and Puga 1998).

There are distinct benefits to adopting Helpman's framework in both theoretical and empirical studies. Helpman's assumptions are perhaps more consistent with the expenditure structures in the United States. As previous consumer expenditure surveys revealed, housing and related services account for a large share of consumer expenditures. Housing cost differentials are the largest single component of interregional cost of living differences (Beeson and Eberts 1989). The non-tradability of housing stocks between regions forces households to evaluate quality-of-life factors when making location decisions. Helpman's modifications, in his own words, 'seem to be closer to standard urban economic models' (1998: 54). Empirical studies adopting his theoretical framework (e.g. Hanson 2005; Redding and Sturm 2008) explain more spatial variation in economic development (Neary 2001).

In this paper, we extend Helpman's framework to investigate the effects of amenities on interregional migrations. In particular, we want to explain why some regions with relatively high levels of amenities remain undeveloped, while other regions with relatively low levels of amenities grow rapidly. Our framework incorporates multimarket equilibriums, in which regional economic activities such as consumption, production, and trade are internally connected. Thus, our framework could be used to evaluate households' location choices with consideration of factors influencing the quality of life and employment opportunities.

We found the effects of amenities on interregional migration are significantly influenced by other factors such as regional economic characteristics, household preferences, and trade barriers. When there are few trade barriers between regions, the immobile housing resource (both land and amenities) plays a dispersing role in influencing the spatial distribution of households, and amenities are a major determinant of the spatial distribution of population and economic activities. Similar to the effects of immobile developable lands, higher amenities induce more households to move out of densely-populated areas to avoid congestion and enjoy a higher quality of life. In the meantime, this dispersed pattern also brings more manufacturing to less-congested regions, provides employment opportunities, and expands local market potentials.

When there is a trade barrier between regions, the situation becomes more complicated. In this case, the effects of amenities on interregional migration and spatial distribution of economic activities are strongly affected by household preferences. If the preference for residential space is relatively low compared with the preference for varieties and amenities, a large city will likely emerge. Agglomeration of economic activities provides more varieties of tradable goods, offers more job opportunities, and increases the overall quality of life. In this situation, the level of amenities has only a limited effect on interregional migration. On the other hand, if the preference for residential space is relatively high compared with the preference for varieties and amenities, a dispersed distribution pattern of population will emerge. In this case, immobile housing factors, including land and amenities, induce people to move out of congested areas, and amenities have a larger effect on the spatial distribution of activity. These results should inform the debate about the effect of amenities on interregional migrations and regional economic development.

Notes

1 The relevant literatures can be found in Graves (1979), Haurin (1980), Blomquist et al. (1988), Graves and Waldman (1991), Linneman and Graves (1983), Rosen (1979), Roback (1980, 1982), Treyz et al. (1993), Banzhaf and Walsh (2008), and others.

2 A higher value of α or ε indicates a lower preference for varieties.

3 For household i, the expenditure will be allocated as

$$p_l l_{i,k} = (1-\theta)\gamma \frac{E_{i,k}}{(1-\theta\gamma)}; \quad \sum_1^{n_k} p_{j,k} x_{i,j,k} + \sum_1^{n_{-k}} (\tau p_{j,-k}) x_{i,j,-k} = (1-\gamma)\frac{E_{i,k}}{(1-\theta\gamma)}.$$

Summing the above results over all individual households in region k gives:

$$p_l l_k = (1-\theta)\gamma \frac{E_k}{(1-\theta\gamma)}; \quad \sum_1^{n_k} p_{j,k} x_{j,k} + \sum_1^{n_{-k}} (\tau p_{j,-k}) x_{j,-k} = (1-\gamma)\frac{E_k}{(1-\theta\gamma)},$$

where $x_{j,k} = N_k x_{i,j,k}$, $x_{j,k} = N_k x_{i,j,-k}$, and $E_k = N_k E_{i,k}$.

4 Notice that except ε, all parameters in equation (16.9) are bounded. Taking the limit of equation (16.9) on both sides as ε approaches $+\infty$ gives:

$$\Rightarrow 1-\theta\gamma = \lim_{\tau\to\infty} \frac{fq^{-\varepsilon}}{fq^{1-\varepsilon}+(1-f)\tau^{1-\varepsilon}}\{(1-\gamma)q+\gamma(1-\theta)[qf+(1-f)]\}$$

$$+\lim_{\tau\to\infty} \frac{(1-f)\tau^{1-\varepsilon}q^{-\varepsilon}}{f\tau^{1-\varepsilon}q^{1-\varepsilon}+(1-f)}\{(1-\gamma)+\gamma(1-\theta)[qf+(1-f)]\}$$

Note that $1-\varepsilon<0$, so $\lim_{\tau\to\infty}\tau^{1-\varepsilon}=0$.

$$\Rightarrow 1-\theta\gamma = \frac{fq^{-\varepsilon}}{fq^{1-\varepsilon}}\{(1-\gamma)q+\gamma(1-\theta)[qf+(1-f)]\}$$

(Notice that this step eliminates $f=0$. In addition, $1-f\neq 0$. Otherwise, $\lim_{\tau\to\infty}\frac{(1-f)\tau^{1-\varepsilon}q^{-\varepsilon}}{f\tau^{1-\varepsilon}q^{1-\varepsilon}+(1-f)}\{(1-\gamma)+\gamma(1-\theta)[qf+(1-f)]\}$ may not equal zero.)

$$\Rightarrow (1-\theta\gamma)q = q-q\gamma+\gamma(1-\theta)qf+\gamma(1-\theta)(1-f)$$

$$\Rightarrow -\theta\gamma q = -\gamma q+(1-\theta)\gamma qf+\gamma(1-\theta)(1-f)$$

$$\Rightarrow (1-\theta)\gamma q(1-f) = \gamma(1-\theta)(1-f)$$

$$\Rightarrow (1-\theta)\gamma(1-q)(1-f)=0$$

As $\gamma\in(0,1)$, $\theta\in(0,1)$, and $f\in(0,1)$, we have $q=1$.

References

Adamson, D.W., Clark, D.E. and Partridge, M.D. (2004) 'Do urban agglomeration effects and household amenities have a skill bias?', *Journal of Regional Science*, 44(2): 201–223.

Anderson, S.T. and West, S.E. (2006) 'Open space, residential property values, and spatial context', *Regional Science and Urban Economics*, 36(6): 773–789.

Banzhaf, H.S. and Walsh, R.P. (2008) 'Do people vote with their feet? An empirical test of Tiebout', *American Economic Review*, 98(3): 843–863.

Beeson, P.E. and Eberts, R.W. (1989) 'Identifying productivity and amenity effects in interurban wage differentials', *Review of Economics and Statistics*, 71(3): 443–452.

Black, D. and Henderson, V. (1999) 'A theory of urban growth', *American Economic Review*, 89(2): 321–327.

Blomquist, G.C., Berger, M.C. and Hoehn, J.P. (1988) 'New estimates of quality of life in urban areas', *American Economic Review*, 78(1): 89–107.

Bolitzer, B. and Netusil, N.R. (2000) 'The impact of open spaces on property values in Portland, Oregon', *Journal of Environmental Management*, 59(3): 185–193.

Brown, G.M., Jr. and Pollakowski, H.O. (1977) 'Economic valuation of shoreline', *Review of Economics and Statistics*, 59(3): 272–278.

Brueckner, K.J., Jacques-François, T. and Yves, Z. (1999) 'Why is central Paris rich and downtown Detroit poor? An amenity-based theory', *European Economic Review*, 43(1): 91–107.

Chen, Y. and Rosenthal S.S. (2008) 'Local amenities and life-cycle migration: Do people move for jobs or fun?', *Journal of Urban Economics*, 64(3): 519–537.

Cheshire, P.C. and Sheppard, S.C. (1998) 'Estimating the demand for housing, land, and neighborhood characteristics', *Oxford Bulletin of Economics and Statistics*, 60(3): 357–382.

Cheshire, P.C. and Sheppard, S.C. (2002) 'The welfare economics of land-use planning', *Journal of Urban Economics*, 52(2): 242–269.

Dixit, A.K. and Stiglitz, J.E. (1977) 'Monopolistic competition and optimum product diversity', *American Economic Review*, 67(3): 297–308.

Doss, C.R. and Taff, S.J. (1996) 'The influence of wetland type and wetland proximity on residential property values', *Journal of Agricultural and Resource Economics*, 21(1): 120–129.

Frey, W. H. (2009) *The Great American Migration Slowdown: Regional and Metropolitan Dimensions*.Washington DC: Brooking. http://www.brookings.edu/~/media/Files/rc/reports/2009/1209_migration_frey/1209_migration_frey.pdf (accessed 25 December 2009).

Fujita, M. and Krugman, P. (2004) 'The New Economic Geography: Past, present and the future', *Papers in Regional Science*, 83: 149–164.

Geoghegan, J., Lynch, L. and Bucholtz, S. (2003) 'Capitalization of open spaces into housing values and the residential property tax revenue impacts of agricultural easement programs', *Agricultural and Resource Economics Review*, 32(1): 33–45.

Glaeser, E.L. and Shapiro, J.M. (2003) 'Urban growth in the 1990s: Is city living back?', *Journal of Regional Science*, 43(1): 139–165.

Glaeser, E.L. and Gyourko, J. (2005) 'Urban decline and durable housing', *Journal of Political Economy*, 113(2): 345–375.

Glaeser, E.L., Gyourko, J., and Saks, R. (2005) 'Why is Manhattan so expensive? Regulation and the rise in housing prices', *Journal of Law and Economics*, 48(2): 331–369.

Graves, P.E. (1979) 'A life-cycle empirical analysis of migration and climate by race', *Journal of Urban Economics*, 6(2): 135–147.

Graves, P.E. and Waldman, D.M. (1991) 'Multimarket amenity compensation and the behavior of the elderly', *American Economic Review*, 81(5): 1374–1381.

Greenwood, M.J. and Hunt, G.L. (1989) 'Job versus amenities in the analysis of metropolitan migration', *Journal of Urban Economics*, 25(1): 1–16.

Greenwood, M.J., Hunt, G.L., Rickman, D.S. and Treyz, G.I. (1991) 'Migration, regional equilibrium, and the estimation of compensating differentials', *American Economic Review*, 81(5): 1382–1390.

Hanson, H.G. (2005) 'Market potential, increasing returns and geographic concentration', *Journal of International Economics*, 67(1): 1–24.

Haurin, D.R. (1980) 'The effect of property taxes on urban areas', *Journal of Urban Economics*, 7(3): 384–396.

Helpman, E. (1998) 'The size of regions'. In: D. Pines, E. Sadka and I. Zilcha (eds), *Topics in Public Economics*, pp. 33–55, New York: Cambridge University Press.

Hunt, G.L. (1993) 'Equilibrium and disequilibrium in migration modeling', *Regional Studies*, 27(4): 341–349.

Irwin, E. (2002) 'The effects of open space on residential property value', *Land Economics*, 78(4): 465–480.

Irwin, E.G. and Bockstael, N.E. (2002) 'Interacting agents, spatial externalities, and the endogenous evolution of residential land-use pattern', *Journal of Economic Geography*, 2(1): 31–54.

Johnson, K.M. and Beale, C.L. (1994) 'The recent revival of widespread population growth in nonmetropolitan areas of the United States', *Rural Sociology*, 59(4): 655–667.

Klaiber, H.A. and Phaneuf, D.J. (2010) 'Valuing open space in a residential sorting model of the Twin Cities', *Journal of Environmental Economics and Management*, 60(2): 57–77.

Kopits, E., McConnell, V. and Walls, M. (2007) 'The tradeoff between private lots and public open space in subdivisions at the urban-rural fringe', *American Journal of Agricultural Economics*, 89(5): 1191–1197.

Krugman, P. (1991) 'Increasing returns and economic geography', *Journal of Political Economy*, 99(3): 483–499.

Lansford, N.H., Jr. and Jones, L.L. (1995) 'Marginal price of lake recreation and aesthetics: An hedonic approach', *Journal of Agricultural and Applied Economics*, 27(1): 212–223.

Leggett, C.G. and Bockstael, N.E. (2000) 'Evidence of the effects of water quality on residential land prices', *Journal of Environmental Economics and Management*, 39(2): 121–144.

Linneman, P. and Graves, P.E. (1983) 'Migration and job change: A multinomial logit approach', *Journal of Urban Economics*, 14(3): 263–279.

Mahan, B.L., Polasky, S. and Adams, R.M. (2000) 'Valuing urban wetlands: A property price approach', *Land Economics*, 76(1): 100–113.

Neary, J.P. (2001) 'Of hype and hyperbolas: Introducing the New Economic Geography', *Journal of Economic Literature*, 39(2): 536–561.

Ottaviano, G.I.P. and Puga, D. (1998) 'Agglomeration in the global economy: A survey of the "New Economic Geography"', *World Economy*, 21(6): 707–731.

Palmquist, R.B. (1992) 'Valuing localized externalities', *Journal of Urban Economics*, 31(1): 59–68.

Polinsky, M.A. (1977) 'The demand for housing: A study in specification and grouping', *Econometrica*, 45(2): 447–461.

Polinsky, M.A. and Shavell, S. (1976) 'Amenities and property values in a model of an urban area', *Journal of Public Economics*, 5(1–2): 119–129.

Pogodzinski, J.M. (1988) 'Amenities in an urban general equilibrium model', *Journal of Urban Economics*, 24(3): 260–278.

Puga, D. (1999) 'The rise and fall of regional inequalities', *European Economic Review*, 43(2): 303–334.

Puga, D. (2002) 'European regional policies in light of recent location theories', *Journal of Economic Geography*, 2(4): 373–406.

Radeloff, V.C., Hammer, R.B. and Stewart, S.I. (2005) 'Rural and suburban sprawl in the U.S. Midwest from 1940 to 2000 and its relation to forest fragmentation', *Conservation Biology*, 19(3): 793–805.

Radeloff, V.C., Hammer, R.B., Voss, P.R., Hagen, A.E., Field, D.R. and Mladenoff, D.J. (2001) 'Human demographic trends and landscape level forest management in the northwest Wisconsin Pine Barrens', *Forest Science*, 47(2): 229–241.

Rasker, R. and Hansen, A.J. (2000) 'Natural amenities and populations growth in the Greater Yellowstone Region', *Human Ecology Review*, 7(2): 30–40.

Rappaport, J. and Sachs, J.D. (2003) 'The United States as a coastal nation', *Journal of Economic Growth*, 8(1): 5–46.

Redding, S.J. and Sturm, D.M. (2008) 'The costs of remoteness: Evidence from German division and reunification', *American Economic Review*, 98(5): 1766–1797.

Roback, J. (1980) *The Value of Local Urban Amenities: Theory and Measurement*, University of Rochester: PhD dissertation.

Roback, J. (1982) 'Wages, rents, and the quality of life', *Journal of Political Economy*, 90(6): 1257–1278.

Rosen, S. (1979) 'Wage-based indices of urban quality of life'. In: P. Mieszkowski and M. Straszheim (eds), *Current Issues in Urban Economics*. Baltimore and London: Johns Hopkins University Press.

Sivitanidou, R. (1995) 'Urban spatial variation in office-commercial rents: The role of spatial amenities and commercial zoning', *Journal of Urban Economics*, 38(1): 23–49.

Smith, B.A. (1978) 'Measuring the value of urban amenities', *Journal of Urban Economics*, 5(3): 370–387.

Smith, V.K., Poulos, C. and Kim, H. (2002) 'Treating open space as an urban amenity', *Resource and Energy Economics*, 24(1–2): 107–29.

Treyz, G.I., Rickman, D.S., Hunt, G.L. and Greenwood, M.J. (1993) 'The dynamics of U.S. internal migration', *Review of Economics and Statistics*, 75(2): 209–214.

Tyrväinen, L. and Miettinen, A. (2000) 'Property prices and urban forest amenities', *Journal of Environmental Economics and Management*, 39(2): 205–23.

Weicher, J.C. and Zerbst, R.H. (1973) 'The externalities of neighborhood parks: An empirical investigation', *Land Economics*, 49(1): 99–105.

Wu, J. (2006) 'Environmental amenities, urban sprawl, and community characteristics', *Journal of Environmental Economics and Management*, 52(2): 527–547.

Wu, J., Xu, W. and Ralph, J.A. (2012) *Optimal Location and Size of Open Space: How Do They Affect Urban Landscapes?*, Oregon State University: AREc Working Paper.

Index